Recent Advances in PMOS Negative Bias Temperature Instability

Souvik Mahapatra
Editor

Recent Advances in PMOS Negative Bias Temperature Instability

Characterization and Modeling of Device Architecture, Material and Process Impact

Springer

Editor
Souvik Mahapatra
Department of Electrical Engineering
IIT Bombay
Mumbai, India

ISBN 978-981-16-6119-8 ISBN 978-981-16-6120-4 (eBook)
https://doi.org/10.1007/978-981-16-6120-4

This Springer imprint is published by the registered company Springer Nature Singapore Pte Ltd.
The registered company address is: 152 Beach Road, #21-01/04 Gateway East, Singapore 189721, Singapore

To my parents

Krishna Chandra Mahapatra and Anjali Mahapatra

Foreword by Muhammad Ashraful Alam

That day in the late afternoon in December 2000, I had just finished a talk to a large audience of engineers at our Lucent Technology manufacturing facility in Florida. The talk has gone well, the topic (TDDB, short for time-dependent dielectric breakdown) was timely and important (we at Bell Labs had just published a paper in Nature predicting that gate oxides can be reliably scaled below 1 nm), the data supporting our new optimistic view was abundant, and an elegant physics of percolation model and interaction between electrons and holes threaded together the results beautifully. After the talk, as people gathered to ask questions and offer suggestions, I felt relaxed and appreciated. On my way out, however, a hesitant-looking young technician caught up with me: "Your talk was all about TDDB, but what about the real problem of NBTI"? Apparently, NBTI (short for Negative Bias Temperature Instability) was holding up the product qualification at the factory. I did not know what he was talking about—flummoxed and flustered, I mumbled a non-answer and decided to look up the topic. I had no idea that afternoon that in 20-years' time, NBTI would turn out to be the single most important technological problem for modern MOSFETs, and it would deserve the detailed and deeply thoughtful book that you are now holding in your hands.

Einstein's 1916 general theory of relativity, fusing space and time and bending lights along curved space–time, transformed astrophysics. In the 1950s, two additional ideas about space and time would arise that would likewise transform solid-state physics. First, Mandelbrot's theory of fractals defined geometrical space with fractional dimensions (space-fractals) and provided the mathematical foundation for percolation, renormalization, phase transition and eventually dielectric breakdown—the topic of my Florida TDDB talk. Similarly, Turing's reaction–diffusion (RD) theory of morphogenesis explained spontaneous pattern formation (i.e., the emergence of order out of disorder), but equally important, it defined for the first time a system of equations with fractional-order time kinetics (time fractals). I was vaguely familiar with the reaction–diffusion model and the promise of fractional time kinetics due to dispersive diffusion in amorphous material. As I read Jeppson and Svensson's somewhat esoteric 1977 paper on the reaction–diffusion (RD) model, I was intrigued by the possibility of finding time fractals hiding in plain sight in the NBTI kinetics. Later on, I had rederived and reframed the RD model solution for NBTI using a simple geometrical argument, which suggests a power law time dependence of interface trap generation at the channel/gate dielectric interface, ~ At^n, where A depends on the technology, voltage, temperature and duty cycle of operation, but the power exponent n—a technology-agnostic rational fraction that would scale NBTI over 7–8 decades of time—does not! The simplicity of the derivation and the robustness of the time-exponent suggested that NBTI degradation is exceptional—a technologically important problem that is also a testbed of one of the deepest theories of nature.

Over the next decade, or so, Prof. Mahapatra and I, along with our students and colleagues, would generalize the simple theory (by adding contributions due to hole trapping into pre-existing and newly generated gate insulator defects) and explore various technology-agnostic and technology-specific aspects of NBTI.

This book contains a deep analysis of both aspects of the problem. The analysis of technology-specific components became increasingly important because by early 2000, the era of classical scaling had ended and that of equivalent scaling started: The gate oxides were being doped with nitrogen, replaced with higher-k alternative, the channel strained increased, the channel material evolved, and the transistors evolved from planar to surround-gate topologies. With company-specific technologies, it became difficult to compare NBTI, let alone define their intrinsic features.

The book reflects Prof. Mahapatra's long journey in re-discovering the essential universality of NBTI hidden within the confusing mess of measured data. Chapters 1–6 focus on topics that are essential for measurement and physical modeling of the NBTI degradation, while Chap. 7–14 focus on the application of the modeling framework to analyse measured NBTI data across different technologies. I do not know if any other book regarding a single degradation phenomenon covers the topic so completely and comprehensively. You may wonder if a reliability phenomenon, whose basic physics can be derived using very simple arguments, deserves a book-length discussion 20-years down the road. Yes, it does, because technology is harder than physics: Calculating the trajectory of a stone projectile requires no more than few lines of algebra, but putting a man on the moon requires a bit more effort!

Still, a single book cannot cover everything and this book is no exception. After reading this book and gaining a deep appreciation of the degradation phenomenon, the reader will continue to find interesting papers in the literature discussing additional topics, such as area-optimized NBTI degradation, circuit-level odometers, coupled NBTI and TDDB degradation, coupled NBTI and hot carrier degradation, NBTI for ferroelectric FET, negative capacitance FET, NBTI for machine learning and neuromorphic systems, alternative interpretations of NBTI and so on. These new topics show that NBTI remains, and will remain, a vibrant topic of research for a long time to come and the style adopted in this book, based on the synthesis of physics, material science, device physics, circuit design would remain a model for how to address these questions. Indeed, I wish I had such a book to read when I was getting started—it would have taught me how to think about big problems in technology with an elegant mix of timely analysis and timeless physics.

June 2021

Muhammad Ashraful Alam
Purdue University
West Lafayette, IN, USA

Foreword by Stephen M. Ramey

I was recently simultaneously amazed and frustrated reading an article about measurements of the muon magnetic precession rate which disagreed from the standard model of physics by 0.0002%.[1] Fascinated that the scientific community could measure such a tiny, obscure quantity with such precision. Frustrated by the comparison to the state of our ability to model MOSFET aging characteristics where we're lucky to match integrated circuit behavior within a few percent. It is a testament to how a seemingly-simple system of "just" the familiar semiconductor and gate oxide has far more complicated physics governing behavior than would be expected at first glance. Closer inspection reveals a highly complicated system comprised of thin, strained layers with material properties that deviate from those observed on larger structures; amorphous gate oxides with a mixture of elements and bond structures; a material interface with still unknown properties and crystal configuration; lattice and bond energies that change with time under different electrical stress conditions; pre-existing trap states with various spatial, energy, and temporal characteristics; and weakly bound hydrogen throughout the system playing various roles in the aging process. With such a complicated system, it is not surprising that the NBTI (Negative Bias Temperature Instability) models have struggled to accurately predict measured behavior even with the expanded in complexity in recent years to capture all the interacting physical mechanisms.

In fact, the semiconductor industry has spent many billions of dollars on research and development over the last 60 or so years yet we still don't fully understand how to completely predict NBTI in integrated circuits. Further, this mechanism can be as important to overall chip performance as all the improvements combined in a new technology node. This is especially apparent in recent technology nodes targeted at high performance CPUs (Central Processing Units) where the voltages have stopped decreasing node over node and in some cases increased in an effort to get higher frequencies. This reverse scaling of the voltage while continuing to shrink transistor features such as gate oxides has compounding exponential implications to the NBTI

[1] Emily Conover, "Muon magnetism could hint at a breakdown of physics' standard model," Science News, April 7, 2021.

and places further demands on the modeling fidelity. Additionally, as recent technologies move further away from the more familiar Silicon-based planar MOSFET with SiO_2 (Silicon Dioxide) gate dielectric, the need for predictive physics understanding and modeling becomes imperative to help proactively guide new technology definition and development.

Gone are the days when simplistic understanding and modeling of NBTI was adequate as the industry continues to push the technology limits and looks for ways to squeeze every last bit of performance without sacrificing the reliability assurance that products will remain healthy in use. Reliability modeling in general has always faced the challenge of accurately predicting long lifetimes through extrapolation of shorter-duration stress tests. This challenge becomes particularly difficult in a mechanism like NBTI where there are multiple mechanisms that contribute to the overall phenomenon, each with its own distinct time, voltage, and temperature dependence. As such, the mechanism that dominates aging during a short duration stress test may not be the dominant mechanism over the lifetime of a chip operating under a variety of workloads. NBTI in particular, unlike most transistor reliability modes, suffers from an additional complication where the traps responsible for aging of the transistor quickly recover upon removal of stress that prevents directly measuring the NBTI characteristics. As such, additional physics-based modeling is needed to translate the measured NBTI aging to the parameters of interest which is generally the NBTI during the stress itself.

Additionally, specialized measurement techniques and data interpretation methodology are needed to map the measured aging behavior to the underlying physics models. Accessing the details of the various trap types and physical mechanisms through electrical test is challenging in MOSFETs since there are only four terminals on the transistor to use for characterization. Even with this limitation, there are a dizzying variety of electrical tests that can be used to tease out the nuances of the various mechanisms. To fully quantify and calibrate the physical models, a combination of DC, AC, and "ultra-fast" stress tests can be run over a range of voltages, times, and temperatures. While none of the techniques can give an absolute direct measurement of a single trap type, the combination of many of these allows a "triangulation" to get insights into each mechanism independently. With the complexity of the measurements needed and the physical measurements, a methodology for interpretation of the measurements in the context of the physical mechanisms is just as important as the physics itself.

The need for research in this book should therefore be apparent. As an industry, the better we can refine the predictive models for aging the better we can create new technologies that are optimized for both performance and reliability. The ability of these model to predict behaviors in various architectures and material systems is critical to enable rapid technology development since technology complexity has outgrown the ability to simply tackle the development strictly empirically. Additionally, the translation of these models into a usable simulation tool is essential to enable circuit design co-optimized for both performance and reliability.

June 2021

Stephen M. Ramey
Manager of transistor reliability
Intel Corporation
Portland, OR, USA

Preface

Although first discovered about 50 years ago, Negative Bias Temperature Instability (NBTI) has become an important reliability concern in p-MOSFETs about 20 years ago, with the introduction of Silicon Oxynitride gate insulator technology. It continues to remain as a serious issue for High-K Metal Gate (HKMG) technology, and impacts various device architectures, e.g., bulk and FDSOI p-MOSFETs, bulk and SOI p-FinFETs, and Gate All Around Stacked Nanosheet (GAA-SNS) FETs, fabricated using either Gate First (GF) or Replacement Metal Gate (RMG) integration scheme.

The transistor parametric drift due to NBTI stress increases with more negative gate bias and higher temperature, but reduces after the removal of stress, resulting in lower degradation for AC compared to DC stress. The degradation is impacted by different process changes related to the gate stack (e.g., Nitrogen), channel material (e.g., Germanium) and mechanical strain (due to changes in layout or device dimensions), and several backend processes (e.g., plasma damage) can also influence if not properly optimized.

NBTI has been extensively studied in the past 20 years. The published reports can be categorized into the following broad sub-topics: (a) development of measurement methods for the characterization of transistor parametric drift and the gate insulator defects responsible for the same, (b) understanding the physical mechanism and development of physical models to calculate the transistor parametric drift, (c) process optimization for reducing NBTI for technology qualification, (d) variability of NBTI in small area devices and (e) development of compact models and circuit simulation. In our earlier book [Fundamentals of Bias Temperature Instability in MOS Transistors, Springer, 2015], the topic (a) is discussed in detail, while the topics (b) and (c) are discussed to some extent. In the past 6 years, a lot of work has happened on NBTI in advanced technologies like FDSOI MOSFETs, FinFETs, GAA-SNS FETs, SiGe channels, and much more insight has been obtained on the physical mechanism of NBTI and related models. These aspects are covered in this book.

Chapter 1 discusses different ultra-fast measurement methods for recovery artifact free parametric characterization and several important features of NBTI during and after DC and AC stress. The impact of stress and recovery biases, temperature, pulse duty cycle, frequency and measurement delay are discussed. Chapter 2 discusses the impact of Nitrogen incorporation in the gate insulator, Germanium incorporation in the channel and mechanical stress effects on NBTI, and variability of NBTI degradation in small area devices.

Chapter 3 discusses the physical mechanism of NBTI, and different characterization techniques for direct estimation of the gate insulator defects (generation of interface and bulk traps, pre-existing bulk traps) that are responsible for the transistor parametric drift. Chapter 4 through Chap. 6 discuss the BTI Analysis Tool (BAT) framework used in this book to model NBTI in different technologies. The Reaction Diffusion (RD) model is discussed in Chap. 4 for the generation of interface traps. The Transient Trap Occupancy Model (TTOM), to obtain the occupancy (charged state) of interface traps calculated from the RD model, is discussed in Chap. 5. Hole trapping in pre-existing (process related) bulk traps is calculated by the Activated Barrier Double Well Thermionic (ABDWT) model also in Chap. 5. The Reaction Diffusion Drift (RDD) model for bulk trap generation is discussed in Chap. 6. These models are independently validated using suitably chosen experiments when only interface trap generation, interface trap generation with trap occupancy, interface trap generation with trap occupancy and either hole trapping or bulk trap generation impact NBTI.

The full BAT framework is used in Chap. 7 through Chap. 14 to model NBTI in devices having different processes, materials and architectures. The time kinetics of measured parametric drift during and after DC, AC and mixed AC-DC stress, for different stress and recovery biases, temperature, AC pulse modes, digital and non-digital AC stress, pulse duty cycle and frequency are modeled.

Chapter 7 models GF HKMG bulk p-MOSFETs and focuses on the impact of gate insulator process changes, *e.g.*, Nitrogen incorporation, interlayer and High-K thickness scaling. Chapter 8 models GF HKMG Si-capped SiGe bulk p-MOSFETs and focuses on the impact of changes in the Si cap and SiGe quantum well thicknesses, and Germanium content in SiGe quantum well. Chapter 9 models Si and SiGe channel based GF HKMG FDSOI p-MOSFETs, and focuses on the impact of Germanium, Nitrogen and layout changes.

Chapter 10 models RMG HKMG SOI p-FinFETs, including the variability observed in few fin devices. Chapter 11 models RMG HKMG Si and SiGe channel bulk p-FinFETs and focuses on the impact of Germanium and Nitrogen. Chapter 12 models RMG HKMG GAA-SNS FETs having different sheet length and width. Chapter 13 models the impact of fin dimension scaling in FinFETs and sheet dimension scaling in GAA-SNS FETs, and analyses the differences between (110) and (100) channel orientations.

The process and material changes discussed above are modeled across different device architectures for DC and AC (digital, Mode-B) stress, at different bias and temperature but at fixed duty cycle and frequency. Chapter 14 models the impact of AC stress modes, digital and non-digital pulse, pulse duty cycle and frequency. The debate regarding the frequency dependence or independence of NBTI degradation is analyzed.

There are several ways by which the material presented in this book can further help future research directions. The BAT framework needs to be implemented in a commercial Technology CAD (TCAD) environment. The RD, ABDWT and RDD frameworks use physical models and those can be suitably adopted, while TTOM, which uses empirical equations in this book, can be modified using ABDWT like framework. This is necessary for full 3-dimensional simulation of modern device structures, with proper electrostatics, quantum and mechanical stress effects. The TCAD implementation would help simulations involving situations when the Self-Heating effect and Hot Carrier Degradation are also present along with NBTI. On the other hand, faster versions of these physical models can help enable activity-aware circuit simulations in a SPICE environment, and would help mitigate some of the challenges faced by simple empirical compact models to handle recovery under random gate excitations. From the physical viewpoint, connections can be made between the hole trapping-detrapping processes during NBTI and Random Telegraph Noise (RTN) observed in small area devices. Similarly, the generation of bulk gate insulator traps during NBTI can be related to the Stress Induced Leakage Current (SILC) and Time Dependent Dielectric Breakdown (TDDB) experiments. The framework can be used to study the tunnel oxide degradation in 3D NAND memories during write-erase cycles, and the degradation of other emerging devices like Negative Capacitance and Ferroelectric FETs, etc.

It is my sincere hope that this book would be helpful to a wide range of readers, from the practicing engineers to graduate students and their supervisors, and provide a comprehensive perspective of NBTI in different state-of-the-art technologies, including a clear picture of the underlying physical mechanisms and model. I am well aware of the alternative viewpoints on the NBTI mechanisms and models, and I hope that independent evaluations will be made to access their capabilities against diverse experimental conditions and devices, as done in this book.

Mumbai, India Souvik Mahapatra
June 2021

Acknowledgements

This book has been made possible by active contribution and support from several people. I am grateful to my former and current students at IIT Bombay, who have directly or indirectly contributed to this book (listed in alphabetical order): Akshay Raj, Ankush Chaudhary, Ayush Ranjan, Bijesh Rajamohanan, Dhanoop Varghese, E. Naresh Kumar, Gautam Kapila, Karansingh Thakor, Kaustubh Joshi, Narendra Parihar, Natasha Goyal, Nilesh Goel, Nilotpal Choudhury, Nirmal Nanaware, P. Bharath Kumar, Ravi Tiwari, Shweta Deora, Subhadeep Mukhopadhyay, Subrat Mishra, Tarun Samadder, Uma Sharma and Vrajesh Maheta.

This book has been made possible from active collaborations with Applied Materials, IBM and ST Microelectronics, and support from Micron and Synopsys. I am grateful to Steven Hung and Johanes Swenberg (Applied Materials), Richard Southwick, Miaomiao Wang, Huimei Zhou and James Stathis (IBM), Vincent Huard and Xavier Federspiel (ST Microelectronics), Chandra Mouli (Micron), Hiu Yung Wong, Steve Motzny, Victor Moroz, Lee Smith and Jamil Kawa (Synopsys) for technical discussions. Satheesh Kuppurao (Applied Materials), Mukesh Khare (IBM) and Terry Ma (Synopsys) are acknowledged for management support. I am also grateful to Jacopo Franco from IMEC for providing the experimental data used in Chap. 8.

I am grateful to Ahmad Ehteshamul Islam and Muhammad Ashraful Alam for fruitful discussions and collaboration in the area of NBTI modeling. I also wish to thank several researchers with whom I had many useful discussions in the past: Amr Haggag, Andreas Kerber, Anand Krishnan, Ben Kaczer, Chetan Prasad, Chittoor Parthasarathy, Eduard Cartier, Eiichi Murakami, Giuseppe La-Rosa, Guido Groeseneken, Hans Reisinger, Hideki Aono, Hussam Amrouch, Sriram Kalpat, Srikanth Krishnan, Sanjay Rangan, Stephen Ramey, Tanya Nigam, Tibor Grasser, Victor van-Santeen and Vijay Reddy. I wish to thank my colleagues at IIT Bombay and friends for their understanding and support.

Last but not the least, I wish to profoundly thank my wife Sahana Mukherjee and daughter Adrija Mahapatra for their constant encouragement and support, which immensely helped me during the course of working on this book.

Mumbai, India
June 2021

Souvik Mahapatra

Contents

About the Editor

Souvik Mahapatra received his Bachelors and Masters degrees in Physics from Jadavpur University, Calcutta, India in 1993 and 1995 respectively, and Ph.D. in Electrical Engineering from IIT Bombay, Mumbai, India in 1999. During 2000–2001, he was with Bell Laboratories, Lucent Technolgies, Murray Hill, NJ, USA. Since 2002 he is with IIT Bombay, and is currently the PK Kelkar Chair Professor in the Department of Electrical Engineering. His primary research interests are in the areas of semiconductor device characterization, modeling and simulation, and in particular, MOS transistor and Flash memory device scaling and reliability. He has interacted closely with major semiconductor industries in the world, and has contributed in several technologically relevant research topics such as MOS gate insulator scaling, Bias Temperature Instability and Hot Carrier Degradation in CMOS devices, CHISEL NOR Flash, SONOS NOR and NAND Flash memory devices. He has authored and co-authored more than 190 papers in peer reviewed journals and conferences and several book chapters, and delivered invited talks and tutorials in major international conferences around the world, including at the IEEE IEDM and IEEE IRPS. He has served as a distinguished lecturer of the IEEE EDS, chair of the IEEE EDS device reliability physics subcommittee, and in paper selection subcommittees and as session chairs in several IEEE conferences. He is a fellow of Institute of Electrical and Electronics Engineers (IEEE), Indian National Science Academy (INSA), Indian National Academy of Engineering (INAE) and Indian Academy of Sciences (IASc).

Chapter 1
Characterization of NBTI Parametric Drift

Check for updates

Souvik Mahapatra, Nilesh Goel, and Narendra Parihar

1.1 Introduction

Negative Bias Temperature Instability (NBTI)-induced parametric drift is a crucial reliability issue for p-channel Metal–Oxide–Semiconductor Field-Effect Transistors (MOSFETs). Although first reported ~50 years ago in Silicon Dioxide (SiO_2) gate insulator devices [1], NBTI became a topic of great interest ~20 years ago with the introduction of Silicon Oxynitride (SiON) gate insulator devices [2–5]. It continues as an important reliability issue for High-K Metal Gate (HKMG) devices having dual layer (SiO_2 or SiON Interlayer (IL) and Hafnium Dioxide (HfO_2) high-K layer) gate insulator stack [6–31]. It impacts different device architectures having HKMG stack, e.g., planar bulk MOSFETs [6–11], Fully Depleted Silicon On Insulator (FDSOI) MOSFETs [12, 13], bulk and SOI FinFETs [13–27], as well as Gate All Around Stack Nanosheet (GAA-SNS) FETs [28–31]. Although the majority of the reports are from devices having conventional Silicon (Si) channel, some of the reports are from devices having HKMG stack on Silicon Germanium (SiGe) channel. The schematic illustration of different device architectures is shown in Fig. 1.1.

A p-channel MOSFET suffers from NBTI when the gate voltage (V_G) becomes negative compared to all other terminals of the device. Figure 1.2 illustrates an example of a complementary MOS (CMOS) inverter, consisting of p- and n- channel MOSFETs connected in series between the power and ground rails. When the input is low ($V_{IN} = 0V$) and the output is high ($V_{OUT} = V_{DD}$, where V_{DD} is the operating power rail bias), the p-MOSFET has $V_G = 0V$ and source (V_S), substrate (V_B) and drain (V_D) biases are at V_{DD}. The gate can be thought of as "effectively" at a negative bias ($V_G = -V_{DD}$) compared to the other terminals of the device, if V_S, V_B and V_D are "assigned" to $0V$ reference. In this condition, positive charges are

S. Mahapatra (✉) · N. Goel · N. Parihar
Department of Electrical Engineering, Indian Institute of Technology Bombay, Powai, Mumbai 400076, India
e-mail: souvik@ee.iitb.ac.in

S. Mahapatra (ed.), *Recent Advances in PMOS Negative Bias Temperature Instability*,
https://doi.org/10.1007/978-981-16-6120-4_1

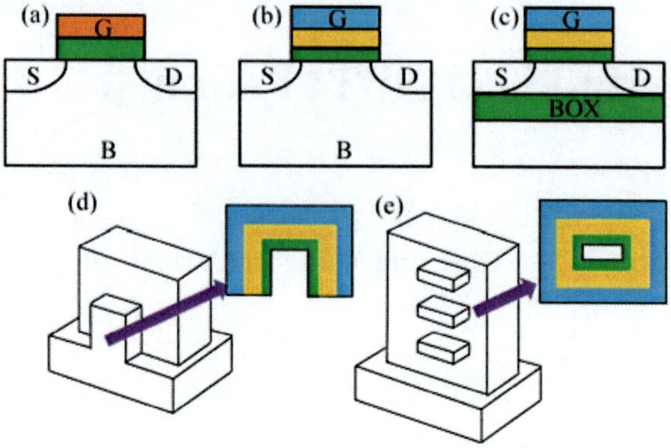

Fig. 1.1 Schematic of different device architectures: (a, b) planar MOSFETs with (a) single-layer SiO_2 or SiON and (b) dual-layer HKMG stacks. Also shown different device architectures having HKMG stack: (c) FDSOI, (d) FinFET and (e) GAA-SNS FET. The gate insulator stacks are separately illustrated for FinFET and GAA-SNS FET. Color coding of the gate insulator stack: green (SiO_2 or SiON) and orange (poly-Si gate) in (a), green (SiO_2 or SiON IL), yellow (high-K) and blue (metal gate) in (b–e)

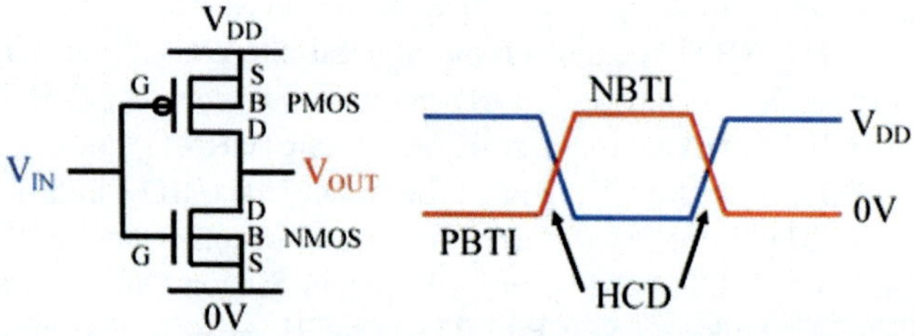

Fig. 1.2 Schematic of a CMOS inverter together with voltage waveforms at the input and output terminals showing NBTI, PBTI, HCD and TDDB degradation phases

accrued in the gate insulator and lead to device parametric drift. However, if the input is subsequently made high ($V_{IN} = V_{DD}$) and so the output goes low ($V_{OUT} = 0V$), the gate of the p-MOSFET is at the same bias as source and substrate (V_G, V_S and V_B are at V_{DD}), and the drain is at $0V$. In this case, the drain can be thought of as "effectively" at a negative bias ($V_D = –V_{DD}$), if V_G, V_S and V_B are "assigned" to $0V$ reference. A fraction of the positive charges reduces in this condition, which results in (partial) recovery of the device parametric drift. Note that AC operation of the inverter (e.g., in a Ring Oscillator (RO) circuit with odd number of serially connected inverters, when oscillating) consists of successive low and high input pulses. Hence, the individual inverter would show lower parametric drift (due to recovery) under AC when compared to a pure DC operation (i.e., in a static RO). NBTI recovery-related relief for AC (dynamic) operation was first reported ~20 years ago [32].

Besides NBTI, the CMOS inverter circuit is impacted by Positive BTI (PBTI) for n-MOSFET, Time Dependent Dielectric Breakdown (TDDB) and Hot Carrier

Degradation (HCD) for both p- and n-MOSFETs as shown in Fig. 1.2. However, NBTI remains the most severe MOSFET reliability concern for latest technology nodes [20] and is the subject of this book.

NBTI impacts different digital circuit blocks (~standard cells), Static Random Access Memory (SRAM) and various analog circuits [33–36]. This book focuses on NBTI characterization and modeling in isolated devices having different architectures, dimensions, gate stack processes and channel materials. The goal of this book is to propose a physical NBTI model and validate it using a wide range of experimental data. This would help develop robust methodologies for NBTI-aware technology qualification and is a suitable framework for circuit simulation.

1.2 NBTI Measurement Methods

Figure 1.3 illustrates the measurement setup for an isolated p-MOSFET, with V_G and V_D waveforms for the two most commonly used methods: (a) Measure–Stress–Measure (MSM) [37, 38] and (b) On-The-Fly (OTF) [39, 40]. In both methods, V_G is set to the stress (V_{GSTR}) and recovery (V_{GREC}) biases during stress and post-stress phases, marked, respectively, as “S” and “R”.

In the MSM method, the magnitude of V_G is changed to the measurement bias (V_{GMEAS}) by periodically interrupting the stress and recovery phases for measurement, marked as “M”. The magnitude of V_{GSTR} is usually higher than V_{DD} to accelerate NBTI, while V_{GREC} is often at $0V$ (as in this illustration) or $0\ V < |V_{GREC}| < |V_{GSTR}|$. Measurement can be done at a single V_G as illustrated in this figure or using a full *I-V* sweep as discussed later. Note that although the V_D is at $0V$ during stress and at $-V_{DD}$ during recovery for the p-MOSFET in a CMOS inverter circuit (with V_S and V_B at $0\ V$ reference), and V_D is held at $0V$ during stress and recovery

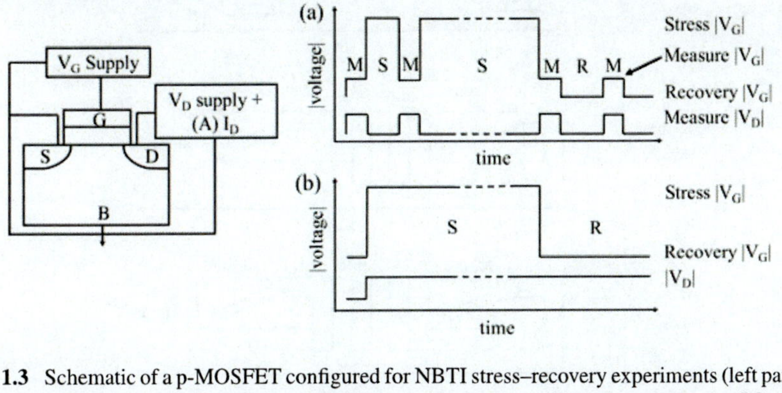

Fig. 1.3 Schematic of a p-MOSFET configured for NBTI stress–recovery experiments (left panel). Schematic of the gate and drain bias pulses for stress (S), recovery (R) and measurement (M) using the (a) MSM and (b) OTF methods. The exact gate waveform in the M phase of MSM method for the full sweep scheme is described in Fig. 1.4. For OTF, M coincides with the S (or R) phase

experiments and only raised (in magnitude) during measurement in an isolated p-MOSFET, the source and substrate terminals always remain grounded. In the OTF method, the measurement is continuously done at stress (or recovery) V_G, with V_D set to measurement condition, source and substrate grounded. Any V_{GSTR} can be used; however, V_{GREC} cannot be $0V$ in this case (since the drain current is continuously sampled).

The terminology "time" denotes the generic stress (or recovery) timescale, and a specific stress (or recovery) value is denoted by t_{STR} (or t_{REC}). The temperature (T) is usually kept identical between stress, recovery and measurement phases.

1.2.1 Full Sweep MSM Method

Figure 1.4 illustrates the V_G waveform for full sweep MSM method for measurement during stress (similar V_G measurement sweep is used during recovery after stress). Linear drain current (I_{DLIN}) versus V_G (at low V_D) transfer characteristics are measured by interrupting stress in log-uniform time intervals. The V_G sweep direction can be made either (a) from high to low (magnitude) or (b) from low to high (magnitude), shown for DC stress.

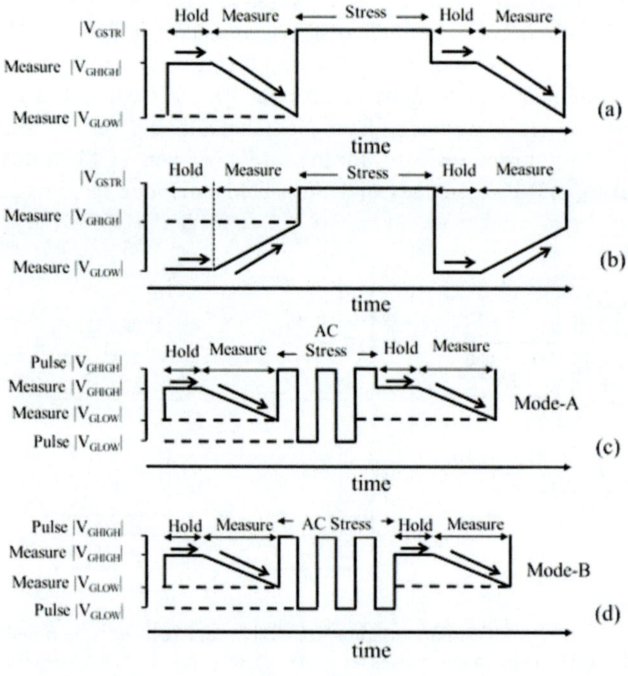

Fig. 1.4 MSM method: gate bias patterns during (a, b) DC, (c) Mode-A and (d) Mode-B AC stress. Sweep is done from high to low V_G in (a, c, d) and low to high V_G in (b)

AC pulse is used for dynamic stress with pulse high (V_{GHIGH}) and low (V_{GLOW}) values. V_{GHIGH} equals V_{GSTR} in this case, and V_{GLOW} is often at $0V$ (digital waveform) or can even have nonzero value. The measurement can be made either (c) after the end of last half cycle (denoted as Mode-A) or (d) after the end of last full cycle (denoted as Mode-B) for AC stress. It is important to keep the sweep direction same for DC, Mode-A and Mode-B AC stress condition (the high to low sweep direction is shown in this example). The Mode-B AC stress is often used in the literature and is also the default throughout this book, unless specifically mentioned otherwise. For AC stress, t_{STR} denotes the total time during which the AC pulse is applied, which includes the pulse high and low phases. Therefore, the actual "stress" would depend on the pulse duty cycle (PDC). The pulse frequency (f) is limited to 1MHz in this book, due to limitations associated with wafer-level device characterization.

The measurement time (~measurement delay) is denoted as t_M and includes the hold time and sweep time. The MOSFET parametric drift, e.g., threshold voltage shift (ΔV_T), is determined using the pre- and post- stress (or recovery) measurements, i.e., $\Delta V_T = V_T$ (post) – V_T (pre). Pre- and post-stress V_T is determined by the peak transconductance (g_m) method. NBTI-induced positive charges result in negative ΔV_T. Note that while $|\Delta V_T|$ is always plotted, it is shown as ΔV_T (without the modulus sign) in all the figures throughout this book.

Figure 1.5 shows the (a) I_{DLIN} and (b) g_m versus V_G traces measured before and after stress. Increase (magnitude) in V_T and reduction in peak g_m (due to mobility degradation because of additional Coulomb scattering) are observed due to positive gate insulator charges accrued during stress.

Figure 1.6 shows the time evolution of ΔV_T for consecutive stress and recovery phases with $V_G = -V_{GSTR}$ and 0 V, respectively [41]. Data are shown for the conventional (slow) and ultra-fast MSM methods, with t_M (= stress off phase) of ~1s and ~100ns, respectively. As ΔV_T recovers when the magnitude of V_G is lowered from V_{GSTR} (more on this in Sect. 1.3), the stress-off time for measurement should be kept as small as possible to increase accuracy. In this example, the ΔV_T buildup and recovery, respectively, during and after stress are significantly underestimated

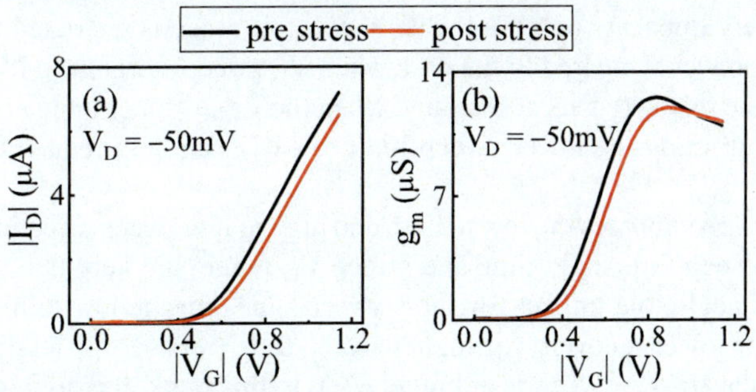

Fig. 1.5 Measured (a) drain current and (b) transconductance versus gate bias characteristics from a HKMG p-MOSFET before and after NBTI stress. The V_T shifts toward more negative values, while peak g_m reduces after NBTI stress. Data from [38]

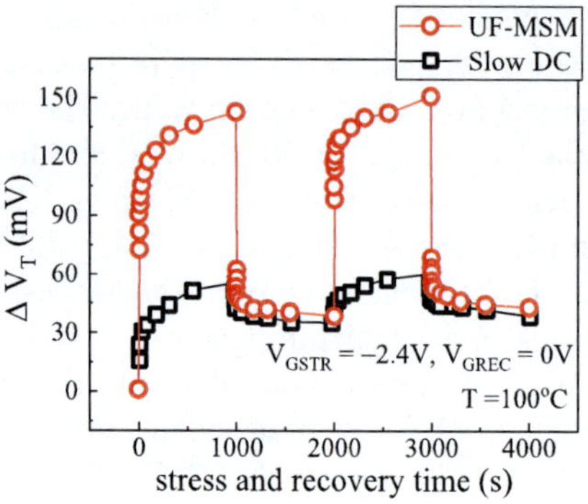

Fig. 1.6 Measured ΔV_T time kinetics during and after stress (two successive stress–recovery cycles are shown) in SiON p-MOSFET using Ultra-Fast MSM (UF-MSM) and conventional slow DC measurement methods, data from [41]

when the slow measurement method is used. Therefore, it is necessary to characterize NBTI kinetics by an ultra-fast method.

Figure 1.7 shows the time evolution of ΔV_T measured by the full sweep method during (a–c) DC and (d) Mode-A and Mode-B AC stress with different measurement delays. Gate first (GF) HKMG Si channel p-MOSFETs are used [10]. The stress time kinetics is usually plotted in a log–log scale and is done throughout the book; a power-law dependence is obtained with slope n. In this book, the value of n is obtained using fit of measured data (plotted in a log–log scale) in the 1s-1Ks time range, unless specifically mentioned otherwise.

Figure 1.7 (a) shows the impact of hold and sweep time; a low to high sweep direction is used. Lower ΔV_T magnitude at a fixed t_{STR} and higher n are obtained when either of the hold or sweep time is large, due to the impact of recovery. As shown in Sect. 1.3, recovery depends on the ratio of t_{REC} $(= t_M)$ to t_{STR}. The impact of recovery becomes smaller at higher t_{STR}, since t_M (= hold + sweep time) is fixed but t_{STR} increases with the passage of stress. Therefore, the difference between the "un-recovered" and "as-measured" ΔV_T reduces at higher t_{STR} and gives rise to an increase in n. This is explained using the schematic illustration above the figure. The hold time can be reduced to reduce t_M without much impact on measured data.

Figure 1.7 (b) shows the impact of sweep V_G range for low to high sweep direction. Lower recovery impact is observed when the sweep range is restricted to the above threshold region as compared to the case when V_G sweep starts from $0V$. As shown in Sect. 1.3, the recovery gets accelerated when the V_{GREC} magnitude is made lower. Therefore, restricting the lower sweep limit to ~V_{T0} can help reduce the recovery artifacts.

Figure 1.7 (c) compares the low to high and high to low sweep directions; all other conditions (sweep time, hold time and sweep V_G range) are kept fixed. The sweep direction has negligible impact for short sweep time, but the low to high direction shows slightly lower recovery for higher sweep time (recovery is less if the longer time part of the I_D–V_G sweep is at higher V_G). In this book, the high to low sweep direction is used for both DC and AC stresses when the full sweep MSM version is used; see Fig. 1.4 (a), (c), (d), with $t_M = 20\mu s$ delay as default.

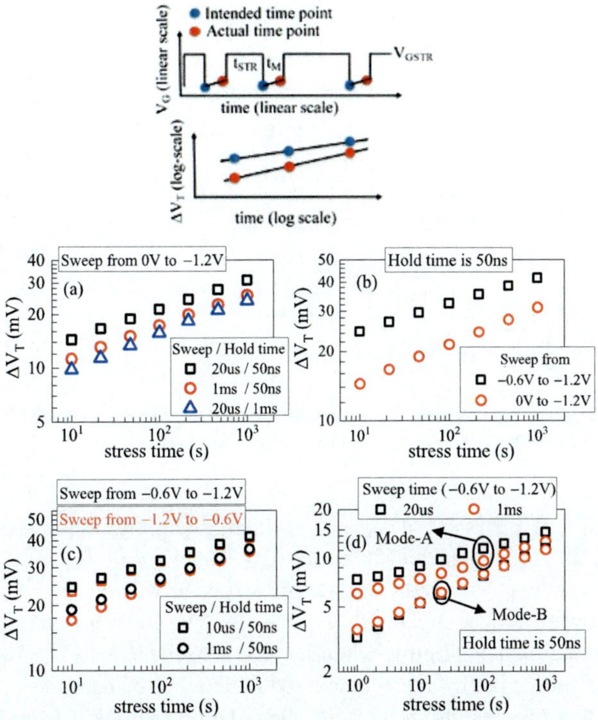

Fig. 1.7 Impact of measurement delay on the time evolution of ΔV_T in MSM method, for variation in (a) sweep and hold time (see Fig. 1.4), (b) range of sweep V_G and (c) sweep direction for Mode-B, and (d) sweep delay for Mode-A and Mode-B AC stress. The top panel explains the reason for increase in slope (n) due to measurement delay. The difference between intended (no delay) and actually measured (with delay) data points is highest for DC stress (as illustrated) and lowest for Mode-B AC stress. Data from [42]

Figure 1.7 (d) shows that the measurement delay (~ sweep time) impacts the Mode-A but not Mode-B AC time kinetics, which is consistent with the recovery kinetics (higher recovery after Mode-A compared to Mode-B AC stress), which is explained later in Sect. 1.3. Note that the impact of measurement delay is highest for DC, moderate for Mode-A AC and lowest for Mode-B AC stress.

1.2.2 One Point Drop Down (OPDD) MSM Method

Figure 1.8 illustrates the V_G waveform for (a) DC, (b) Mode-A and (c) Mode-B AC stress for the OPDD method, where (d) full I_{DLIN} versus V_G sweep is measured only before stress to estimate V_{T0} (the post-stress *I-V* curve is shown for illustration only, and it is not actually measured). The stress is periodically interrupted in log-uniform

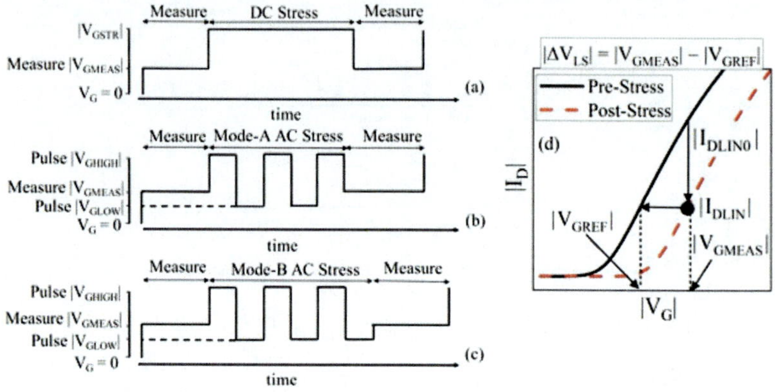

Fig. 1.8 OPDD MSM method: pulse waveforms for (a) DC, (b) Mode-A and (c) Mode-B AC stress. (d) Schematic illustrating the lateral shift method; see text

time intervals, and V_G is set at V_{GMEAS}, which is slightly above V_{T0} to measure I_{DLIN} (shown as the "black dot" in panel (d)). The V_G (= V_{GREF}) corresponding to the post-stress I_{DLIN} is noted from the pre-stress curve, and the lateral (V_G-axis) difference between V_{GREF} and V_{GMEAS} equals to ΔV_T [43]. It is presumed that the mobility degradation is negligible when I_{DLIN} is sensed close to V_{T0}, which is a fair assumption as shown below and is discussed in detail elsewhere [42].

Since NBTI stress results in mobility degradation (peak g_m degradation shown in Fig. 1.5 (b), it is important to verify that the lateral shift method used in OPDD version results in correct ΔV_T. Note that the I_{DLIN} versus V_G traces before and after stress should be parallel to each other if mobility degradation is negligible. Therefore, the lateral shift (at iso-I_{DLIN}) is ΔV_T; see Fig. 1.8 (d). By noting a simple relationship as $\Delta I_{DLIN}/I_{DLIN0} = -\Delta V_T/(V_{GMEAS} - V_{T0})$, where V_{GMEAS} is the gate bias at which I_{DLIN} is measured, ΔV_T can be determined if mobility degradation is absent. This can be defined as the vertical shift method.

Figure 1.9 shows the time evolution of ΔV_T obtained from the (a) lateral and (b) vertical shift methods for different values of V_{GMEAS} (hence different values of the pre-stress I_{DLIN} reference) for the device shown in Fig. 1.5. The reference ΔV_T kinetics from the peak g_m method (from full sweep I_{DLIN} versus V_G data) is also shown. It is important to remark that ΔV_T from the lateral shift, vertical shift and peak g_m methods are identical only for low V_{GMEAS} (~V_{T0}). Due to non-negligible mobility degradation, the lateral or vertical shift methods overestimate ΔV_T when higher V_{GMEAS} values are used.

The MSM method is used throughout this book, with the OPDD version having $t_M = 10\mu s$ delay as default, but in few cases the full sweep version is used instead, which is explicitly mentioned. The interested reader can refer to [42] for further details of these methods.

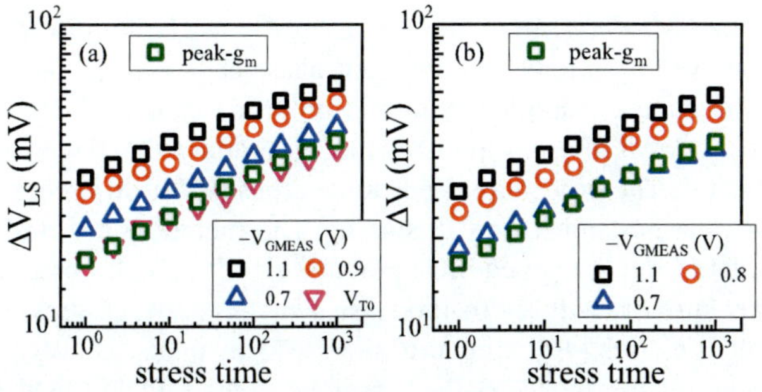

Fig. 1.9 Comparison of lateral and vertical voltage shifts (see text) at different measurement biases, and comparison with ΔV_T from peak g_m method. Data from [42]

1.2.3 Other Measurement Methods

Figure 1.10 illustrates the applied V_G waveform and corresponding I_{DLIN} for OTF stress measurements. I_{DLIN} (at low V_D) is continuously sampled in a log-uniform timescale during stress at $V_G = V_{GSTR}$. I_{DLIN} reduces during stress due to buildup of positive charges because of mobility degradation and V_T shift. The first data point is measured with a time delay t_0 from the application of V_{GSTR} and is denoted as I_{DLIN0}. Note that there are no recovery issues during stress, as V_G is not reduced (in magnitude) below V_{GSTR} for measurement. However, I_{DLIN0} is assumed to be unstressed to estimate the parametric drift, $\Delta I_{DLIN} = I_{DLIN} - I_{DLIN0}$, and the error in I_{DLIN0} (deviation from the actual pre-stress value) impacts the fractional degradation ($\Delta I_{DLIN}/I_{DLIN0}$) [40]. The ultra-fast version uses a t_0 delay of $1\mu s$, which was found to be sufficient [40, 42]. However, as I_{DLIN} degradation is due to both mobility degradation (evident from g_m degradation) and V_T shift, a complex post-processing mobility correction procedure is needed to obtain ΔV_T [42, 44]. This method is not generally used in this book (except in some cases which is explicitly mentioned); the interested reader can refer to [42] for details.

Fig. 1.10 Schematic to explain the concept of time-zero delay in OTF method. The measured I_{DLIN0} is closer to the true value for lower t_0 delay

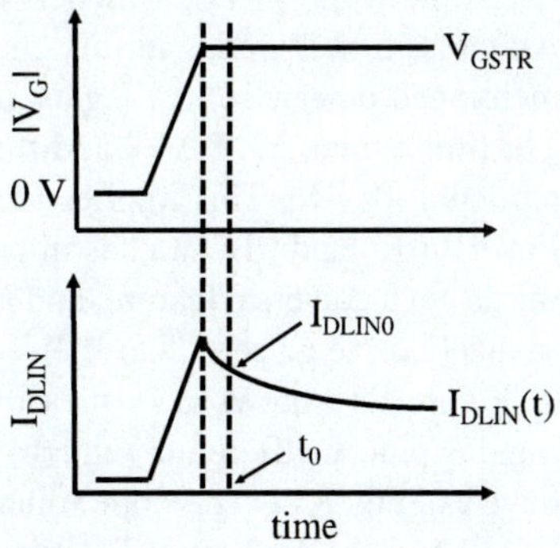

In the extended-MSM (e-MSM) method, a stress pulse of duration t_{STR} and at $V_G = V_{GSTR}$ is followed by a long measurement phase at $V_G \sim V_{T0}$, and I_{DLIN} (at $V_D = 50mV$) is continuously sampled in log-uniform time scale [45]. No measurement is done during the stress phase, and the method is similar to OTF recovery measurements; see Fig. 1.3. The procedure needs to be repeated for different t_{STR} to obtain the stress characteristics and is time consuming. This method is not used in this book, except where data from Si capped SiGe p-MOSFETs [8] are discussed.

The constant current feedback method also switches between stress (or recovery) and measurement phases [46], similar to the MSM methods. However, rather than measuring I_{DLIN} at constant V_G ($\sim V_{T0}$), as done in the OPDD-MSM method, this method forces a constant I_{DLIN} and measures change in V_G. The reference I_{DLIN} is chosen at $V_G \sim V_{T0}$ from prestress measurements. The shift in V_G during stress (or recovery) can therefore be directly linked to ΔV_T, as mobility degradation can be ignored. This is similar to the V_G-axis shift technique illustrated in Fig. 1.8 for the OPDD-MSM method, except that the shift is naturally obtained. This method is not used in this book.

1.3 Basic Features of Measured Data

All data shown in this section are measured by the MSM full sweep method, unless specifically mentioned otherwise, in GF HKMG Si channel p-MOSFETs [10].

1.3.1 Static (DC) Stress

Figure 1.11 (a) and (b) shows the time evolution of ΔV_T measured during DC stress at different V_{GSTR} and T. The time kinetics is plotted in a log–log scale, and ΔV_T rapidly increases at the onset of stress, which is visible when data are plotted from short to long time as in Fig. 1.11 (a). However, power-law time dependence is observed at long time ($t_{STR} \geq 1s$), which is clarified by replotting the data in Fig. 1.11 (b). All stress time kinetics in this chapter is plotted for $t_{STR} \geq 1s$, unless specifically mentioned otherwise. ΔV_T gets accelerated at more negative V_{GSTR} and higher T. The time evolution of ΔV_T at different V_{GSTR} and T is parallel to each other for this particular device. This implies identical n across V_{GSTR} and T, which is shown in Fig. 1.11 (c) and (d), and has important implications as discussed below. However, this is not a universal feature, and n can indeed vary with V_{GSTR} and T as shown later in this book (e.g., see Chap. 10).

Figure 1.12 shows ΔV_T measured at a fixed t_{STR} as a function of (a) V_{GSTR} (in a log–log plot, at different T) and (b) $1/kT$ (in a semilog plot, at different V_{GSTR}), for the device of Fig. 1.11. The slope values denote (a) power-law voltage acceleration factor (VAF) and (b) Arrhenius T activation energy (E_A), and these parameters quantify, respectively, the impact of V_{GSTR} and T on ΔV_T. Although any stress time can

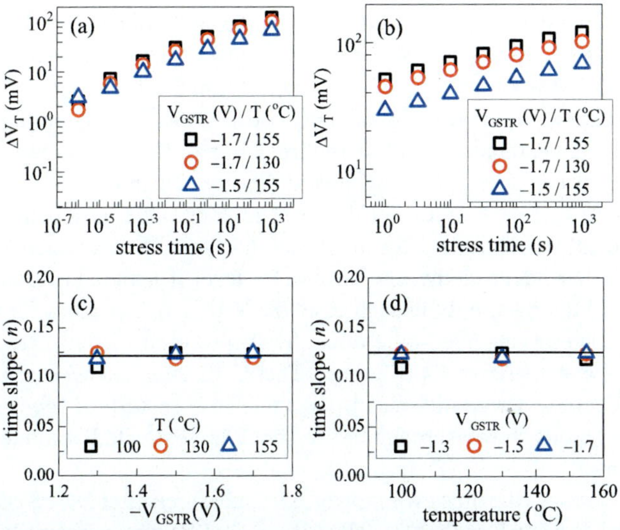

Fig. 1.11 Time evolution of measured ΔV_{T} at different V_{GSTR} and T, plotted (a) from short to long time and (b) re-plotted only for long time stress. Extracted power-law time slope (n) as a function of (c) stress voltage and (d) temperature. Data from [10]

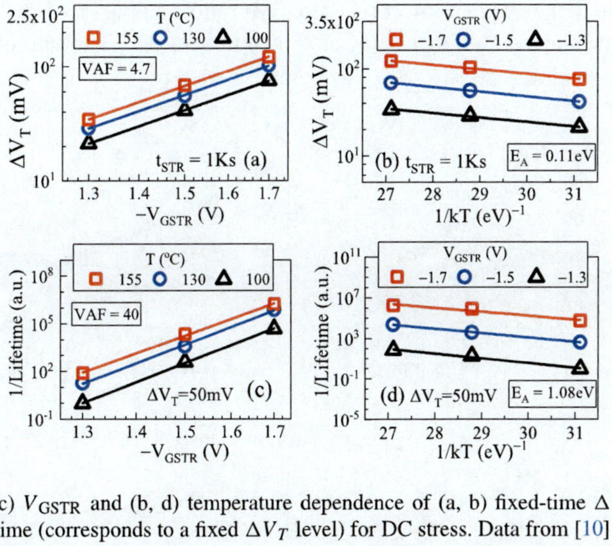

Fig. 1.12 (a, c) V_{GSTR} and (b, d) temperature dependence of (a, b) fixed-time ΔV_T and (c, d) inverse of lifetime (corresponds to a fixed ΔV_T level) for DC stress. Data from [10]

be used, t_{STR} of 1Ks is used to extract VAF and E_{A} throughout this book, unless specifically mentioned otherwise. Note that when n is identical across V_{GSTR} and T, the VAF and E_{A} values remain identical for all values of t_{STR}. However, n can indeed vary with V_{GSTR} and T as shown later in this book. In such a case, the VAF and E_{A} values would become dependent on the value of t_{STR} at which they are extracted.

Moreover, identical VAF is obtained across different T (hence identical E_A across different V_{GSTR}) for this device. This is also not a universal feature, and VAF (and E_A) can indeed vary with T (and V_{GSTR}), as shown later in this book (e.g., see Chap. 11).

Another method of determining the impact of V_G and T is by noting the time to reach a particular degradation level (= lifetime). Figure 1.12 also shows 1/lifetime as a function of (c) V_{GSTR} (in a log–log plot, at different T) and (d) $1/kT$ (in a semilog plot, at different V_{GSTR}), for the same device. The slope values now denote (c) power-law VAF and (d) Arrhenius E_A for the device lifetime. The extracted VAF and E_A would stay independent of the reference ΔV_T level if n remains identical across V_{GSTR} and T. Moreover, only in such case, the VAF and E_A values for the lifetime plots can be related to the fixed-time values by the power-law slope n, i.e., VAF or E_A (at fixed t_{STR}) = n * VAF or E_A (lifetime). Therefore, identical VAF across different T (and identical E_A across different V_{GSTR}) is also observed for lifetime. Note, all VAF and E_A values shown throughout this book are extracted at a fixed t_{STR} (and not for lifetime).

Figure 1.13 shows the time evolution of ΔV_T measured after DC stress at different (a) V_{GSTR} and T, as well as (b) t_{STR} (all at $V_{GREC} = 0V$), and different V_{GREC} for (c) longer and (d) shorter t_{STR}, for the device of Fig. 1.11. Usually, the recovery kinetics is plotted in a semilog scale. The ΔV_T recovery starts immediately after the stoppage of stress at $V_{GREC} = 0V$, and it proceeds over several decades of time, which is seen at different V_{GSTR}, T and t_{STR}; see Fig. 1.13 (a) and (b). For a given t_{REC}, recovery reduces for higher t_{STR}, i.e., lower t_{REC} to t_{STR} ratio; see Fig. 1.13 (b). This causes increased time slope n at higher delay as shown in Sect. 1.2. The start of recovery is

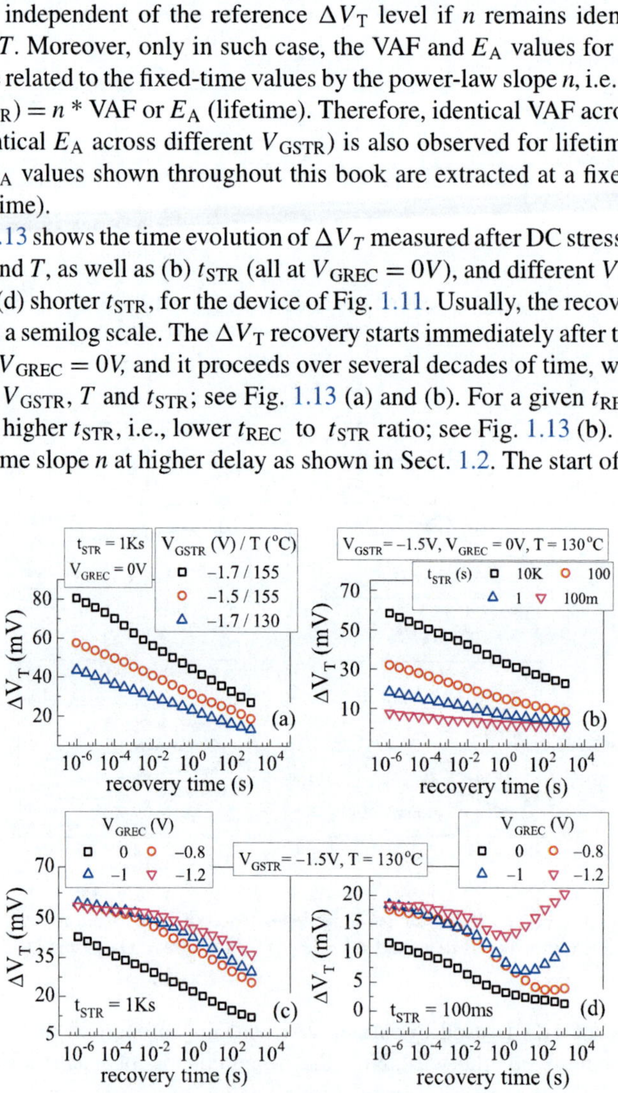

Fig. 1.13 Time evolution of measured ΔV_T recovery after stress for variation in (a) V_{GSTR} and T, (b) stress time and (c, d) V_{GREC} for (c) long and (d) short t_{STR} values. Data from [10]

delayed for $|V_{GREC}| > 0$ V, and the delay gets larger for smaller difference between V_{GSTR} and V_{GREC}; see Fig. 1.13 (c) and (d). Moreover, the device gets re-stressed at longer t_{REC} following stress at shorter t_{STR} and for small V_{GSTR} to V_{GREC} difference; see Fig. 1.13 (d). However, no re-stressing is seen (within the measurement time window) during recovery following stress at long t_{STR}, irrespective of V_{GSTR} and V_{GREC} difference; see Fig. 1.13 (c).

1.3.2 Dynamic (AC) Stress

Figure 1.14 shows the time evolution of ΔV_T measured (a) during and (b) after DC as well as Mode-A and Mode-B AC stress at identical V_{GSTR} and T, for the device of Fig. 1.11. Data are plotted in a log–log scale and show power-law time dependence. For a fixed t_{STR}, the buildup of ΔV_T is highest for DC and lowest for Mode-B AC stress. The power-law time dependence has highest n for Mode-B AC and lowest n for DC stress. The recovery is shown in a semilog plot. The start of recovery is immediately after the stoppage of stress for DC but is significantly delayed for Mode-B AC stress. Therefore, for a fixed t_{REC}, the recovery of ΔV_T is highest for DC and lowest for Mode-B AC stress. The stress and recovery kinetics of Mode-A AC stress is in between that of DC and Mode-B AC stress. The difference in recovery kinetics for Mode-A and Mode-B AC stress explains the difference in delay impact shown in Sect. 1.2. Due to very slow start of recovery, the delay impact is negligible for Mode-B AC stress; see Fig. 1.7 (d).

Figure 1.15 shows the time evolution of ΔV_T measured for different (a) V_{GHIGH} (= V_{GSTR}) and T, (b) PDC, (c) frequency and (d) V_{GLOW} (except the experimental parameters explicitly mentioned in each panel, all other conditions are kept fixed) under Mode-B AC stress in the same device of Fig. 1.11. Power-law time dependence is observed with identical n ~ 1/6 in all cases. The ΔV_T magnitude increases at (a) larger magnitude of V_{GHIGH} and T, (b) higher PDC and (d) higher (nonzero) magnitude of V_{GLOW}, while it remains invariant with (c) changes in f (frequency

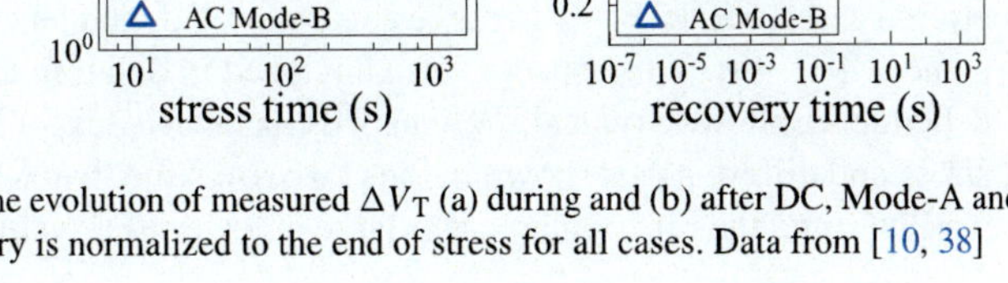

Fig. 1.14 Time evolution of measured ΔV_T (a) during and (b) after DC, Mode-A and Mode-B AC stress. Recovery is normalized to the end of stress for all cases. Data from [10, 38]

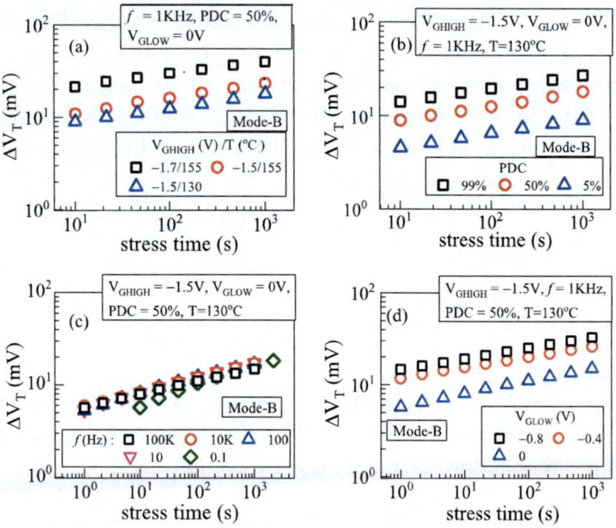

Fig. 1.15 Time evolution of measured ΔV_T for Mode-B AC stress for different (a) V_{GSTR} and T, (b) PDC, (c) frequency and (d) V_{GLOW} values of the gate pulse. Data from [10, 38]

independence). The $n \sim 1/6$ time kinetics under Mode-B AC stress is a very significant result and will be analyzed in detail in Chaps. 3 and 4.

The impact of PDC and f can be further accessed by plotting the measured ΔV_T at fixed t_{STR} for AC stress. Figure 1.16 shows the impact of (a) f at different V_{GLOW} and (b) PDC at different f during Mode-A and Mode-B AC stress. Figure 1.16 also shows the PDC dependence of Mode-B AC stress for different (c) V_{GLOW} and (d) measurement delay. The reference DC value is shown, obtained at $V_{GSTR} = V_{GHIGH}$ and at identical t_{STR} as AC stress.

Note that ΔV_T is always larger for higher (nonzero) V_{GLOW} magnitude, which remains valid for both Mode-A and Mode-B AC stress and different f; see Fig. 1.16 (a), and for different PDC, see Fig. 1.16 (c). Moreover, Mode-A shows larger ΔV_T than Mode-B AC stress at lower f, Mode-A shows f dependence particularly at lower f, while Mode-B shows f independence for all f; see Fig. 1.16 (a). Importantly, ΔV_T for Mode-A reduces at larger f and merges with Mode-B AC data, which is seen at different V_{GLOW}. Note that, for practical circuit operation at high f, measured ΔV_T is identical between Mode-A and Mode-B AC stresses and shows f independence. The f independence of NBTI was verified up to ~2 GHz pulse using specialized on-chip circuits [47, 48]. The f independence of Mode-B AC stress kinetics is another key result, which is analyzed in detail later in Chap. 14. However, in some exceptional cases, Mode-B AC stress can also show f dependence, *especially* at high V_{GSTR} and T [27, 49]. This aspect is also analyzed in detail in Chap. 14.

The PDC dependence shows a typical, "*S*"-shaped characteristics for Mode-A and Mode-B AC stress and different f as shown in Fig. 1.16 (b). Note that ΔV_T increases with increase in PDC, and the rate of increase is large at lower PDC values, becomes

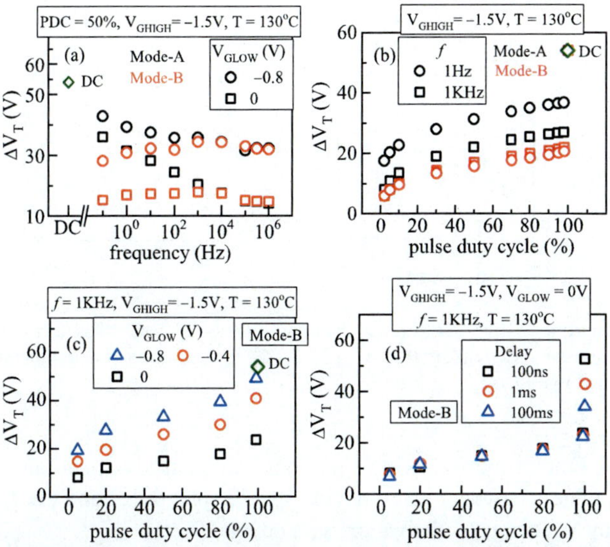

Fig. 1.16 Frequency and (b) PDC dependence of fixed-time ΔV_T for Mode-A and Mode-B stress. The PDC dependence of ΔV_T for Mode-B stress at different (c) V_{GLOW} and (d) measurement delay. All data are shown at $t_{STR} = 1$Ks. Data from [10, 38]

less sensitive at moderate PDC values, increases sharply and shows a "kink" or jump at large PDC values close to DC. Larger jump near DC is observed for Mode-B compared to Mode-A stress, while the difference in the jump between Mode-A and Mode-B is more prominent at lower f.

For a particular PDC, ΔV_T increases with increase in the magnitude of V_{GLOW} (for nonzero V_{GLOW}) as shown in Fig. 1.16 (c), while the "kink" near DC is lower for higher V_{GLOW} magnitude and vice versa. Interestingly, the measurement delay has no impact (for the range of values used here) on Mode-B AC ΔV_T across PDC (even for very large PDC), and only the DC value gets impacted as shown in Fig. 1.16 (d). This is as expected since the recovery following Mode-B AC stress is much slower than that for DC stress as shown in Fig. 1.14. The PDC and f dependencies will be analyzed later in this book (see Chap. 14).

1.4 Estimation of EOL Degradation

The conventional, empirical fit-based method for the determination of extrapolated degradation at end of life (EOL) under use condition ($V_G = V_{DD}$) is explained using Fig. 1.17 (a) and (b). The measured ΔV_T time kinetics at accelerated V_{GSTR} is fitted by a power-law time dependence and extrapolated to EOL, Fig. 1.17 (a). This exercise is repeated at different V_{GSTR}, and the extrapolated EOL ΔV_T is fitted using either an exponential or a power-law V_{GSTR} dependence and extrapolated to use condition,

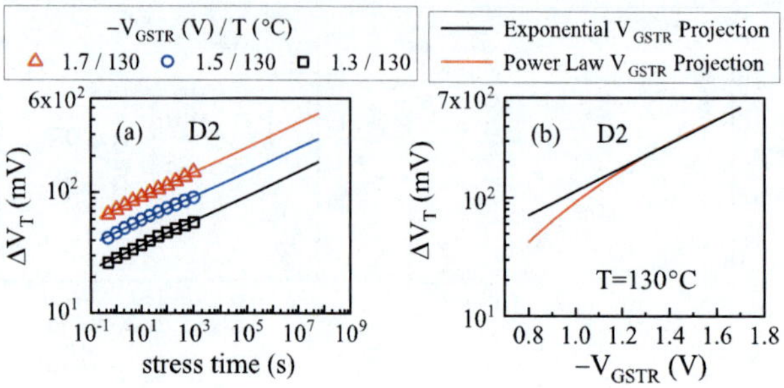

Fig. 1.17 Estimation of EOL ΔV_{T} at use condition, using regression-based (a) extrapolation of measured data in time to EOL at different V_{GSTR} and (b) extrapolation of extrapolated ΔV_{T} at EOL in gate bias to use V_{DD}. Data from [10]

Fig. 1.17 (b) . One of the following equations can be used depending on the choice of V_{GSTR} dependence, where A is a device-specific pre-factor and the T dependence is governed by Arrhenius T activation with energy E_{A}:

$$\Delta V_T = A * e^{\Gamma_V * V_{\mathrm{GSTR}}} * e^{-\left(E_A/kT\right)} * t^n \tag{1.1}$$

$$\Delta V_T = A * V_{GSTR}^{\Gamma} * e^{-\left(E_A/kT\right)} * t^n \tag{1.2}$$

The EOL degradation depends on the stress bias and time range used for fitting the measured data. This aspect is discussed in detail later in Chap. 6.

1.5 Summary

NBTI recovery necessitates the use of ultra-fast methods to characterize the time kinetics of device parametric shift without any recovery artifacts. Ultra-fast MSM is the method of choice and can be used for DC, Mode-A and Mode-B AC stress. The full sweep and OPDD versions are shown to be equivalent, and the latter is generally faster as only one sense V_{G} is used.

Measured ΔV_{T} kinetics shows power-law time dependence during stress, whose magnitude gets accelerated at higher V_{GSTR} and T. Several quantities are of interest, e.g., the ΔV_{T} magnitude, time slope n, VAF, E_{A}, T dependence of VAF and hence V_{GSTR} dependence of E_{A}. Lower ΔV_{T} is seen for AC compared to DC stress, and the AC-to-DC ratio depends on AC stress mode (A or B), PDC, f and V_{GLOW}. The ΔV_{T} recovery after stress proceeds over several decades in logarithmic timescale. The start of recovery after stoppage of stress and also the rate of recovery depend on DC or AC stress, AC stress mode, as well as the stress and recovery conditions (V_{GSTR}, T, t_{STR} and V_{GREC}).

The NBTI stress and recovery kinetics are strongly impacted by gate insulator process, channel material, device architecture and mechanical strain in the channel. These aspects are discussed next in Chap. 2.

Acknowledgements All experimental data presented in this chapter are re-plotted from previously published reports. The authors acknowledge Applied Materials for providing planar MOSFETs. Ankush Chaudhury is acknowledged for data in Fig. 1.8. Karansingh Thakor is acknowledged for help with manuscript preparation.

References

1. B.E. Deal, M. Sklar, A.S. Grove, E.H. Snow, J. Electrochem. Soc. **114**, 266 (1967)
2. N. Kimizuka, K. Yamaguchi, K. Imai, T. Iizuka, C.T. Liu, R.C. Keller, T. Horiuchi, in Symposium on VLSI Technology Digest of Technical Papers (2000), p. 92
3. V. Huard, M. Denais, F. Perrier, N. Revil, C. Parthasarathy, A. Bravaix, E. Vincent, Microelectron. Reliab. **45**, 83 (2005)
4. S. Mahapatra, K. Ahmed, D. Varghese, A.E. Islam, G. Gupta, L. Madhav, D. Saha, M.A. Alam, in IEEE International Reliability Physics Symposium Proceedings (2007), p. 1
5. Y. Mitani, H. Satake, A. Toriumi, IEEE Trans. Device Mater. Reliab. **8**, 6 (2008)
6. S. Pae, M. Agostinelli, M. Brazier, R. Chau, G. Dewey, T. Ghani, M. Hattendorf, J. Hicks, J. Kavalieros, K. Kuhn, M. Kuhn, J. Maiz, M. Metz, K. Mistry, C. Prasad, S. Ramey, A. Roskowski, J. Sandford, C. Thomas, J. Thomas, C. Wiegand, J. Wiedemer, in IEEE International Reliability Physics Symposium Proceedings (2008), p. 352
7. K. Joshi, S. Hung, S. Mukhopadhyay, V. Chaudhary, N. Nanaware, B. Rajamohanan, T. Sato, M. Bevan, A. Wei, A. Noori, B. McDougal, C. Ni, G. Saheli, C. Lazik, P. Liu, D. Chu, L. Date, S. Datta, A. Brand, J. Swenberg, S. Mahapatra, in IEEE International Reliability Physics Symposium Proceedings (2013), p. 4C.2.1
8. J. Franco, B. Kaczer, P.J. Roussel, J. Mitard, M. Cho, L. Witters, T. Grasser, G. Groeseneken, IEEE Trans. Electron Devices **60**, 396 (2013)
9. P. Srinivasan, J. Fronheiser, K. Akarvardar, A. Kerber, L.F. Edge, R.G. Southwick, E. Cartier, H. Kothari, in IEEE International Reliability Physics Symposium Proceedings (2014), p. 6A.3.1.
10. N. Parihar, N. Goel, S. Mukhopadhyay, S. Mahapatra, IEEE Trans. Electron Devices **65**, 392 (2018)
11. N. Parihar, S. Mahapatra, in IEEE International Reliability Physics Symposium Proceedings (2018), p. TX.5.1
12. V. Huard, C. Ndiaye, M. Arabi, N. Parihar, X. Federspiel, S. Mhira, S. Mahapatra, A. Bravaix, in IEEE International Reliability Physics Symposium Proceedings (2018), p. TX.4.1
13. N. Parihar, R. Tiwari, C. Ndiaye, M. Arabi, S. Mhira, H. Wong, S. Motzny, V. Moroz, V. Huard, S. Mahapatra, in International Conference on Simulation of Semiconductor Processes and Devices (2018), p. 167
14. S. Ramey, A. Ashutosh, C. Auth, J. Clifford, M. Hattendorf, J. Hicks, R. James, A. Rahman, V. Sharma, A.St. Amour, C. Wiegand, in IEEE International Reliability Physics Symposium Proceedings (2013), p. 4C.5.1
15. K.T. Lee, W. Kang, E.-A. Chung, G. Kim, H. Shim, H. Lee, H. Kim, M. Choe, N.-I. Lee, A. Patel, J. Park, J. Park, in IEEE International Reliability Physics Symposium Proceedings (2013), p. 2D.1.1
16. J. Franco, B. Kaczer, A. Chasin, H. Mertens, L.-A. Ragnarsson, R. Ritzenthaler, S. Mukhopadhyay, H. Arimura, P.J. Roussel, E. Bury, N. Horiguchi, D. Linten, G. Groeseneken, A. Thean, in IEEE International Reliability Physics Symposium Proceedings (2016), p. 4B.2.1

17. G. Jiao, M. Toledano-Luque, K.-J. Nam, N. Toshiro, S.-H. Lee, J.-S. Kim, T. Kauerauf, E. Chung, D. Bae, G. Bae, D.-W. Kim, K. Hwang, in IEEE International Electron Devices Meeting Technical Digest (2016), p. 31.2.1
18. N. Parihar, R.G. Southwick, U. Sharma, M. Wang, J.H. Stathis, S. Mahapatra, in IEEE International Reliability Physics Symposium Proceedings (2017), p. 2D.4.1
19. N. Parihar, R. Southwick, M. Wang, J.H. Stathis, S. Mahapatra, in IEEE International Electron Devices Meeting Technical Digest (2017), p. 7.3.1
20. A. Rahman, J. Dacuna, P. Nayak, G. Leatherman, S. Ramey, in IEEE International Reliability Physics Symposium Proceedings (2018), p. 6F.4.1.
21. N. Parihar, U. Sharma, R.G. Southwick, M. Wang, J.H. Stathis, S. Mahapatra, IEEE Trans. Electron. Devices **65**, 23 (2018)
22. N. Parihar, R.G. Southwick, M. Wang, J.H. Stathis, S. Mahapatra, IEEE Trans. Electron. Devices **65**, 1699 (2018)
23. N. Parihar, R.G. Southwick, M. Wang, J.H. Stathis, S. Mahapatra, IEEE Trans. Electron. Devices **65**, 1707 (2018)
24. N. Parihar, R. Tiwari, S. Mahapatra, in International Conference on Simulation of Semiconductor Processes and Devices (2018), p. 176
25. R. Tiwari, N. Parihar, K. Thakor, H.Y. Wong, S. Motzny, M. Choi, V. Moroz, S. Mahapatra, IEEE Trans. Electron. Devices **66**, 2086 (2019)
26. R. Tiwari, N. Parihar, K. Thakor, H.Y. Wong, S. Motzny, M. Choi, V. Moroz, S. Mahapatra, IEEE Trans. Electron. Devices **66**, 2093 (2019)
27. N. Parihar, U. Sharma, R.G. Southwick, M. Wang, J.H. Stathis, S. Mahapatra, in IEEE International Reliability Physics Symposium Proceedings (2019)
28. M. Wang, J. Zhang, H. Zhou, R.G. Southwick, R. Hsin, K. Chao, X. Miao, V.S. Basker, T. Yamashita, D. Guo, G. Karve, H. Bu, in IEEE International Reliability Physics Symposium Proceedings (2019)
29. N. Choudhury, U. Sharma, H. Zhou, R.G. Southwick, M. Wang, S. Mahapatra, in IEEE International Reliability Physics Symposium Proceedings (2020)
30. H. Zhou, M. Wang, J. Zhang, K. Watanabe, C. Durfee, S. Mochizuki, R. Bao, R. Southwick, M. Bhuiyan, B. Veeraraghavan, IEEE International Reliability Physics Symposium Proceedings (2020)
31. N. Choudhury, T. Samadder, R. Tiwari, H. Zhou, R.G. Southwick, M. Wang, S. Mahapatra, in IEEE International Reliability Physics Symposium Proceedings (2021)
32. G. Chen, M.F. Li, C.H. Ang, J.Z. Zheng, D.L. Kwong, IEEE Electron. Device Lett. **23**, 734 (2002)
33. S. Mahapatra, V. Huard, A. Kerber, V. Reddy, S. Kalpat, A. Haggag, in IEEE International Reliability Physics Symposium Proceedings (2014), p. 3B.1.1
34. N. Goel, P. Dubey, J. Kawa, S. Mahapatra, in IEEE International Reliability Physics Symposium Proceedings (2015), p. CA.5.1
35. S. Mishra, H. Amrouch, J. Joe, C.K. Dhabi, K. Thakor, Y.S. Chauhan, J. Henkel, S. Mahapatra, IEEE Trans. Electron. Devices **66**, 271 (2019)
36. A. Thirunavukkarasu, H. Amrouch, J. Joe, N. Goel, N. Parihar, S. Mishra, C.K. Dhabi, Y.S. Chauhan, J. Henkel, S. Mahapatra, IEEE Trans. Electron. Devices **66**, 316 (2019)
37. C. Shen, M.-F. Li, C.E. Foo, T. Yang, D.M. Huang, A. Yap, G.S. Samudra, Y.-C. Yeo, in IEEE International Electron Devices Meeting Technical Digest (2006), p. 333
38. N. Goel, N. Nanaware, S. Mahapatra, IEEE Electron. Device Lett. **34**, 1476 (2013)
39. S. Rangan, N. Mielke, E.C.C. Yeh, in IEEE International Electron Devices Meeting Technical Digest (2003), p. 14.3.1
40. E.N. Kumar, V.D. Maheta, S. Purawat, A.E. Islam, C. Olsen, K. Ahmed, M.A. Alam, S. Mahapatra, in IEEE International Electron Devices Meeting Technical Digest (2007), p. 809
41. T. Yang, M.F. Li, C. Shen, C.H. Ang, C. Zhu, Y.C. Yeo, G. Samudra, S.C. Rustagi, M.B. Yu, D.L. Kwong, in Symposium on VLSI Technology Digest of Technical Papers (2005), p. 92
42. S. Mahapatra, N. Goel, A. Chaudhary, K. Joshi, S. Mukhopadhyay, in Fundamentals of Bias Temperature Instability in MOS Transistors (Springer India, 2015), pp. 43–92

43. B. Kaczer, V. Arkbipov, R. Degraeve, N. Collaert, G. Groeseneken, M. Goodwin, in IEEE International Reliability Physics Symposium Proceedings (2005), p. 381
44. A.E. Islam, V.D. Maheta, H. Das, S. Mahapatra, M.A. Alam, in IEEE International Reliability Physics Symposium Proceedings (2008), p. 87
45. B. Kaczer, T. Grasser, Ph.J. Roussel, J. Martin Martinez, R. O'Connor, B. O'Sullivan, G. Groeseneken, in IEEE International Reliability Physics Symposium Proceedings (2008), p. 20
46. H. Reisinger, O. Blank, W. Heinrigs, A. Muhlhoff, W. Gustin, C. Schlunder, in IEEE International Reliability Physics Symposium Proceedings (2006), p. 448
47. R. Fernandez, B. Kaczer, A. Nackaerts, S. Demuynck, R. Rodrigues, M. Nafaria, G. Groeseneken, in International Electron Devices Meeting Technical Digest (2006)
48. V. Huard, R. Chevallier, C. Parthasarathy, A. Mishra, N. R-Amador, F. Persin, V. Robert, A. Chimeno, E. Pion, N. Planes, D. Ney, F. Cacho, N. Kapoor, V. Kulshrestha, S. Chopra, N. Vialle, in IEEE International Reliability Physics Symposium Proceedings (2010), p. 5E.3.1
49. L. Zhou, Q. Zhang, H. Yang, Z. Ji, Z. Zhang, Q. Liu, H. Xu, B. Tang, E. Simoen, X. Ma, X. Wang, Y. Li, H. Yin, J. Luo, C. Zhao, W. Wang, IEEE Electron. Device Lett. **41**, 965 (2020)

Chapter 2
Device Architecture, Material and Process Dependencies of NBTI Parametric Drift

Souvik Mahapatra and Narendra Parihar

2.1 Introduction

Negative Bias Temperature Instability (NBTI) induced parametric drift is a crucial reliability issue for p-MOSFETs. As discussed in Chap. 1, NBTI became important with the migration from Silicon Dioxide (SiO_2) to Silicon Oxynitride (SiON) gate insulator devices [1–4]. It continues to remain a concern for High-K Metal Gate (HKMG) devices featuring dual layer (SiO_2 or SiON Interlayer (IL) and Hafnium Dioxide (HfO_2) High-K layer) gate insulator stack, e.g., planar bulk MOSFETs [5–10], Fully Depleted Silicon On Insulator (FDSOI) MOSFETs [11, 12], bulk and SOI FinFETs [12–26], as well as Gate All Around Stack Nano sheet (GAA-SNS) FETs [27–30], with either Silicon (Si) or Silicon Germanium (SiGe) channel. The schematic of different device architectures is illustrated in Fig. 2.1.

As discussed in Chap. 1, Sect. 1.3, NBTI results in the accumulation of positive charges in the gate insulator of a p-MOSFET, when the gate bias (V_G) is held at a negative value w.r.t the other terminals of the device. This causes a shift in device parameters, e.g., threshold voltage shift (ΔV_T), over time. ΔV_T accelerates at higher magnitude of V_G during stress ($V_G = V_{GSTR}$) and higher temperature (T), governed by the Voltage Acceleration Factor (VAF) and Arrhenius T activation energy (E_A) respectively. The parametric shift accrued during stress partially recovers after stress when the magnitude of V_G is reduced or removed ($V_G = V_{GREC}$ or 0V), and therefore, AC stress results in lower ΔV_T than DC. The AC to DC ratio depends on the Pulse Duty Cycle (PDC), however, it may or may not depend on frequency (f) of the gate pulse (depends on AC stress mode). However, NBTI recovery necessitates the use of ultra-fast methods for error-free characterization, which is discussed in Chap. 1,

S. Mahapatra (✉) · N. Parihar
Department of Electrical Engineering, Indian Institute of Technology Bombay, Powai, Mumbai 400076, India
e-mail: souvik@ee.iitb.ac.in

S. Mahapatra (ed.), *Recent Advances in PMOS Negative Bias Temperature Instability*,
https://doi.org/10.1007/978-981-16-6120-4_2

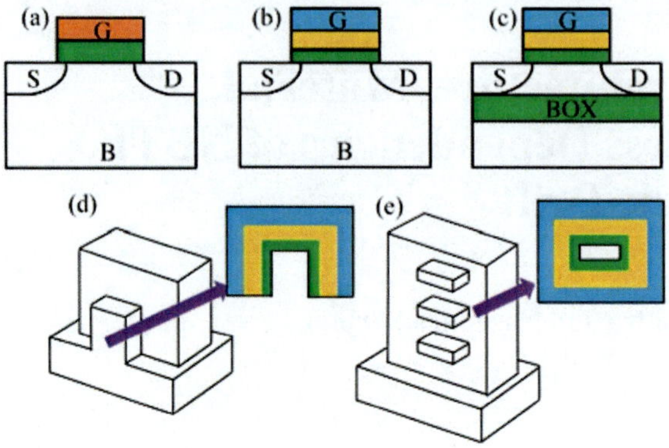

Fig. 2.1 Schematic of different device architectures: (a), (b) planar MOSFETs with (a) single layer SiO_2 or SiON and (b) dual layer HKMG gate stacks. Also shown are different device architectures having HKMG gate stack: (c) FDSOI, (d) FinFET, and (e) GAA-SNS FET. The gate insulator stacks are separately illustrated for FinFET and GAA-SNS FET. Color coding of the gate insulator stack: Green (SiO_2 or SiON) and Orange (Poly-Si gate) in (a), Green (SiO_2 or SiON IL), Yellow (High-K), and Blue (Metal Gate) in (b)–(e). Reproduced from Chap. 1.1

Sect. 1.2. The One Point Drop Down (OPDD) version of the ultra-fast Measure-Stress-Measure (MSM) method is used in this chapter unless specifically mentioned otherwise.

The process dependence for the conventional SiO_2 and SiON gate insulator devices has been discussed in [31]. In summary, NBTI improves (lower ΔV_T) when dry oxidation is used instead of wet oxidation for growing the gate insulator [1, 2], Fluorine (F) is incorporated, either directly into the gate insulator or indirectly by diffusion from implanted junctions into the gate insulator [2, 32], and Deuterium (D_2) is used instead of Hydrogen (H_2) for the Post Metallization Anneal [1, 2]. However, NBTI worsens (higher ΔV_T) with the introduction of Nitrogen (N) in the gate insulator [1–4]. The exact method of N incorporation plays an important role, as it is the N density near the channel/gate insulator interface (not the overall N content in the gate stack) that controls NBTI [4, 33–36]. The Remote Plasma Nitridation (RPN) process shows lowest NBTI amongst different methods when compared at identical N dose, since it places the majority of N near the gate insulator/poly-Si gate interface, away from the channel. The Post Nitridation Anneal (PNA) plays an important role to cure the plasma damage created during the RPN process [37, 38]. In addition, several back ends of line processes [39] and plasma damage [40] also affect NBTI.

For dual layer HKMG stack, NBTI improves if the IL is thermally treated, but it becomes worse with the introduction of N especially into the IL, which is shown for Si channel bulk planar MOSFETs [6, 9]. Similar impact of N is also reported in HKMG based SiGe channel FDSOI MOSFETs [11, 12] and FinFETs [12, 17, 18, 21, 22]. The impact of N% on the time kinetics of NBTI is analyzed and modeled in detail in Chap. 7 (Si channel bulk MOSFETs), Chap. 9 (SiGe channel FDSOI MOSFETs), and Chap. 11 (SiGe channel FinFETs).

NBTI improves when SiGe is used as the channel material instead of Si, which is universally demonstrated for HKMG bulk [7, 8] and FDSOI [11, 12] planar MOSFETs and FinFETs [12, 15–18, 21, 22, 24]. The impact of Ge% on NBTI time kinetics is analyzed and modeled in detail in Chap. 8 (Si-capped SiGe MOSFETs), Chap. 9 (SiGe channel FDSOI MOSFETs), and Chap. 11 (SiGe channel FinFETs).

NBTI improves when the spacing between STI (Shallow Trench Isolation) to the channel is increased, which is shown for FDSOI MOSFETs [11, 12]. NBTI improves with the scaling of fin length and width in FinFETs and scaling of sheet length in GAA-SNS FETs, but it worsens with the scaling of sheet width in GAA-SNS FETs [23, 25, 29, 30]. These aspects are analyzed and modeled in detail in Chaps. 9, 12, and 13 in devices having different architectures.

A brief summary of key process impact is discussed below, only DC stress and recovery data are shown. As mentioned above, these aspects will be further analyzed in detail in later chapters of this book.

2.2 Incorporation of Nitrogen

Figure 2.2 shows the time evolution of ΔV_T measured (a, b) during and (c, d) after stress in (a, c) Gate First (GF) HKMG Si planar p-MOSFETs and (b, d) Replacement Metal Gate (RMG) HKMG SiGe p-FinFETs having different Nitrogen content (N%) in the gate stack (the comparison of Si versus SiGe channel is shown in the next section). Data are measured by ultra-fast MSM (a, b) full sweep and (c, d) OPDD methods. In each panel, the stress (or recovery) V_G and T conditions are kept

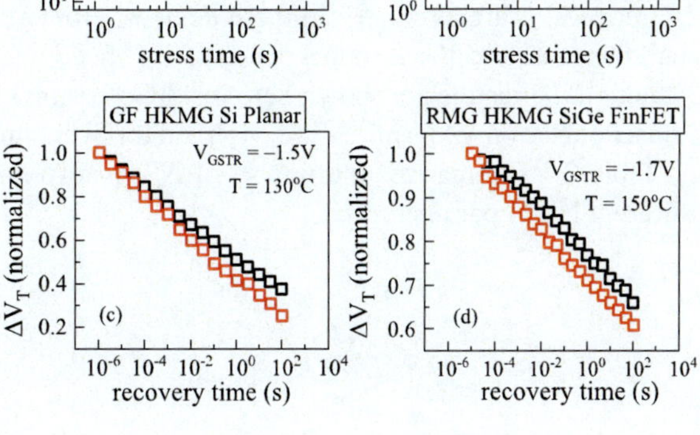

Fig. 2.2 Time evolution of ΔV_T (a, b) during and (c, d) after stress in (a, c) GF HKMG Si p-MOSFETs and (c, d) RMG HKMG SiGe p-FinFETs with different N% in the gate stack. Recovery is normalized to end of stress. Data from [9, 21]

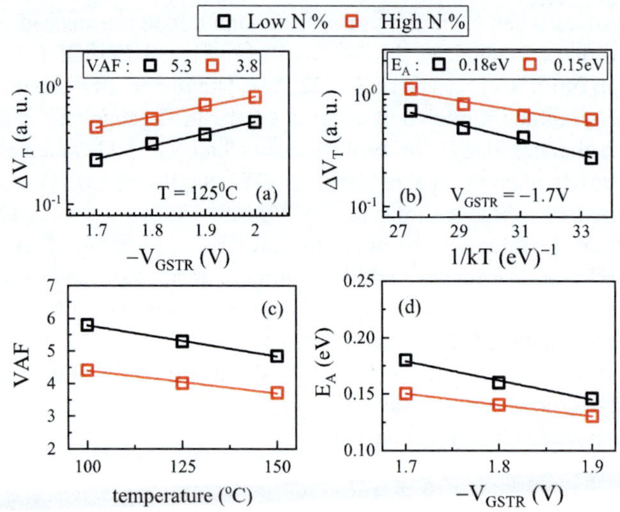

Fig. 2.3 Fixed-time ($t_{STR} = 1$Ks) ΔV_T as a function of (a) V_{GSTR} and (b) T, and the corresponding, (c) T dependence of VAF and (d) V_{GSTR} dependence of E_A, obtained during stress in RMG HKMG SiGe p-FinFETs having different N% in the gate stack. Data from [21]

identical between devices having different N%. Note that power-law time dependence is observed during stress. The magnitude of ΔV_T increases but the time slope n reduces during stress for devices having higher N%. Moreover, recovery becomes faster and the Fraction Remaining (FR) reduces (FR is defined as the ratio of ΔV_T at $t = t_{REC}$ during recovery to that at $t = t_{STR}$ at the end of stress) for higher N% devices. As shown, the above features are universally observed irrespective of the channel material and gate insulator process.

Figure 2.3 shows ΔV_T measured at a fixed t_{STR} during stress as a function of (a) V_{GSTR} (at fixed T) and (b) T (at fixed V_{GSTR}) in SiGe FinFETs with different N% in the gate stack. The magnitude of ΔV_T (at fixed V_{GSTR} and T) increases, however, the power-law VAF (at fixed T) and Arrhenius E_A (at fixed V_{GSTR}) reduce with higher N%. The T dependence of VAF and V_{GSTR} dependence of E_A at different N% are shown in Fig. 2.3(c) and (d) respectively. Note, the VAF reduces at higher T and E_A reduces at higher V_{GSTR} in these devices. However, the T dependence of VAF (and therefore the V_{GSTR} dependence of E_A) becomes less sensitive at higher N%. Although the results are shown here for SiGe channel FinFETs, similar data on the impact of N% on VAF and E_A are observed for Si channel devices (e.g., see Chap. 7).

Table 2.1 summarizes the impact of N% incorporation in the gate insulator on different NBTI parameters.

Table 2.1 Impact of variation in Ge% in the channel and N% in the gate stack on different NBTI parameters

	Process	
Parameters	Increase in Ge%	Increase in N%
ΔV_T magnitude	Lower	Higher
Long-time slope (n)	Higher	Lower
VAF	Higher	Lower
Tactivation E_A	Higher	Lower
T dependence of VAF	Higher	Lower
Recovery	Lower	Higher

2.3 Si Versus SiGe Channel

Figure 2.4 shows the time evolution of ΔV_T measured in Si and SiGe (different Ge%) p-channel (a) GF HKMG Si-capped SiGe planar, (b) GF HKMG FDSOI as well as (c) RMG HKMG FinFET devices during stress, and measured using (a) e-MSM, (b) full sweep MSM and (c) OPDD MSM methods. The ΔV_T recovery kinetics is shown only for FinFETs in Fig. 2.4(d). In each panel, the stress (or recovery) V_G and T conditions are kept identical between devices having different Ge%. Note that power-law time dependence is observed during stress. The ΔV_T magnitude reduces but the time slope n increases with higher Ge% in the channel, and these features are

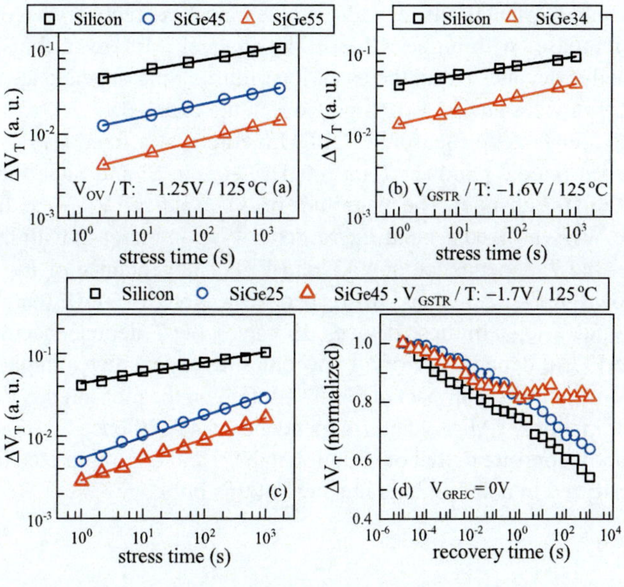

Fig. 2.4 Time evolution of ΔV_T (a–c) during and (d) after stress in p-channel (a) Si-capped SiGe planar MOSFET [7], (b) FDSOI MOSFET [12] and (c, d) FinFET [21] devices having different Ge% in the channel. Recovery is normalized to end of stress

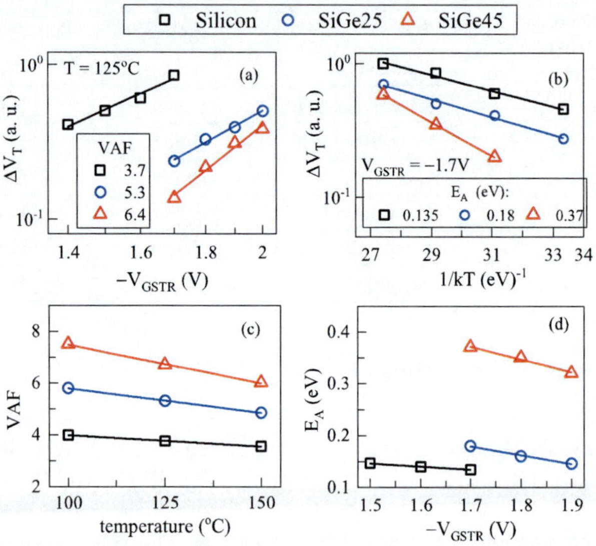

Fig. 2.5 Fixed-time ($t_{STR} = 1$Ks) ΔV_T as a function of (a) V_{GSTR} and (b) T, and the corresponding (c) T dependence of VAF and (d) V_{GSTR} dependence of E_A, obtained during stress in RMG HKMG SiGe p-FinFETs having different Ge% in the channel. Data from [21]

universally observed in planar bulk and FDSOI MOSFETs with (100) and FinFET with (110) channel orientations. In addition, the long-time recovery becomes slower and the FR increases with higher Ge% in the channel. Moreover, the shape of the recovery kinetics deviates (from the usual logarithmic time dependence observed in Si devices) for devices having very high Ge% in the channel.

Figure 2.5 shows ΔV_T measured (OPDD method) at a fixed t_{STR} as a function of (a) V_{GSTR} (at fixed T) and (b) T (at fixed V_{GSTR}) in Si and SiGe FinFETs with different Ge% in the channel. The magnitude of ΔV_T (at fixed V_{GSTR} and T) reduces, however, the VAF (at fixed T) and E_A (at fixed V_{GSTR}) increase with higher Ge%. Furthermore, the T dependence of VAF and V_{GSTR} dependence of E_A at different Ge% are shown in Fig. 2.5(c) and (d) respectively. Note, the VAF (and E_A) reduce at higher T (and V_{GSTR}) in these devices. However, the T dependence of VAF (and therefore the V_{GSTR} dependence of E_A) becomes more sensitive at higher Ge%.

It is noteworthy that the impact of Ge% variation on the measured ΔV_T stress and recovery time kinetics, VAF, E_A, and T dependence of VAF (hence V_{GSTR} dependence of E_A) is exactly opposite to that of variation in N%, and is summarized in Table 2.1. These are analyzed in detail in later chapters of this book.

2.4 Impact of Mechanical Strain

The fin dimensions (length, width, and height) of a FinFET are usually not varied purposefully for a particular class (e.g., core or Input–Output (I/O)) of device, although they can vary due to process-induced variability, which is discussed in the next section. However, purposeful and large variations in fin dimensions are done for different classes of devices in a particular technology, or when the technology is scaled.

Figure 2.6 shows ΔV_T measured at fixed t_{STR} as a function of V_{GSTR} in RMG HKMG SiGe channel p-FinFETs having different fin (a) length and (b) width. The magnitude of ΔV_T reduces while VAF increases with reduction in fin dimensions, for both fin length and width scaling. Such results are obtained for Si channel SOI FinFETs (length scaling) as well. The impact of fin dimension scaling on NBTI is explained by changes in the channel strain due to compressive mechanical stress in Chap. 13.

Figure 2.7 shows ΔV_T measured at fixed t_{STR} as a function of V_{GSTR} in RMG HKMG Si channel GAA-SNS p-FETs having different sheet (a) length and (b) width. Note that contrary to the FinFET data shown above, the magnitude of ΔV_T and VAF reduce with reduction in sheet length, while ΔV_T increases but VAF reduces with the reduction in sheet width. The impact of sheet dimension scaling on NBTI is also explained by changes in channel strain due to compressive mechanical stress in Chaps. 12 and 13. It is important to remark that the mechanical strain impact is different between (110) and (100) surfaces that are relevant respectively for FinFETs and GAA-SNS FETs.

The spacing (SA) between the transistor channel region and STI edge, illustrated using Fig. 2.8(a) for a planar MOSFET, is an important design parameter. Due to strain relaxation near the STI edge [11], devices having lower SA experience lower

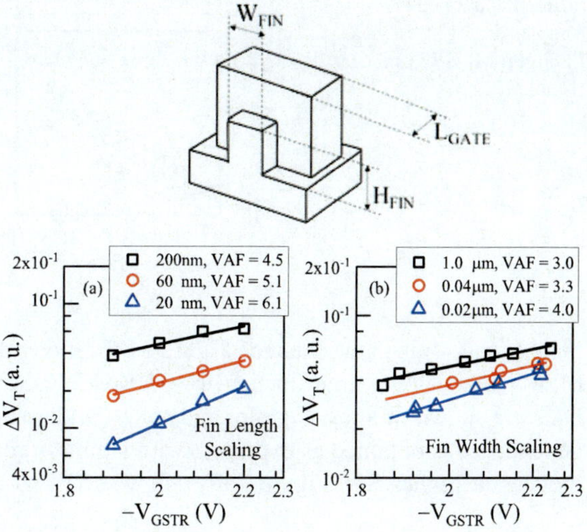

Fig. 2.6 Fixed-time (t_{STR} = 1Ks) ΔV_T as a function of V_{GSTR} during stress in SiGe channel p-FinFETs having different fin (a) length and (b) width. The VAF values are from power-law fits of measured data. The FinFET schematic is illustrated at the top. Data from [25]

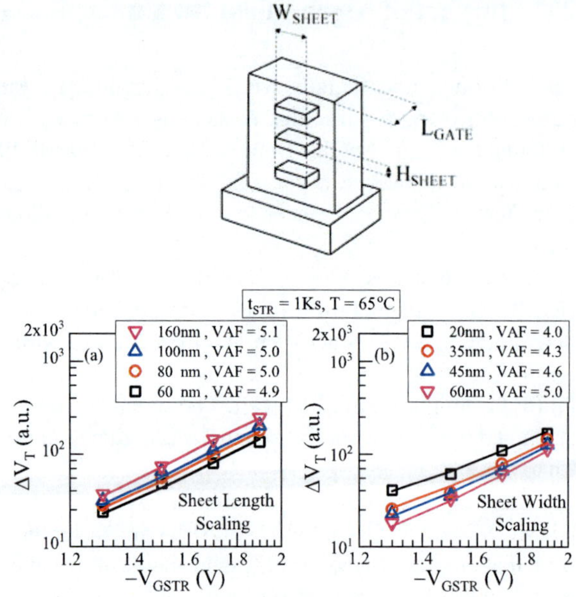

Fig. 2.7 Fixed-time ($t_{\mathrm{STR}} =$ 1Ks) ΔV_{T} as a function of V_{GSTR} during stress in Si channel GAA-SNS p-FETs having different sheet (a) length and (b) width. The VAF values are from power-law fits of measured data. The GAA-SNS FET schematic is illustrated at the top. Data from [30]

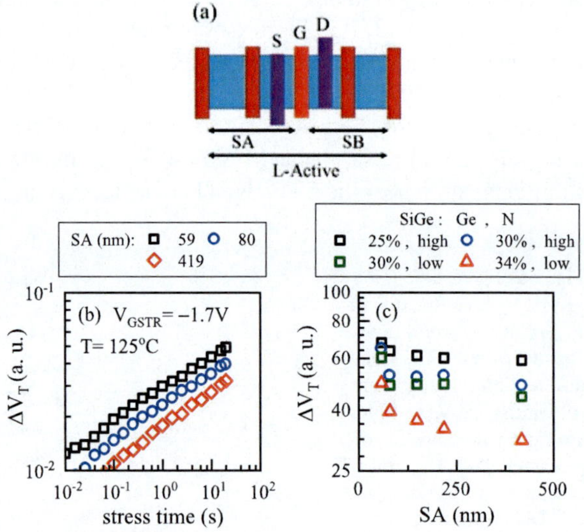

Fig. 2.8 (a) Schematic to illustrate the spacing (SA) between STI edge and channel (S, G, and D represent source, gate, and drain, L-active is total length of active). Impact of SA variation on (b) time kinetics of ΔV_{T} and (c) fixed-time ($t_{\mathrm{STR}} = 1$Ks) ΔV_{T} for different Ge% and N% samples during stress in FDSOI p-MOSFETs. Data from [11, 12]

mechanical strain in the channel. Figure 2.8(b) shows the measured (using full sweep MSM method) ΔV_{T} time kinetics in GF HKMG FDSOI devices having different SA. Data are shown in a log–log plot and over a wide range of t_{STR}. A power-law time dependence is obtained as expected, with similar n for different SA, although ΔV_{T} reduces at higher SA. Figure 2.8(c) shows ΔV_{T} at a fixed t_{STR} as a function of

SA in devices having different Ge% and N%, the V_{GSTR} and T are kept identical across different devices. Note that ΔV_T reduces with an increase in SA for all (Ge%, N%) devices, although the usual impact of increased Ge% (reduction in ΔV_T) and increased N% (increase in ΔV_T) is observed for all SA devices. This is discussed in more detail in Chap. 9.

2.5 Variability in Small Area Devices

The as-fabricated transistor performance has device-to-device variability in small area devices. This can be classified as variability across the wafer because of processing variation [41], and variability due to stochastic fluctuations of different entities such as doping, metal gate work function (for HKMG gate stack), and device dimensions [41–47]. Furthermore, the stochastic nature of charge buildup in the gate insulator during NBTI stress results in variation in device performance degradation across different small area devices [41–46, 48–54], which, however, gets "averaged out" when measurements are done in large area devices (e.g., data shown in the earlier sections of this chapter). The process and NBTI variability are briefly reviewed in this section.

Figure 2.9 shows the distribution of pre-stress V_T (V_{T0}) due to process variability and stochastic effects in a production quality 14 nm node FinFET technology [47]. Data from over 30 million FinFETs can be modeled by a Normal distribution as shown, whose variance increases with reduction in the area of the devices (it is inversely proportional to the square of area, not explicitly shown). Similar results are also reported from planar [41, 42], and FDSOI [44] MOSFETs.

Figure 2.10 shows the post-stress ΔV_T distribution measured in over 3100 FinFETs from a 22 nm production quality technology, (a) at different t_{STR} but fixed fin dimension (designed), and (b) for different designed fin dimensions but at fixed t_{STR} [51]. In Fig. 2.10(b), V_{GSTR} and T are kept identical across different devices. Note that both the mean and variance of the distribution increase at larger t_{STR}, see Fig. 2.10(a), and for a particular t_{STR}, larger variance is observed for devices with

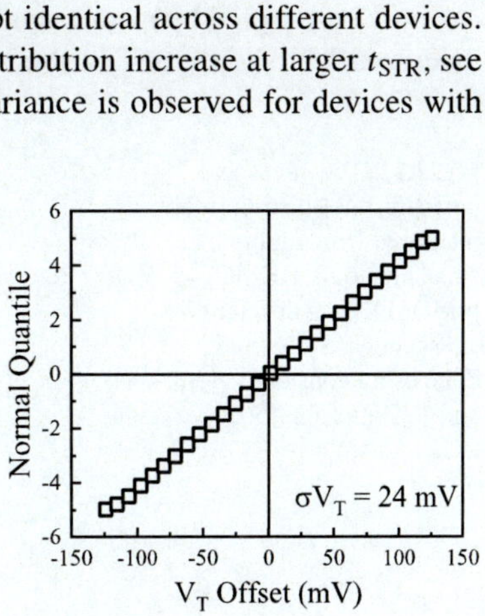

Fig. 2.9 Pre-stress V_{T0} distribution measured in 30 million few-fin p-FinFETs from the 14 nm technology node. Data from [47]

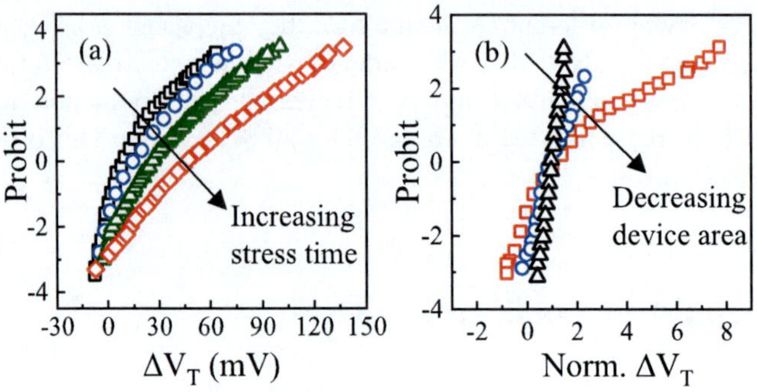

Fig. 2.10 Post-stress ΔV_T distribution measured in over 3100 few-fin p-FinFETs from the 22 nm technology node (a) at different stress times and (b) at fixed t_{STR} but from devices with different numbers of fins. Data from [51]

smaller fin area (because of smaller number of fins), see Fig. 2.10(b). Similar data are also shown in other reports [43–45, 54].

The mean of ΔV_T extracted from multiple small areas FinFETs remains identical across FinFETs having different numbers of fins, see Fig. 2.10(b), which is also reported elsewhere [52, 54]. In particular, the stress and recovery time kinetics of the mean of ΔV_T from multiple few-fin FinFETs remain identical to the ΔV_T time kinetics of the corresponding multi-fin device, when the designed fin dimensions are kept fixed [54]. This enables modeling the mean of ΔV_T from multiple few-fin FinFETs and ΔV_T from multi-fin FinFETs using the same framework, which will be demonstrated in Chap. 10 [54].

However, the shape of the post-stress ΔV_T distribution remains debated, since it has been modeled by different distributions: Dispersive Skellam [48], Exponential Poisson [49, 51], Normal [43, 45, 50] and Gamma [54–56]. Note, the choice of ΔV_T distribution would impact the projected ΔV_T at higher percentiles, and this aspect can only be clarified by modeling a statistically large measured dataset.

Figure 2.11 shows the distributions of pre-stress and post-stress V_T at different

Fig. 2.11 Pre-stress and post-stress V_T distributions measured from multiple small area bulk HKMG p-MOSFETs at different stress times. Stressing is done using constant V_{GSTR} and T. Data from [50]

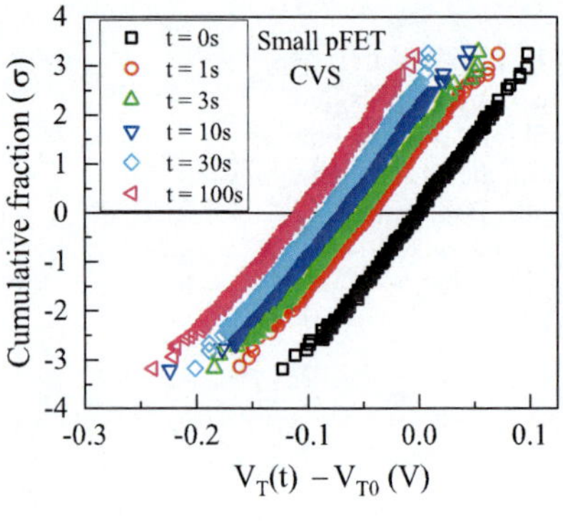

t_{STR}, measured from multiple small area bulk MOSFETs stressed under constant V_{GSTR} and T [50]. The pre-stress and post-stress V_T show Normal distribution and remain parallel to each other. This implies that although the mean V_T shifts (negative V_T shift due to positive gate insulator charges) at higher t_{STR}, no change is observed in the variance of the distribution, which is also reported elsewhere [54].

The variance of V_T distribution does not change after stress (for the t_{STR} range studied in this experiment) due to the dominance of the time-zero variability over NBTI variation. It is important to remark that under constant overdrive stress ($V_{OV} = V_{GSTR} - V_{T0}$), the post-stress V_T distribution stays Normal but both the mean and variance shift at higher t_{STR} [49, 50]. The time-zero variability is "taken care of" when V_{GSTR} is varied according to V_{T0} to maintain identical V_{OV}. However, this is not a relevant stress condition, as devices in a circuit are exposed to constant V_G and not constant V_{OV} during operation. Therefore, from a practical standpoint, the exact nature of ΔV_T distribution is less important, as the post-stress V_T distribution can be modeled by only mean shift, with identical variance as the pre-stress device. However, in some cases, the variance associated with stochastic NBTI can become larger than that of process variability, especially so in devices having low process variability and stressed over very large t_{STR}. In this situation, the exact nature of ΔV_T distribution would become important, and the variance of post-stress V_T distribution would also change for constant V_{GSTR} stress.

Figure 2.12 shows the fixed time measured ΔV_T (using OPDD method) as a function of pre-stress V_T from multiple SOI FinFETs (2 fins per device). All devices are stressed at identical V_{GSTR} and T. No meaningful correlation is observed, which is consistent with most other reports. Note that no correlation between the process variability related V_{T0} distribution and NBTI induced post-stress ΔV_T distribution is reported in [42, 44, 46, 53], although a weak correlation has been reported in [45]. In the absence of a correlation, the pre-stress V_{T0} and post-stress ΔV_T distributions can be presumed as independent Poisson processes, and during modeling, they can be added (randomly) to obtain the post-stress V_T distribution as shown in [55, 56].

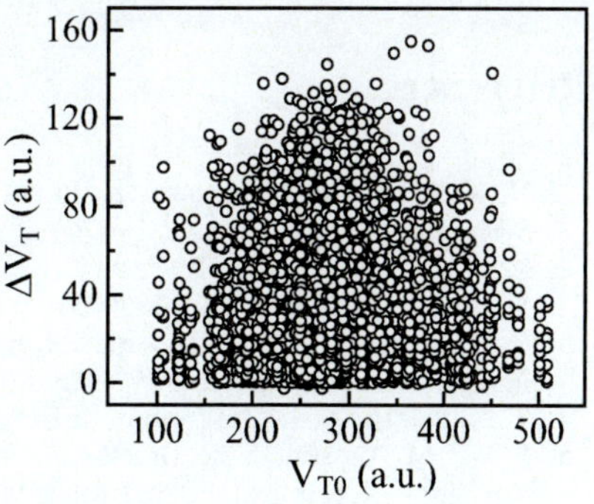

Fig. 2.12 Post-stress ΔV_T versus pre-stress V_T from multiple RMG HKMG SOI p-FinFETs, showing the absence of any meaningful correlation. Identical V_{GSTR} and T are used across different devices. Data from [54]

2.6 Summary

The ultra-fast measured ΔV_T kinetics shows power-law time dependence during NBTI stress, whose magnitude gets accelerated at higher V_{GSTR} and T. The magnitude of ΔV_T, as well as its time slope n, VAF, E_A, T dependence of VAF and hence V_{GSTR} dependence of E_A, depend on the gate insulator quality (N%), channel material (Si or SiGe), device architecture, device dimension, and layout. The ΔV_T recovery after stress proceeds over several decades in logarithmic timescale. The start of recovery after stoppage of stress and the rate of recovery also depends on N% and Ge%.

NBTI in small area devices shows variation in measured ΔV_T, the mean and variance of ΔV_T distribution increase at higher t_{STR}. The mean of multiple small area devices and large area devices show identical stress-recovery kinetics. No correlation is observed between time-zero variation in V_{T0} and NBTI variability in ΔV_T, and they can be assumed as independent Poisson processes. However, due to the dominance of time-zero variation (related to process variation) over NBTI variability, only the mean of V_T distribution shifts at higher t_{STR} during stress, while no change is observed in variance. Therefore, in most practical situations, the modeling of NBTI variability can be handled by modeling the mean ΔV_T.

The NBTI physical mechanism and framework needed to model the measured data are described in Chap. 3 through Chap. 6.

Acknowledgements All experimental data presented in this chapter are re-plotted from previously published reports. The authors acknowledge Applied Materials for providing planar MOSFETs, IBM for providing measurement facilities, FinFETs and GAA-SNS FETs, IMEC and ST Microelectronics respectively for providing data from Si-capped SiGe and FDSOI MOSFETs. Nilesh Goel and Nilotpal Choudhury are acknowledged for data in Figs. 2.2 and 2.7. One of the authors (Narendra Parihar) acknowledges Richard Southwick, Miaomiao Wang and James Stathis for help with measurement facilities. Karansingh Thakor is acknowledged for help with manuscript preparation.

References

1. N. Kimizuka, K. Yamaguchi, K. Imai, T. Iizuka, C. T. Liu, R. C. Keller, T. Horiuchi, in *Symposium on VLSI Technology Digest of Technical Papers*, 92 (2000)
2. V. Huard, M. Denais, F. Perrier, N. Revil, C. Parthasarathy, A. Bravaix, E. Vincent, Microelectron. Reliab. **45**, 83 (2005)
3. S. Mahapatra, K. Ahmed, D. Varghese, A.E. Islam, G. Gupta, L. Madhav, D. Saha, M.A. Alam, in *IEEE International Reliability Physics Symposium Proceedings*, 1 (2007)
4. Y. Mitani, H. Satake, A. Toriumi, IEEE Trans. Dev. Mater. Reliab. **8**, 6 (2008)
5. S. Pae, M. Agostinelli, M. Brazier, R. Chau, G. Dewey, T. Ghani, M. Hattendorf, J. Hicks, J. Kavalieros, K. Kuhn, M. Kuhn, J. Maiz, M. Metz, K. Mistry, C. Prasad, S. Ramey, A. Roskowski, J. Sandford, C. Thomas, J. Thomas, C. Wiegand, J. Wiedemer, in *IEEE International Reliability Physics Symposium Proceedings*, 352 (2008)
6. K. Joshi, S. Hung, S. Mukhopadhyay, V. Chaudhary, N. Nanaware, B. Rajamohanan, T. Sato, M. Bevan, A. Wei, A. Noori, B. McDougal, C. Ni, G. Saheli, C. Lazik, P. Liu, D. Chu, L.

Date, S. Datta, A. Brand, J. Swenberg, S. Mahapatra, in *IEEE International Reliability Physics Symposium Proceedings*, 4C.2.1 (2013)

7. J. Franco, B. Kaczer, P.J. Roussel, J. Mitard, M. Cho, L. Witters, T. Grasser, G. Groeseneken, IEEE Trans. Electron Devices **60**, 396 (2013)
8. P. Srinivasan, J. Fronheiser, K. Akarvardar, A. Kerber, L.F. Edge, R.G. Southwick, E. Cartier, H. Kothari, in *IEEE International Reliability Physics Symposium Proceedings*, 6A.3.1 (2014)
9. N. Parihar, N. Goel, S. Mukhopadhyay, S. Mahapatra, IEEE Trans. Electron Devices **65**, 392 (2018)
10. N. Parihar, S. Mahapatra, in *IEEE International Reliability Physics Symposium Proceedings*, TX.5.1 (2018)
11. V. Huard, C. Ndiaye, M. Arabi, N. Parihar, X. Federspiel, S. Mhira, S. Mahapatra, A. Bravaix, in *IEEE International Reliability Physics Symposium Proceedings*, TX.4.1 (2018)
12. N. Parihar, R. Tiwari, C. Ndiaye, M. Arabi, S. Mhira, H. Wong, S. Motzny, V. Moroz, V. Huard, S. Mahapatra, in *International Conference on Simulation of Semiconductor Processes and Devices*, 167 (2018)
13. S. Ramey, A. Ashutosh, C. Auth, J. Clifford, M. Hattendorf, J. Hicks, R. James, A. Rahman, V. Sharma, A.St. Amour, C. Wiegand, in *IEEE International Reliability Physics Symposium Proceedings*, 4C.5.1 (2013)
14. K.T. Lee, W. Kang, E.-A. Chung, G. Kim, H. Shim, H. Lee, H. Kim, M. Choe, N.-I. Lee, A. Patel, J. Park, J. Park, in *IEEE International Reliability Physics Symposium Proceedings*, 2D.1.1 (2013)
15. J. Franco, B. Kaczer, A. Chasin, H. Mertens, L.-A. Ragnarsson, R. Ritzenthaler, S. Mukhopadhyay, H. Arimura, P. J. Roussel, E. Bury, N. Horiguchi, D. Linten, G. Groeseneken, A. Thean, in *IEEE International Reliability Physics Symposium Proceedings*, 4B.2.1 (2016)
16. G. Jiao, M. Toledano-Luque, K.-J. Nam, N. Toshiro, S.-H. Lee, J.-S. Kim, T. Kauerauf, E. Chung, D. Bae, G. Bae, D.-W. Kim, K. Hwang, in *IEEE International Electron Devices Meeting Technical Digest*, 31.2.1 (2016)
17. N. Parihar, R.G. Southwick, U. Sharma, M. Wang, J.H. Stathis, S. Mahapatra, in *IEEE International Reliability Physics Symposium Proceedings*, 2D.4.1 (2017)
18. N. Parihar, R. Southwick, M. Wang, J.H. Stathis, S. Mahapatra, in *IEEE International Electron Devices Meeting Technical Digest*, 7.3.1 (2017)
19. A. Rahman, J. Dacuna, P. Nayak, G. Leatherman, S. Ramey, in *IEEE International Reliability Physics Symposium Proceedings*, 6F.4.1 (2018)
20. N. Parihar, U. Sharma, R.G. Southwick, M. Wang, J.H. Stathis, S. Mahapatra, IEEE Trans. Electron Devices **65**, 23 (2018)
21. N. Parihar, R.G. Southwick, M. Wang, J.H. Stathis, S. Mahapatra, IEEE Trans. Electron Devices **65**, 1699 (2018)
22. N. Parihar, R.G. Southwick, M. Wang, J.H. Stathis, S. Mahapatra, IEEE Trans. Electron Devices **65**, 1707 (2018)
23. N. Parihar, R. Tiwari, S. Mahapatra, in *International Conference on Simulation of Semiconductor Processes and Devices*, 176 (2018)
24. R. Tiwari, N. Parihar, K. Thakor, H.Y. Wong, S. Motzny, M. Choi, V. Moroz, S. Mahapatra, IEEE Trans. Electron Devices **66**, 2086 (2019)
25. R. Tiwari, N. Parihar, K. Thakor, H.Y. Wong, S. Motzny, M. Choi, V. Moroz, S. Mahapatra, IEEE Trans. Electron Devices **66**, 2093 (2019)
26. N. Parihar, U. Sharma, R.G. Southwick, M. Wang, J.H. Stathis, S. Mahapatra, in *IEEE International Reliability Physics Symposium Proceedings* (2019)
27. M. Wang, J. Zhang, H. Zhou, R.G. Southwick, R. Hsin, K. Chao, X. Miao, V.S. Basker, T. Yamashita, D. Guo, G. Karve, H. Bu, in *IEEE International Reliability Physics Symposium Proceedings* (2019)
28. N. Choudhury, U. Sharma, H. Zhou, R.G. Southwick, M. Wang, S. Mahapatra, in *IEEE International Reliability Physics Symposium Proceedings* (2020)
29. H. Zhou, M. Wang, J. Zhang, K. Watanabe, C. Durfee, S. Mochizuki, R. Bao, R. Southwick, M. Bhuiyan, B. Veeraraghavan, *IEEE International Reliability Physics Symposium Proceedings* (2020)

30. N. Choudhury, T. Samadder, R. Tiwari, H. Zhou, R.G. Southwick, M. Wang, S. Mahapatra, in *IEEE International Reliability Physics Symposium Proceedings* (2021)
31. S. Mahapatra, N. Goel, S. Mukhopadhyay, in *Fundamentals of Bias Temperature Instability in MOS Transistors* (Springer India, 2015), pp. 1–42
32. Y. Mitani, T. Yamaguchi, H. Satake, A. Toriumi, in *IEEE International Reliability Physics Symposium Proceedings*, 226 (2007)
33. C.H. Liu, M.T. Lee, Chih-Yung Lin, J. Chen, K. Schruefer, J. Brighten, N. Rovedo, T.B. Hook, M.V. Khare, Shih-Fen Huang, C. Wann, Tze-Chiang Chen, T.H. Ning, in *International Electron Devices Meeting Technical Digest*, 39.2.1 (2001)
34. T. Sasaki, K. Kuwazawa, K. Tanaka, J. Kato, D.-L. Kwong, IEEE Electron Device Lett. **24**, 150 (2003)
35. M. Terai, K. Watanabe, S. Fujieda, IEEE Trans. Electron Devices **54**, 1658 (2007)
36. V.D. Maheta, C. Olsen, K. Ahmed, S. Mahapatra, IEEE Trans. Electron Devices **55**, 1630 (2008)
37. K. Sakuma, D. Matsushita, K. Muraoka, Y. Mitani, in *2006 IEEE International Reliability Physics Symposium Proceedings*, 454 (2006)
38. V.D. Maheta, C. Olsen, K. Ahmed, S. Mahapatra, in *International Symposium on the Physical and Failure Analysis of Integrated Circuits* (2008)
39. A. Suzuki, K. Tabuchi, H. Kimura, T. Hasegawa, S. Kadomura, in *Symposium on VLSI Technology Digest of Technical Papers*, 216 (2002)
40. A.T. Krishnan, V. Reddy, S. Krishnan, in *International Electron Devices Meeting Technical Digest*, 39.3.1 (2001)
41. A. Kerber, IEEE Electron Device Lett. **35**, 294 (2014)
42. T. Tsunomura, J. Nishimura, A. Kumar, A. Nishida, S. Inaba, K. Takeuchi, T. Hiramoto, T. Mogami, in *Symposium on VLSI Technology Digest of Technical Papers*, 150 (2015)
43. A. Kerber, T. Nigam, in *IEEE International Reliability Physics Symposium Proceedings*, 2D.4.1 (2013)
44. D. Angot, V. Huard, L. Rahhal, A. Cros, X. Federspiel, A. Bajolet, Y. Carminati, M. Saliva, E. Pion, F. Cacho, A. Bravaix, in *IEEE International Electron Devices Meeting Technical Digest*, 15.4.1 (2013)
45. A. Kerber, T. Nigam, in *IEEE International Reliability Physics Symposium Proceedings*, 6A.6.1 (2014)
46. S. Ramey, M. Chahal, P. Nayak, S. Novak, C. Prasad, J. Hicks, in *IEEE International Reliability Physics Symposium Proceedings*, 3B.2.1 (2015)
47. M.D. Giles, N. Arkali Radhakrishna, D. Becher, A. Kornfeld, K. Maurice, S. Mudanai, S. Natarajan, P. Newman, P. Packan, T. Rakshit, in *Symposium on VLSI Technology Digest of Technical Papers*, T150 (2015)
48. S.E. Rauch, IEEE Trans. Device Mater. Reliab. **7**, 524 (2007)
49. B. Kaczer, T. Grasser, P. J. Roussel, J. Franco, R. Degraeve, L.-A. Ragnarsson, E. Simoen, G. Groeseneken, H. Reisinger, in *IEEE International Reliability Physics Symposium Proceedings*, 26 (2010)
50. A. Kerber, P. Srinivasan, IEEE Electron Device Lett. **35**, 431 (2014)
51. C. Prasad, M. Agostinelli, J. Hicks, S. Ramey, C. Auth, K. Mistry, S. Natarajan, P. Packan, I. Post, S. Bodapati, M. Giles, S. Gupta, S. Mudanai, K. Kuhn, in *IEEE International Reliability Physics Symposium Proceedings*, 6A.5.1 (2014)
52. C. Liu, H. Nam, K. Kim, S. Choo, H. Kim, H. Kim, Y. Kim, S. Lee, S. Yoon, J. Kim, J.J. Kim, L. Hwang, S. Ha, M.-J. Jin, H.C. Sagong, J.-K. Park, S. Pae, J. Park, in *IEEE International Electron Devices Meeting Technical Digest*, 11.3.1 (2015)
53. A. Chaudhary, B. Kaczer, P. J. Roussel, T. Chiarella, N. Horiguchi, S. Mahapatra, in *IEEE International Reliability Physics Symposium Proceedings*, 3B.4.1 (2015)

54. N. Parihar, R. Anandkrishnan, A. Chaudhary, S. Mahapatra, IEEE Trans. Electron Dev. **66**, 3273 (2019)
55. T. Naphade, K. Roy, S. Mahapatra, in *IEEE International Electron Devices Meeting Technical Digest*, 33.6.1 (2013)
56. N. Goel, P. Dubey, J. Kawa, S. Mahapatra, in *IEEE International Reliability Physics Symposium Proceedings*, CA.5.1 (2015)

Chapter 3
Physical Mechanism of NBTI Parametric Drift

Souvik Mahapatra, Narendra Parihar, Subhadeep Mukhopadhyay, and Nilesh Goel

3.1 Introduction

As shown in the previous chapters, Negative Bias Temperature Instability (NBTI) became an important reliability concern during the migration from Silicon Dioxide (SiO_2) to Silicon Oxynitride (SiON) gate insulator MOSFETs [1–7], and continues to remain so in dual layer (SiO_2 or SiON Interlayer (IL) and Hafnium Dioxide (HfO_2) High-K layer) High-K Metal Gate (HKMG) gate insulator planar bulk [8–14] and Fully Depleted Silicon On Insulator (FDSOI) [15, 16] MOSFETs, bulk and SOI FinFETs [16–29], and Gate All Around Stacked Nanosheet FETs [30–33], with either Silicon (Si) or Silicon Germanium (SiGe) channel.

As described in Chap. 1, Sect. 1.3, NBTI results in a gradual buildup of positive gate insulator charges in a p-MOSFET under the application of a negative gate bias (V_G), and shifts different device parameters, e.g., threshold voltage shift (ΔV_T), over time. ΔV_T accelerates at higher magnitude of V_G during stress ($V_G = V_{GSTR}$) and higher temperature (T), governed by the Voltage Acceleration Factor (VAF) and Arrhenius T activation energy (E_A) respectively. The parametric shift accrued during stress partially recovers when the magnitude of V_G is reduced or removed ($V_G = V_{GREC}$ or 0 V) after stress. As a consequence, AC stress results in lower ΔV_T than DC stress. The AC to DC ratio depends on the Pulse Duty Cycle (PDC) and pulse low value (V_{GLOW}), however, it may or may not depend on the frequency (f) of the gate pulse, as it depends on pulse high (V_{GHIGH}) and AC stress mode. On the other hand, NBTI recovery necessitates the use of ultra-fast methods for recovery artifact free measurements. As described in Chap. 1, Sect. 1.2, the ultra-fast Measure Stress Measure (MSM) method is used throughout this book.

S. Mahapatra (✉) · N. Parihar · S. Mukhopadhyay · N. Goel
Department of Electrical Engineering, Indian Institute of Technology Bombay, Powai, Mumbai 400076, India
e-mail: souvik@ee.iitb.ac.in

S. Mahapatra (ed.), *Recent Advances in PMOS Negative Bias Temperature Instability*,
https://doi.org/10.1007/978-981-16-6120-4_3

It is described in Chap. 2 that the time kinetics of measured ΔV_T and related parameters, such as the power-law slope (n) at longer stress time (t_{STR}), VAF, E_A, T dependence of VAF during stress and Fraction Remaining (FR) during recovery after stress (FR is defined as ΔV_T at $t = t_{REC}$ after stress to that at $t = t_{STR}$ at the end of stress) depend on different transistor processes. In modern HKMG gate insulator-based p-MOSFETs, some of the key processes that impact NBTI are Nitrogen content (N%) in the gate insulator and Germanium content (Ge%) in the channel. The magnitude of ΔV_T increases, while the associated n, VAF, E_A, T dependence of VAF and FR reduce with higher N%. On the other hand, the magnitude of ΔV_T reduces, while the associated n, VAF, E_A, T dependence of VAF and FR increase with higher Ge%. Moreover, NBTI reduces with fin length and width scaling in FinFETs, sheet length scaling in GAA-SNS FETs and for larger spacing (SA) between the Shallow Trench Isolation (STI) and device active in FDSOI MOSFETs, however, it increases with sheet width scaling in GAA-SNS FETs.

Note that a practical and technologically relevant modeling framework should be able to explain the following experimental features (see Chaps. 1 and 2):

- Time evolution of ΔV_T during DC stress at different V_{GSTR} and T, including the T dependence of VAF and V_{GSTR} dependence of E_A.
- Time evolution of ΔV_T during Mode-A and Mode-B AC stress at different pulse high bias (V_{GHIGH}), PDC, f, and V_{GLOW}.
- Time evolution of recovery after DC, Mode-A, and Mode-B AC stress, as well as the impact of t_{STR} and V_{GREC} on recovery kinetics.
- Impact of measurement delay (t_M) on DC and AC stress kinetics.
- Explanation of process dependence, especially that related to N% in the gate insulator stack and Ge% in the channel, for different device architectures or technologies.
- Explanation of fin/sheet dimensions and layout effects as applicable for a particular technology.
- Variability of ΔV_T in small area devices.

These aspects will be analyzed and modeled in this book, using the BTI Analysis Tool (BAT) framework explained in Chap. 4 through Chap. 6. The BAT framework suggests NBTI parametric drift to be due to uncorrelated contributions from generated interface traps (density ΔN_{IT}), hole trapping in pre-existing bulk gate insulator traps (ΔN_{HT}), and generated bulk gate insulator traps (ΔN_{OT}), as illustrated in Fig. 3.1. Charges associated with these traps give rise to the underlying subcomponents (ΔV_{IT}, ΔV_{OT}, and ΔV_{HT}) of overall ΔV_T. In this chapter, independent experimental evidences are provided to justify the above hypothesis.

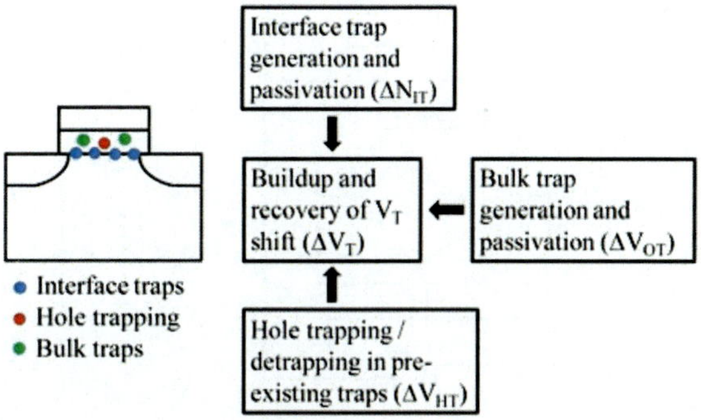

Fig. 3.1 Measured NBTI parametric drift is modeled by uncorrelated contributions from ΔN_{IT}, ΔN_{HT}, and ΔN_{OT} subcomponents in this book

3.2 Evidence of Interface Trap Contribution

As shown in Chap. 1, Sect. 1.2, positive charges accrued during NBTI stress result in transconductance degradation (Δg_m, because of additional Coulomb scattering induced mobility degradation) and ΔV_T. However, it is difficult to identify if these charges are due to generated traps or hole trapping in pre-existing defects.

The generation and passivation of interface trap respectively during and after NBTI stress have been reported using multiple direct measurement methods:

- Change in subthreshold slope (ΔSS) for linear drain current (I_{DLIN}) versus V_G sweep [34], since SS is due to the capacitance associated with interface traps [35].
- Change in current (ΔI_{CP}) from Charge Pumping (CP) method [2, 4–7, 36], as I_{CP} is due to interface trap assisted electron-hole recombination when the gate is repeatedly pulsed between inversion and accumulation, I_{CP} is measured at the substrate with the source and drain grounded [37].
- Change in current (ΔI_{DCIV}) from Direct-Current I-V (DCIV) or Gated Diode method [10, 13, 24, 25, 38–43], as I_{DCIV} is due to interface trap assisted electron-hole recombination when the gate is scanned from inversion to accumulation with the source-drain to substrate junctions in forward bias, I_{DCIV} is measured at the substrate [44].
- Change in gate leakage (ΔI_G) from Low-Voltage Stress Induced Leakage Current (LV-SILC) method [1, 3, 39], as LV-SILC is due to interface trap assisted gate tunneling between the substrate and gate when sensed in accumulation [45].

The CP and DCIV are the most commonly used methods and are discussed below. Interested readers can refer to [46] for further details on these measurement methods.

3.2.1 Charge Pumping (CP) Method

Figure 3.2(a) illustrates a p-MOSFET configured for CP measurement. The gate is repetitively pulsed between inversion and accumulation, the source and drain

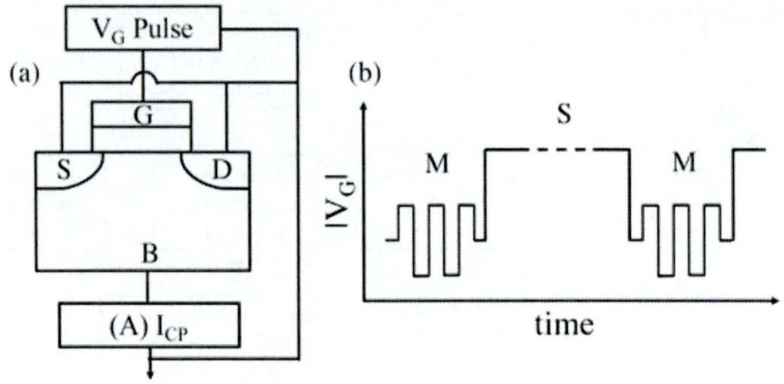

Fig. 3.2 Charge pumping method: (a) measurement setup and (b) gate pulses for implementation in the Measure-Stress-Measure (MSM) mode to characterize NBTI stress

terminals are tied together and grounded, and the substrate current is measured. During inversion, the MOS channel is flooded with holes from the source drain, some of which get trapped in the interface traps. During accumulation, the channel is flooded with electrons from the substrate, and some of these electrons would recombine with previously trapped holes at the interface traps. The resulting trap-assisted electron-hole recombination current is measured at the substrate (I_{CP}) as the gate is continuously pulsed, which is given by the following expression [37]:

$$I_{CP} = qf(WL)N_{IT} \tag{3.1}$$

where q is the electronic charge, f is the frequency of the gate pulse, W and L are the width and length of the MOSFET respectively (WL = gate area) and N_{IT} is the interface trap density (/cm^2). CP measurements are performed before and during (by interrupting) NBTI stress as shown in Fig. 3.2(b), and the generated interface traps (ΔN_{IT}) can be determined using Eq. 3.1. Similar considerations apply for CP measurements during the post-stress recovery phase.

Figure 3.3 shows the time evolution of ΔV_T and ΔI_{CP} measured respectively by slow I_{DLIN}-V_G and CP methods in p-MOSFETs having different SiON gate insulator processes (PNO: Plasma Nitrided Oxide, RTNO: Rapid Thermal Nitrided Oxide) and poly-Si gate, (a) during and (b) after NBTI stress [7]. The stress kinetics is plotted on a log–log scale, but recovery is plotted on a linear scale. The buildup and recovery of ΔV_T and ΔI_{CP} (and hence ΔN_{IT}) are observed respectively during and after stress. Power-law time dependence is observed during stress for ΔV_T and ΔI_{CP}. However, it is difficult to compare their time slopes (n) due to different delays associated with these measurement probes (the impact of measurement delay in Measure-Stress-Measure (MSM) method is discussed in Chap. 1, Sect. 1.2). However, ΔI_{CP} does reproduce the higher ΔV_T buildup during stress and higher ΔV_T fractional recovery after stress for the RTNO device as compared to the PNO device (note, the density of N near the channel/gate insulator interface is higher for RTNO as compared to the

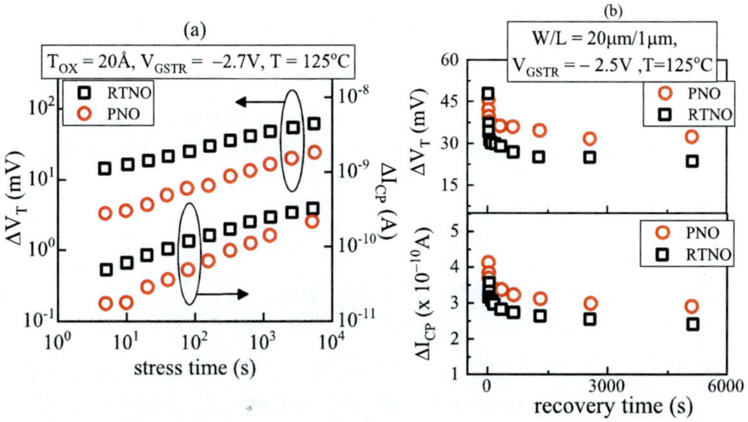

Fig. 3.3 Time evolution of ΔV_T and ΔI_{CP} respectively from slow MSM I-V and CP methods (a) during and (b) after stress in differently processed SiON p-MOSFETs. Data from [7]

PNO process, see Chap. 1, Sect. 1.4). Therefore, Fig. 3.3 suggests that at least a part of ΔV_T buildup and recovery is contributed by ΔN_{IT} in these devices.

As mentioned in Chap. 2, Sect. 2.1, incorporation of Fluorine (F), either directly into the gate insulator or indirectly by diffusion from implanted junctions into the gate insulator reduces NBTI. Figure 3.4 shows the time evolution of (a) ΔV_T and ΔN_{IT} as well as (b) ΔV_T and ΔI_{CP} for NBTI stress in SiON p-MOSFETs without and with F in the gate stack (ΔV_T is measured by the slow I_{DLIN}-V_G and ΔN_{IT} or ΔI_{CP} by CP methods) [2, 5]. In panel (a), Boron (B) and Difluoroboron (BF_2) implanted source-drain junctions are used and F incorporation into the gate insulator is present for the latter (BF_2) device [2], while in panel (b), F is directly implanted into the gate insulator [5]. Data are plotted in a log–log scale and power-law time

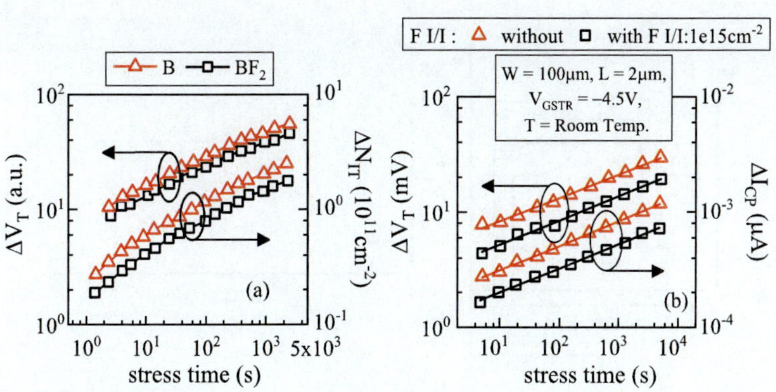

Fig. 3.4 Time evolution of ΔV_T and ΔN_{IT} (or ΔI_{CP}) respectively from slow MSM I-V and CP methods during stress in p-MOSFETs without and with F incorporation in the gate insulator. Data from (a) [2] and (b) [5]

dependence is observed for both measurement methods. The reduction in NBTI due to F incorporation is visible by both methods, and therefore, once again, at least a part of ΔV_T is contributed by ΔN_{IT} in these devices.

CP measurement is difficult in devices having thin gate insulators due to corruption by high gate leakage. DCIV measurements are used in such cases, which are discussed next.

3.2.2 *Direct-Current IV (DCIV) Method*

Figure 3.5(a) illustrates a p-MOSFET configured for DCIV measurement. The gate is swept from accumulation to inversion, the source and drain terminals are tied together and a forward bias (V_F) is applied (V_F should be well below the junction cut-in voltage to avoid leakage), and the interface trap assisted electron-hole recombination current (I_{DCIV}) is measured at the substrate. The gate waveform during stress and DCIV measurement is illustrated using Fig. 3.5(b). Either DC or AC stress can be used. The same waveform is also used for measurement during recovery after stress.

Figure 3.6 shows the measured I_{DCIV} versus V_G characteristics before stress, as well as (a) during and (b) after NBTI stress (by interrupting stress and recovery for a duration t_M) in a p-MOSFET having Gate First (GF) HKMG gate insulator stack [46]. The measured I_{DCIV} versus V_G curves show a peak that increases during stress and reduces after stress, and respectively indicates generation and passivation of interface traps. Furthermore, the location of the peak shifts towards more negative V_G during stress and towards more positive V_G after stress, and respectively indicate buildup and removal of positive charges.

The difference between the peak ($I_{DCIV,P}$) and base ($I_{DCIV,B}$) of the I_{DCIV} versus V_G characteristics is used to estimate N_{IT} using the following expression [44]:

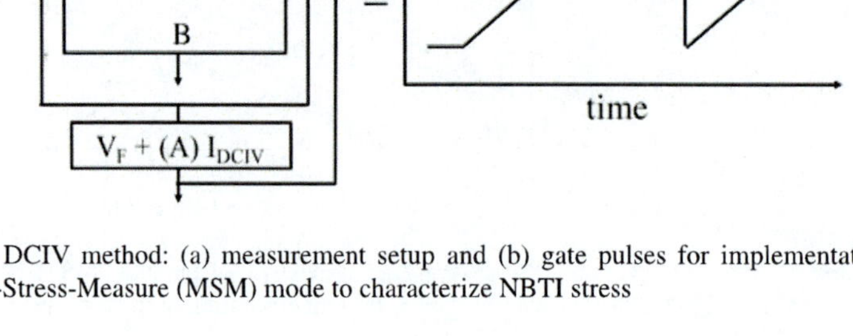

Fig. 3.5 DCIV method: (a) measurement setup and (b) gate pulses for implementation in the Measure-Stress-Measure (MSM) mode to characterize NBTI stress

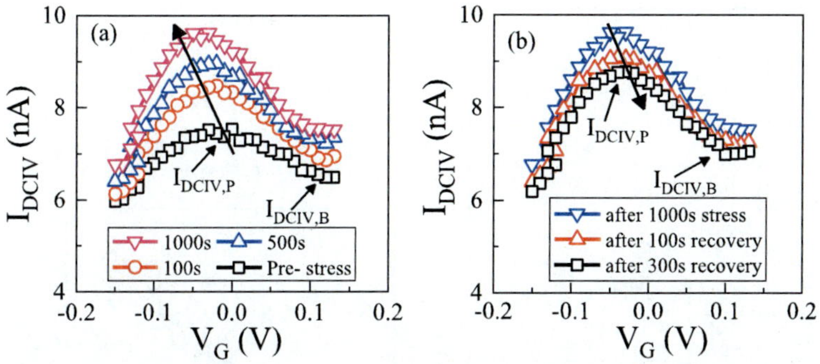

Fig. 3.6 Measured DCIV current–voltage characteristics (a) during and (b) after stress for different stress-recovery time. Data from [46]

$$I_{\mathrm{DCIV,P}} - I_{\mathrm{DCIV,B}} = qn_{\mathrm{i}}v_{\mathrm{th}}\frac{\pi}{4}\sqrt{\sigma_n\sigma_p}(WL)N_{\mathrm{IT}}e^{qV_{\mathrm{F}}/2kT} \tag{3.2}$$

where q is the electronic charge, n_{i} is the intrinsic carrier concentration, v_{th} is the thermal velocity, σ_{n} and σ_{p} are the electron and hole capture cross sections of the interface traps, W and L are the device width and length respectively. The capture cross sections of Eq. 3.2 can be determined by comparing the pre-stress N_{IT} from DCIV and SS methods. The change in N_{IT} due to NBTI is proportional to change in $(I_{\mathrm{DCIV,P}} - I_{\mathrm{DCIV,B}})$, and it is presumed that σ_{n} and σ_{p} do not change after stress.

Strictly speaking, Eq. 3.2 is valid for an interface trap located at a single energy level in the energy bandgap, and the expression corresponding to continuous trap distribution is quite complicated. However, Eq. 3.2 can still be used to estimate the change in average interface trap density (ΔN_{IT}) due to stress [43, 44]. A pragmatic approach is to use an equivalent expression, $N_{\mathrm{IT}} = K\left(I_{\mathrm{DCIV,P}} - I_{\mathrm{DCIV,B}}\right)$, determine K in pre stress by comparing N_{IT} from the DCIV and SS methods, and by assuming no change in K (i.e., no change in σ_{n} and σ_{p}) after stress.

Figure 3.7 shows the time evolution of (a) ΔI_{DCIV}, (b) ΔV_{T}, and (c) Δg_{m} during multiple NBTI stress-recovery cycles in a p-MOSFET having SiON gate insulator and poly-Si gate [43]. DCIV is used for ΔN_{IT} and slow I_{DLIN}-V_{G} method is used for ΔV_{T} and Δg_{m}. Data from all the measurement probes increase during stress and reduce after stress, and are correlated to each other as shown in Fig. 3.7(d). Therefore, once again, at least a part of the I-V parametric degradation can be attributed to generated interface traps.

Figure 3.8 shows the oxide electric field (E_{OX}) dependence of (a) ΔV_{T} and (b) fractional ΔI_{DCIV} during fixed-time NBTI stress in p-MOSFETs with GF HKMG gate insulator and different IL processes [10]. The MSM-OPDD method explained in Chap. 1, Sect. 1.2 is used to measure ΔV_{T} with t_{M} of 1ms, while DCIV measurement has t_{M} of ~5s. Note that the conventional chemical oxide-based IL is a low temperature process and is of inferior quality compared to the thermal IL process (details in [10]). Therefore, the thermal IL device shows lower ΔV_{T}, Fig. 3.8(a),

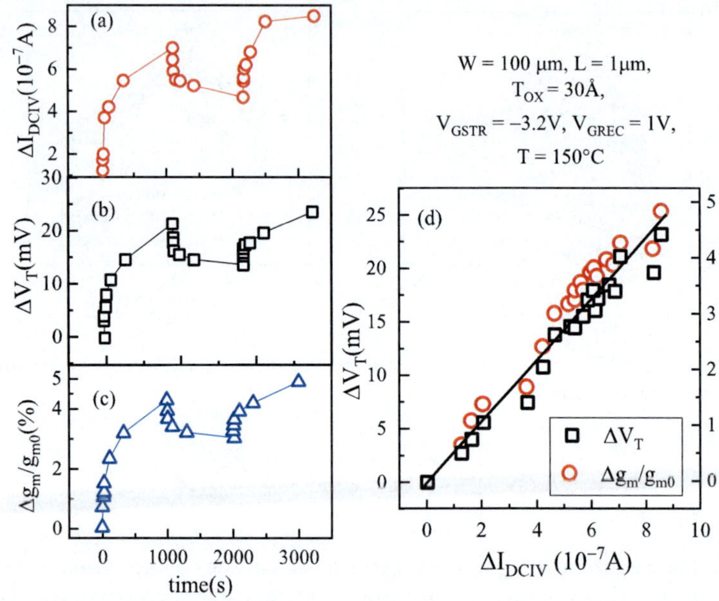

Fig. 3.7 Measured shifts in (a) I_{DCIV}, (b) V_{T} and (c) $\mathrm{g_m}$ during and after stress, and (d) correlation of V_{T} and g_m to I_{DCIV} shift. Data from [38]

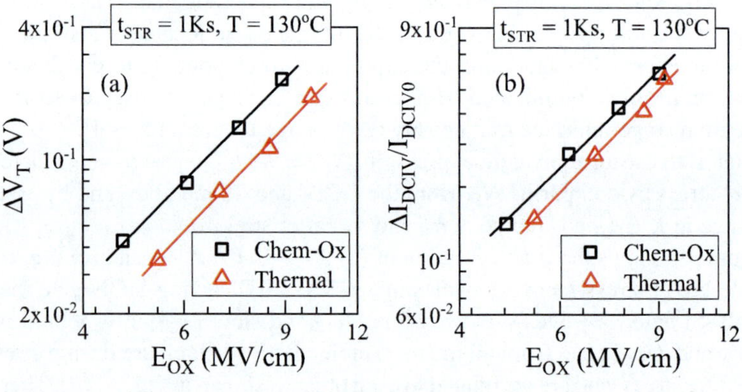

Fig. 3.8 Oxide field dependence of fixed time (a) ΔV_{T} and (b) ΔI_{DCIV} during stress in dual layer (IL/High-K) HKMG p-MOSFETs with conventional chemical oxide (Chem-Ox) and thermal interlayer (IL) processes (see [10] for details). The HKMG gate stack is illustrated in Fig. 1.1 of Chap. 1

which is partially attributed to lower ΔN_{IT} (~ΔI_{DCIV}), Fig. 3.8(b). However, due to some inherent differences between both methods, it is difficult to make a direct quantification of the contribution of ΔN_{IT} to ΔV_{T} for different IL processes. Such a comparison needs certain corrections to the as-measured DCIV data, which is discussed next.

3.2.3 Corrections for Measurement Delay and Bandgap

DCIV is implemented in a slow MSM mode and suffers from measurement delay artifacts (see Chap. 1, Sect. 1.2 for measurement delay effect on measured data), as ultra-fast measurement of the I_{DCIV} versus V_G characteristics is difficult due to low level of measured I_{DCIV}. Therefore, the as-measured data need to be corrected for measurement delay before the ΔN_{IT} time kinetics can be modeled.

Figure 3.9(a) shows the time evolution of ΔN_{IT} measured by DCIV using two different measurement delays in GF HKMG p-MOSFETs [13]. Lower ΔN_{IT} magnitude and higher slope n are observed for larger t_M due to recovery artifacts. Note that ΔN_{IT} recovery is universal as shown in Fig. 3.9(b), where measured (using default t_M) ΔN_{IT} at a certain time after the end of stress is compared to ΔN_{IT} measured immediately after stress in p-MOSFETs having differently processed (changes in IL and/or High-K thickness, N%) GF HKMG gate insulator stacks [10].

The following expression can be used to correct for measurement delay and obtain "recovery-free" data [47]:

$$\Delta N_{IT}(t_{STR}, t_M) = \frac{\Delta N_{IT}(t_{STR})}{1 + B\left(\frac{t_M}{t_{STR}}\right)^{\beta}} \tag{3.3}$$

where t_M is the measurement delay, t_{STR} is the cumulative stress time before the onset of measurement, B and β are the device specific fitting constants. Once the fitting constants are obtained using recovery measurements, Eq. 3.3 can be used to convert ΔN_{IT} (t_{STR}, t_M) measured after t_{STR} duration of stress and with t_M delay to obtain the "delay free" ΔN_{IT} (t_{STR}) after the end of stress.

Figure 3.9(a) also shows the corrected ΔN_{IT} kinetics. The accuracy of the fitting constants B and $\beta$is verified since the measured ΔN_{IT} kinetics using different t_M results in identical ΔN_{IT} kinetics after delay correction. Note that the delay corrected ΔN_{IT} time kinetics shows a power-law slope of n ~1/6 for different devices, which is observed across different V_{GSTR}, Fig. 3.9(c) and T, Fig. 3.9(d) for DC stress, and versus PDC, Fig. 3.9(e) and Fig. 3.9(f) for AC stress (also for different V_{GSTR} and T for AC stress, not explicitly shown).

Note that the energy zone scanned by the DCIV method is ~$q * V_F$ (typically, V_F is ~0.3–0.4 V to remain much below the junction cut-in voltage and avoid junction leakage from corrupting the measurements) and is centered around the energy bandgap of the Si substrate. Therefore, the delay corrected DCIV data also need to be corrected for bandgap before the ΔN_{IT} time kinetics can be compared to that of ΔV_T obtained using ultra-fast I_{DLIN}-V_G methods. As I_{DLIN} measurement scans the

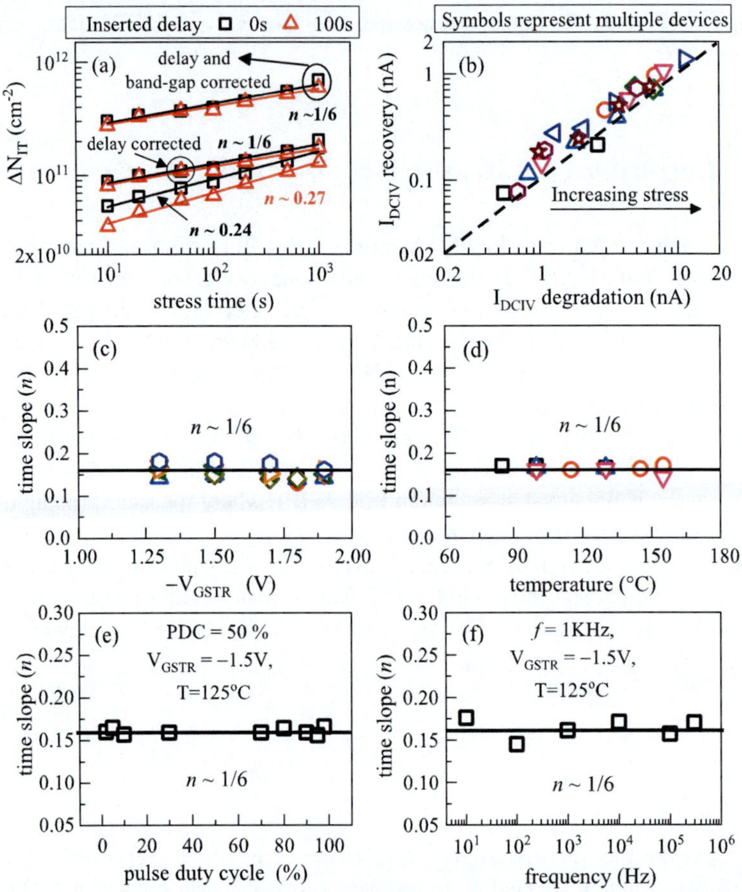

Fig. 3.9 (a) As-measured time evolution of ΔN_{IT} for two different measurement delays by the DCIV method (inserted delay is additional on top of default delay of ~5s). Impact of delay and bandgap corrections are shown. (b) Correlation of DCIV current measured just after stress and after a certain recovery time on different devices. (c)–(f) Extracted power-law time slope n as a function of (c) V_{GSTR}, (d) temperature, (e) PDC, and (f) frequency for different devices. Each symbol type represents a particular device in panels (b) through (d). Data from [13, 43]

entire bandgap (E_G), the delay corrected DCIV data are multiplied by the ratio of $E_G/(q * V_F)$ to obtain delay and bandgap corrected data [43]. It is assumed that the generated traps are uniformly distributed throughout the bandgap for this correction. Correction for bandgap, shown in Fig. 3.9(a), only changes the ΔN_{IT} magnitude (increases by ~3X) and not time slope n.

Note, similar delay correction is also needed for the modeling of CP measured ΔN_{IT} kinetics (the time kinetics again shows n ~1/6 after correction) [6, 46]. As discussed in Chap. 4, the Reaction Diffusion (RD) model with defect assisted dimerization results in n ~1/6 power-law time slope for ΔN_{IT} kinetics during DC and AC stress, consistent with measured data.

3.3 Evidence of Hole Trapping Contribution

It is now well-accepted that hole trapping into (and detrapping out of) process induced pre-existing gate insulator defects (density ΔN_{HT}) also contributes (ΔV_{HT}) to the buildup (and recovery) of overall ΔV_T during (and after) NBTI stress. Most of the past reports have suggested the saturation of ΔV_{HT} for stress time higher than ~1 s [2, 6, 10, 13 21–25, 31, 33, 42, 48–51]. Some of these reports have also shown lower E_A of ΔV_{HT} as compared to that of ΔV_{IT} [2, 6, 10, 13 21–25, 31, 33, 42, 51].

Figure 3.10 illustrates the impact of uncorrelated contributions from ΔV_{IT} and ΔV_{HT} on the (a) time kinetics and (b) T dependence (at fixed t_{STR}) of overall ΔV_T. Note, ΔV_{IT} shows n ~1/6 power-law time dependence at longer t_{STR} and relatively higher E_A. However, ΔV_{HT} shows saturated time kinetics (n ~0 in a log–log plot) at longer t_{STR} and relatively lower E_A. If ΔV_{HT} is not negligible, the resultant ΔV_T is a sum of ΔV_{IT} and ΔV_{HT} and would show higher magnitude but lower n and E_A compared to the case when ΔV_{IT} dominates overall ΔV_T (negligible ΔV_{HT}). These presumptions are experimentally verified in this section. The ultra-fast full sweep MSM method (see Chap. 1, Sect. 1.2) is used for ΔV_T measurements in all cases unless specifically mentioned otherwise.

3.3.1 *Flicker Noise Measurement*

Hole trapping occurs in gate insulator process related pre-existing defects, which can be independently characterized by the flicker noise method [46]. Figure 3.11(a)

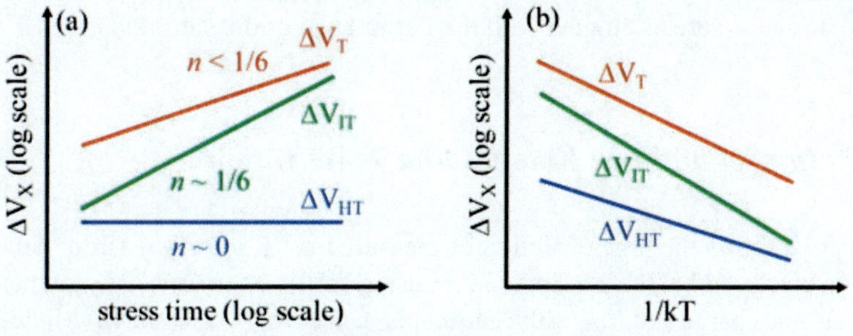

Fig. 3.10 Schematic illustration of the impact of uncorrelated contributions from ΔV_{IT} and ΔV_{HT} on the (a) time kinetics and (b) T activation (at fixed t_{STR}) of overall ΔV_T

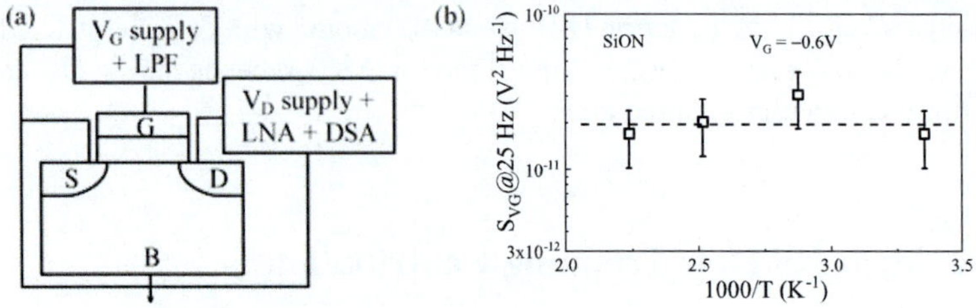

Fig. 3.11 (a) Schematic of flicker noise setup. (b) Measured S_{VG} as a function of T from a p-MOSFET having SiON gate insulator. Data from [52]

shows the measurement setup. The gate is biased close to threshold ($V_G \sim V_{T0}$) using a power supply via a Low Pass Filter (LPF) to reduce noise. The power spectral density of drain current noise (S_{ID}) is measured in the frequency domain at the drain using a Digital Spectrum Analyzer (DSA), a Low Noise Amplifier (LNA) is used to boost the measurement signal. The input referred noise (S_{VG}) is obtained from S_{ID} using the relation $S_{VG} = S_{ID}/{g_m}^2$, where g_m is the measured transconductance at V_G ($\sim V_{T0}$). Figure 3.11(b) plots the measured S_{VG} as a function of T, showing negligible T dependence [52].

Drain current noise has been related to fluctuations in channel carrier density, channel mobility, or both, refer to [46] for details. The combined number-mobility fluctuation model [53] can be used to determine the density of pre-existing defects from measured S_{VG}:

$$S_{VG} = \frac{kTq^2}{\gamma f W L C_{OX}^2}(1 + \alpha\mu N_C) N_T(E_{FN}) \tag{3.4}$$

where kT is the thermal energy, q is the electronic charge, γ is the attenuation factor of hole wavefunction into the gate insulator (can be calculated from tunneling probability), f is the measurement frequency, W and L are the width and length of the device respectively, C_{OX} is the gate capacitance, α is the scattering coefficient, μ is the inversion layer mobility, $N_C = C_{OX}\,(V_G - V_{T0})$ and $N_T\,(E_{FN})$ is the density of gate insulator defects aligned with the Fermi level of the substrate (E_{FN}).

3.3.2 *Impact on Time Kinetics and T Activation*

Figure 3.12 shows the time evolution of measured ΔV_T and ΔV_{IT} (from delay and bandgap corrected DCIV, see Sect. 3.2.3, using $\Delta V_{IT} = q * \Delta N_{IT}/C_{OX}$, where q is the electronic charge and C_{OX} is the gate capacitance) for (a) DC and (b) Mode-B AC stress (see Chap. 1, Sect. 1.2 for different modes of AC stress) in a p-MOSFET having GF HKMG gate insulator stack with low Nitrogen content (N%). The difference

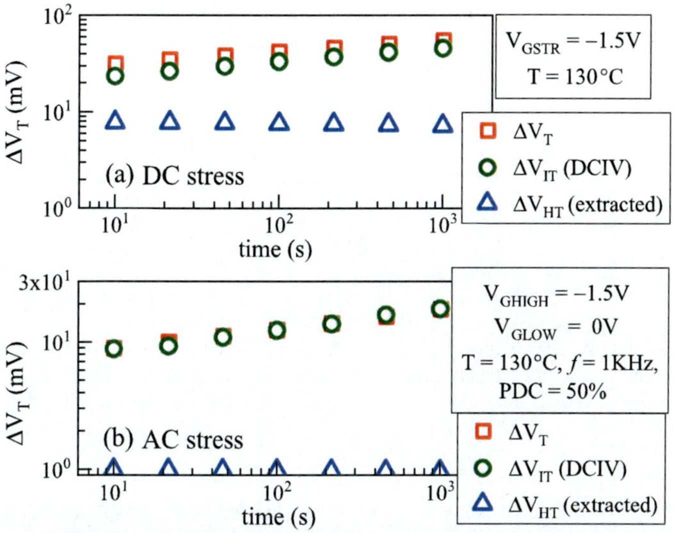

Fig. 3.12 Time evolution of ultra-fast measured ΔV_T and DCIV measured (delay and band gap corrected) ΔV_{IT} during (a) DC and (b) mode-B AC stress in the D1 device having low N%. The extracted difference (ΔV_{HT}) is also shown. Data from [13]

between ΔV_T and ΔV_{IT} is ΔV_{HT}, as ΔV_{OT} contribution is almost negligible for this stress condition in this device (this is explored in more detail in Chap. 7).

Note that ΔV_{HT} is negligible for Mode-B AC stress, which is consistent with data shown in Chap. 1, Fig. 1.14. Both ΔV_{IT} and ΔV_T show power-law time kinetics with slope of n ~1/6 for Mode-B AC stress. However, ΔV_{HT} is not negligible (but small) for DC stress. Therefore, the power-law slope of ΔV_T (n ~0.13) is slightly smaller than that of ΔV_{IT} (n ~1/6) due to additional contribution from the saturated ΔV_{HT} component during DC stress.

Figure 3.13 shows the correlation between measured ΔV_T and ΔV_{IT} in devices

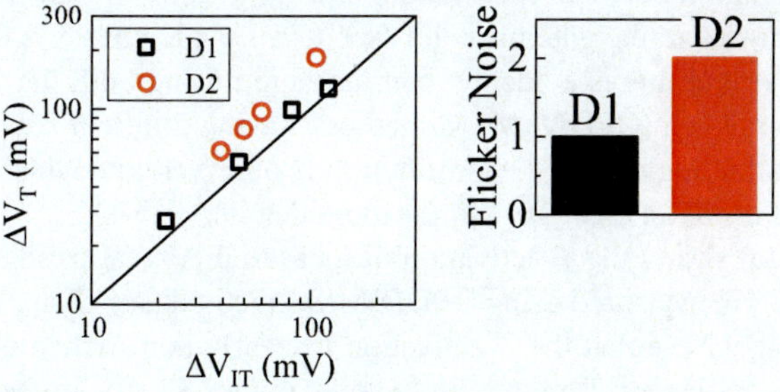

Fig. 3.13 Correlation of ultra-fast measured ΔV_T and DCIV measured (delay and band gap corrected) ΔV_{IT} for the D1 and D2 devices respectively having low and high N%. The pre-stress flicker (or 1/f) noise signals are also shown for these devices. Data from [13]

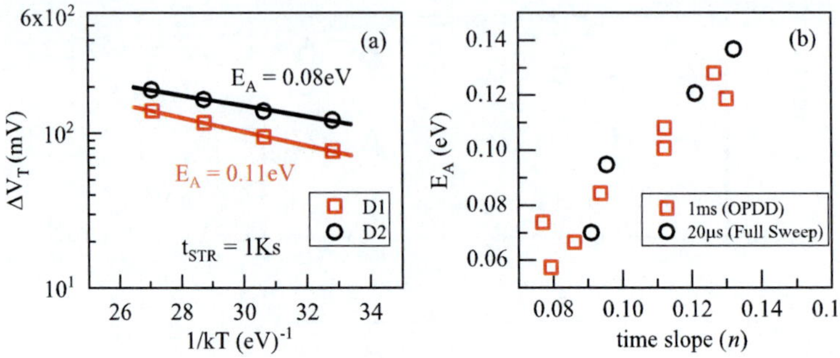

Fig. 3.14 (a) T activation of ultra-fast measured ΔV_T at fixed t_{STR} for the D1 and D2 devices respectively having low and high N%. (b) Correlation of the activation energy (E_A) and power-law time slope (n) for different GF HKMG devices having different N% in the gate insulator stack. Data from [10, 13]

having low (D1) and high (D2) N% based GF HKMG gate stack. Note that the D2 device shows higher ΔV_T for a particular ΔV_{IT} (and hence higher ΔV_{HT}) compared to the D1 device. As mentioned before, the contribution from ΔV_{OT} is almost negligible in both these devices for this stress condition. The pre-stress flicker noise measurements result in higher signal for the D2 device and indicate higher density of pre-existing gate insulator defects compared to the D1 device (i.e., the defect density is proportional to N%), which is also shown in [9, 54]. Therefore, it is expected that higher N% in the gate insulator would result in higher ΔV_{HT} contribution and hence lower time slope n and lower E_A for DC stress.

Figure 3.14(a) shows the T activation of measured ΔV_T at fixed t_{STR} for the D1 and D2 devices of Fig. 3.13. The D2 device with higher N% shows higher ΔV_T but lower E_A compared to the D1 device. Note, higher N% results in higher ΔV_{IT} contribution and lower E_A for the defect dissociation (ΔN_{IT} buildup) process, which is discussed later in Chap. 4. However, higher N% also results in higher ΔV_{HT} contribution, see Fig. 3.13, which also has lower E_A. Hence, the combined effect of ΔV_{IT} and ΔV_{HT} can explain the measured data (more on this in Chap. 7). Figure 3.14(b) shows the correlation between measured n and E_A in different GF HKMG devices having different N% in the gate insulator (see Chap. 7). Higher N% results in lower n and lower E_A, and this is evident when measured using both the full sweep and One Point Drop Down (OPDD) MSM methods having different t_M. Therefore, it is clearly evident that higher ΔV_{HT} contribution is observed in devices having higher N% in the gate insulator, see [10, 54] for more details.

Figure 3.15(a) shows the T activation of measured ΔV_T at fixed t_{STR} for the D2 device of Fig. 3.13, measured using OPDD-MSM method (see Chap. 1, Sect. 1.2) for extended T range. Note that the T activation is clearly non Arrhenius if considered over the entire, extended T range, and shows higher E_A for higher T and lower E_A for lower T regions. The difference in E_A between the ΔV_{IT} (relatively higher E_A) and ΔV_{HT} (relatively lower E_A) subcomponents results in non-Arrhenius T activation of ΔV_T over the extended T range. This is because of increased relative

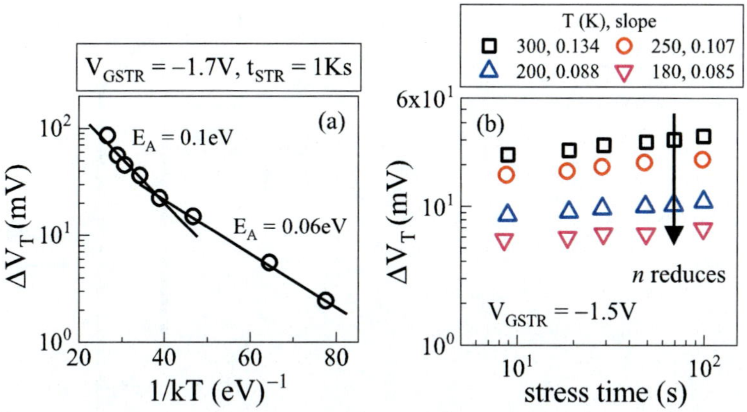

Fig. 3.15 (a) T activation of ΔV_T at fixed t_{STR} and (b) time evolution of ΔV_T at different T for the D2 device having high N%. Ultra-fast OPDD measurements are done over extended low T range to highlight the impact of ΔV_{HT}. Data from [13]

ΔV_{IT} contribution at higher T and increased relative ΔV_{HT} contribution at lower T (since E_A is high, ΔV_{IT} would become less than ΔV_{HT} at lower T). Figure 3.15(b) shows the time evolution of ΔV_T for the same device measured over extended T range. The slope n reduces at lower T due to increased relative ΔV_{HT} contribution and is consistent with the data shown in Fig. 3.15(a).

The ΔV_{HT} time kinetics during and after stress is modeled using the Activated Barrier Double Well Thermionic (ABDWT) model in Chap. 5.

3.4 Evidence of Bulk Gate Insulator Trap Generation Contribution

NBTI stress at relatively higher V_{GSTR} and T is similar to the Time Dependent Dielectric Breakdown (TDDB) experiments, which result in the generation of bulk gate insulator traps (ΔN_{OT}) [55–57]. Charges associated with these traps contribute (ΔV_{OT}) to overall ΔV_T, while Trap Assisted Tunneling (TAT) via these traps gives rise to Stress Induced Leakage Current (SILC) or increase in the gate leakage current (ΔI_G) [58, 59]. Typically, TDDB experiments focus on the formation of a percolation path between the channel and gate caused by these generated traps leading to breakdown of the gate insulator. However, for NBTI, their contribution to ΔV_T before the device breaks is of interest.

Figure 3.16 illustrates SILC due to TAT of electrons from the gate to substrate in a p-MOSFET when measured in inversion. An electron tunnels from the gate to the trap and gets captured, the trap relaxes in energy by the amount E_R, and subsequently, the electron is emitted from the trap to the substrate. Generated bulk trap density can be obtained from the following expression [46]:

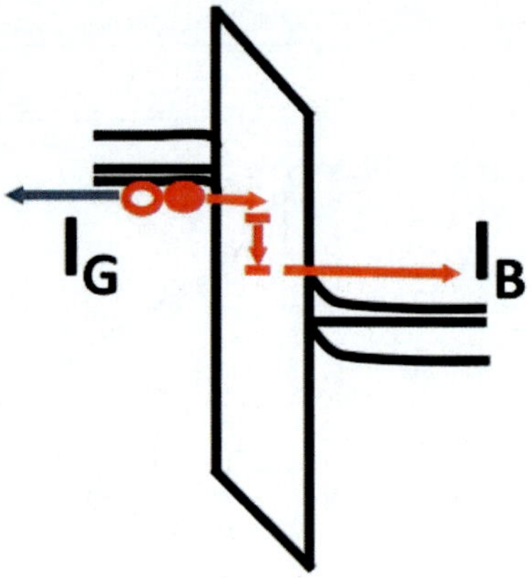

Fig. 3.16 Schematic representation of p-MOSFET energy band diagram showing TAT during SILC measurements

$$\Delta I_{\mathrm{G}} = \iint q\sigma_n v_{\mathrm{th}} N_C S_N f_{\mathrm{CA}} WLN_{\mathrm{T}}(x, E) \frac{T_{\mathrm{CT}}(E)T_{\mathrm{TA}}(E - E_{\mathrm{R}})}{T_{\mathrm{CT}}(E) + T_{\mathrm{TA}}(E - E_{\mathrm{R}})} \mathrm{d}x\mathrm{d}E \quad (3.5)$$

where q is the electronic charge, α_n is the capture cross section of electrons in the bulk traps, v_{th} is the thermal velocity, N_{C} is the density of electrons in the conduction band edge, S_{N} is the supply function in the cathode (gate), f_{CA} is the Fermi energy difference between the cathode and anode (substrate), W and L are the device width and length respectively, N_{T} (x, E) is the density of traps at a spatial location x and energy location E in the gate insulator (see Fig. 3.16), T_{CT} and T_{TA} are the electron tunneling probabilities from the cathode to trap and trap to anode. The interested reader may refer to [46] for details of SILC characterization and analysis.

Figure 3.17 shows the time evolution of (a, c) ΔV_{T} and (b, d) ΔI_{G} measured in a p-MOSFET with relatively thicker SiO_2 gate insulator, for changes in (a, c) V_{GSTR} and (b, d) reverse substrate bias (V_{B}). Both ΔV_{T} and ΔI_{G} measurements are done using the slow MSM method [60]. Note, ΔV_{T} shows power-law time dependence with a single slope n when V_{GSTR} is low or $V_{\mathrm{B}} = 0$ V and no SILC is observed. However, the magnitude of ΔV_{T} and the corresponding slope n increase at longer t_{STR} for stress at higher V_{GSTR} or with V_{B} >0 V (reverse bias), and SILC is also observed. Although the as-measured ΔV_{T} (and n) is somewhat lower (and higher) than the "actual" value since slow MSM method is used (see Chap. 1, Sect. 1.2), the clear "break" from monotonicity in the time kinetics of ΔV_{T} and increase in n at higher t_{STR} is not a delay artifact, and the appearance of SILC rather suggests additional contribution due to ΔV_{OT} at higher V_{GSTR} and V_{B} >0 V stress. Similar results of higher time slope n at higher V_{GSTR} are also reported in [61].

Figure 3.18 shows the time evolution of ΔV_{T} measured (a) during and (b) after stress in a p-MOSFET with slightly thinner (than that of Fig. 3.17) SiO_2 gate insulator. The stress is done without and with reverse V_{B}, and measurements are done using slow MSM method [62]. The difference in measured ΔV_{T} for stress with and without reverse V_{B} is shown in panels (a) and (b). SILC is measured during stress using a slow MSM method, which is also plotted in Fig. 3.18(a). Consistent with Fig. 3.17, ΔV_{T} time kinetics shows higher magnitude and n at longer t_{STR} for V_{B} >0 V stress,

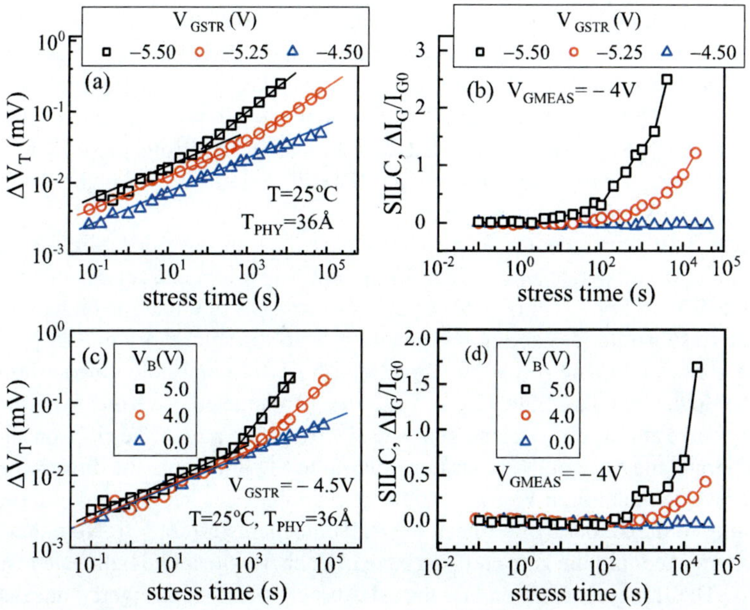

Fig. 3.17 Time evolution of measured (a, c) ΔV_T and (b, d) SILC during stress (a, b) without and (c, d) with reverse V_B. Data from [60]

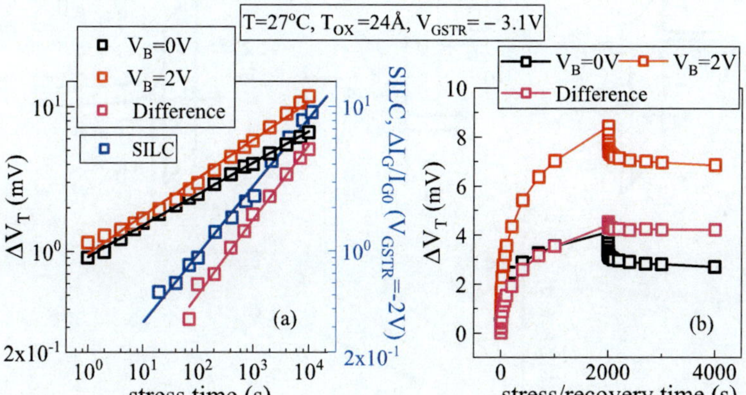

Fig. 3.18 Time evolution of measured ΔV_T (a) during and (b) after stress without and with reverse V_B, and their difference. SILC data are also measured for stress with reverse V_B and are shown in (a). Data from [62]

Fig. 3.18(a), which coincides with the appearance of SILC. Moreover, the difference in ΔV_T (between V_B >0 V and $V_B = 0$ V stress) and SILC (V_B >0 V stress only, no SILC is observed for $V_B = 0$ V stress) show power-law time kinetics with identical slope n ~0.5. Note, the "additional" ΔV_T under V_B >0 stress shows no recovery (and no recovery is also observed in SILC in this case). Therefore, the ΔV_T recovery fraction is lower after V_B >0 V stress, since the end of stress ΔV_T is higher compared to the $V_B = 0$ V case.

Figure 3.19 illustrates the energy band diagrams during p-MOSFET inversion stress under (a) moderate and (b) high V_{GSTR} and $V_B = 0$ V, and (c) moderate V_{GSTR} but $V_B > 0$ V. In Fig. 3.19(a), gate current (I_G) consists of electrons tunneling from the gate to substrate (I_B) and holes tunneling from the source-drain (SD) junctions via the substrate to gate (I_{SD}). The direction of electron and hole flow is shown by arrows. However, under high V_{GSTR}, Fig. 3.19(b), tunneled electrons from the gate impact ionize at the substrate and generate electron-hole pairs. The electrons flow to the substrate, majority of the holes flow out of the SD junctions (the direction of I_{SD} would reverse under high V_{GSTR}), and the remaining holes get injected into the gate according to the Anode Hole Injection (AHI) mechanism [56, 63]. Note that holes that are injected into the gate and also have high energy generate bulk gate insulator defects. The impact ionization and the AHI process also get triggered at relatively lower V_{GSTR} under V_B >0 V stress, Fig. 3.19(c).

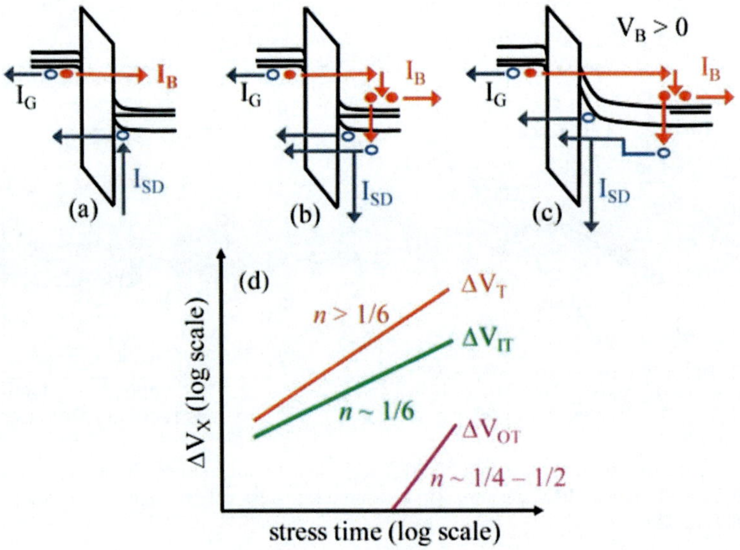

Fig. 3.19 Energy band diagram of (a) p-MOSFET during inversion stress under (a) moderate and (b) high V_{GSTR} but $V_B = 0$ V, and (c) moderate V_{GSTR} but V_B >0 V. The gate tunneling current and the underlying source drain and substrate components are shown. In (b, c) the AHI process is illustrated. (d) Schematic illustration of the impact of uncorrelated contributions from ΔV_{IT} and ΔV_{OT} on the time kinetics of overall ΔV_T

Note that bulk trap generation shows power-law time dependence with reported n ~0.25–0.5 from SILC (see Fig. 3.17, and also [43, 59, 64–69]). Therefore, when present (for stress conditions shown in Figs. 3.17 and 3.18), the addition of ΔV_{OT} to ΔV_{IT} would increase the magnitude and n of overall ΔV_{T} as illustrated in Fig. 3.19(d). Although the impact of ΔV_{OT} on overall ΔV_{T} may not be always as drastic as shown in Figs. 3.17 and 3.18, they can still be present during stress and hence needs to be modeled [13–16, 21–26, 26, 29, 31, 33]. The Reaction Diffusion Drift (RDD) model is used for the ΔV_{OT} time kinetics, which is explained in Chap. 6.

3.5 Summary

Interface trap generation is always present during NBTI stress, for different devices and stress conditions. The time kinetics of ΔV_{IT} shows power-law dependence during stress, with n ~1/6 slope under different experimental conditions. The impact of hole trapping is more prominent for devices having higher density of pre-existing (process related) bulk gate insulator defects. The impact of bulk trap generation is more prominent during stress under high V_{GSTR} and T. The time kinetics of ΔV_{HT} saturates at longer time during stress, and therefore, the addition of ΔV_{HT} to ΔV_{IT} reduces the slope n of overall ΔV_{T} at longer time. Due to the lower T activation E_{A} of ΔV_{HT} compared to ΔV_{IT}, the presence of ΔV_{HT} also reduces the E_{A} of overall ΔV_{T}. Finally, ΔV_{OT} shows power-law time dependence with n ~1/3 slope at longer stress time. Therefore, the addition of ΔV_{OT} to ΔV_{IT} increases the slope n of overall ΔV_{T} at longer time. The E_{A} of overall ΔV_{T} also increases in this case, due to higher E_{A} of the ΔV_{OT} compared to ΔV_{IT} subcomponent, which is discussed in detail in later chapters of this book.

The interface trap contribution is modeled in Chap. 4. Contributions due to hole trapping and bulk trap generation are modeled in Chap. 5 and Chap. 6 respectively.

Acknowledgements All experimental data presented in this chapter are re-plotted from previously published reports. The authors acknowledge Applied Materials for providing planar MOSFETs. Kaustubh Joshi is acknowledged for data in Fig. 3.8. Karansingh Thakor is acknowledged for help with manuscript preparation.

References

1. N. Kimizuka, K. Yamaguchi, K. Imai, T. Iizuka, C.T. Liu, R.C. Keller, T. Horiuchi, in *Symposium on VLSI Technology Digest of Technical Papers*, 92 (2000)
2. V. Huard, M. Denais, F. Perrier, N. Revil, C. Parthasarathy, A. Bravaix, E. Vincent, Microelectron. Reliab. **45**, 83 (2005)
3. A.T. Krishnan, C. Chancellor, S. Chakravarthi, P.E. Nicollian, V. Reddy, A. Varghese, R.B. Khamankar, S. Krishnan, in *IEEE International Electron Devices Meeting Technical Digest*, 688 (2005)

4. D. Varghese, D. Saha, S. Mahapatra, K. Ahmed, F. Nouri, M. Alam, in *IEEE International Electron Devices Meeting Technical Digest*, 684 (2005)
5. Y. Mitani, T. Yamaguchi, H. Satake, A. Toriumi, in *IEEE International Reliability Physics Symposium Proceedings*, 226 (2007)
6. S. Mahapatra, K. Ahmed, D. Varghese, A.E. Islam, G. Gupta, L. Madhav, D. Saha, M.A. Alam, in *IEEE International Reliability Physics Symposium Proceedings*, 1 (2007)
7. Y. Mitani, H. Satake, A. Toriumi, IEEE Trans. Device Mater. Reliab. **8**, 6 (2008)
8. S. Pae, M. Agostinelli, M. Brazier, R. Chau, G. Dewey, T. Ghani, M. Hattendorf, J. Hicks, J. Kavalieros, K. Kuhn, M. Kuhn, J. Maiz, M. Metz, K. Mistry, C. Prasad, S. Ramey, A. Roskowski, J. Sandford, C. Thomas, J. Thomas, C. Wiegand, J. Wiedemer, in *IEEE International Reliability Physics Symposium Proceedings*, 352 (2008)
9. S. Krishnan, U. Kwon, N. Moumen, M.W. Stoker, E.C.T. Harley, S. Bedell, D. Nair, B. Greene, W. Henson, M. Chowdhury, D.P. Prakash, E. Wu, D. Ioannou, E. Cartier, M.-H. Na, S. Inumiya, K. Mcstay, L. Edge, R. Iijima, J. Cai, M. Frank, M. Hargrove, D. Guo, A. Kerber, H. Jagannathan, T. Ando, J. Shepard, S. Siddiqui, M. Dai, H. Bu, J. Schaeffer, D. Jaeger, K. Barla, T. Wallner, S. Uchimura, Y. Lee, G. Karve, S. Zafar, D. Schepis, Y. Wang, R. Donaton, S. Saroop, P. Montanini, Y. Liang, J. Stathis, R. Carter, R. Pal, V. Paruchuri, H. Yamasaki, J.-H. Lee, M. Ostermayr, J.-P. Han, Y. Hu, M. Gribelyuk, D.-G. Park, X. Chen, S. Samavedam, S. Narasimha, P. Agnello, M. Khare, R. Divakaruni, V. Narayanan, M. Chudzik, in *IEEE International Electron Devices Meeting Technical Digest*, 28.1.1 (2011)
10. K. Joshi, S. Hung, S. Mukhopadhyay, V. Chaudhary, N. Nanaware, B. Rajamohanan, T. Sato, M. Bevan, A. Wei, A. Noori, B. McDougal, C. Ni, G. Saheli, C. Lazik, P. Liu, D. Chu, L. Date, S. Datta, A. Brand, J. Swenberg, S. Mahapatra, in *IEEE International Reliability Physics Symposium Proceedings*, 4C.2.1 (2013)
11. J. Franco, B. Kaczer, P.J. Roussel, J. Mitard, M. Cho, L. Witters, T. Grasser, G. Groeseneken, IEEE Trans. Electron Devices **60**, 396 (2013)
12. P. Srinivasan, J. Fronheiser, K. Akarvardar, A. Kerber, L.F. Edge, R.G. Southwick, E. Cartier, H. Kothari, in *IEEE International Reliability Physics Symposium Proceedings*, 6A.3.1 (2014)
13. N. Parihar, N. Goel, S. Mukhopadhyay, S. Mahapatra, IEEE Trans. Electron Devices **65**, 392 (2018)
14. N. Parihar, S. Mahapatra, in *IEEE International Reliability Physics Symposium Proceedings*, TX.5.1 (2018)
15. V. Huard, C. Ndiaye, M. Arabi, N. Parihar, X. Federspiel, S. Mhira, S. Mahapatra, A. Bravaix, in *IEEE International Reliability Physics Symposium Proceedings*, TX.4.1 (2018)
16. N. Parihar, R. Tiwari, C. Ndiaye, M. Arabi, S. Mhira, H. Wong, S. Motzny, V. Moroz, V. Huard, S. Mahapatra, in *International Conference on Simulation of Semiconductor Processes and Devices*, 167 (2018)
17. S. Ramey, A. Ashutosh, C. Auth, J. Clifford, M. Hattendorf, J. Hicks, R. James, A. Rahman, V. Sharma, A.St. Amour, C. Wiegand, in *IEEE International Reliability Physics Symposium Proceedings*, 4C.5.1 (2013)
18. K.T. Lee, W. Kang, E.-A. Chung, G. Kim, H. Shim, H. Lee, H. Kim, M. Choe, N.-I. Lee, A. Patel, J. Park, J. Park, in *IEEE International Reliability Physics Symposium Proceedings*, 2D.1.1 (2013)
19. J. Franco, B. Kaczer, A. Chasin, H. Mertens, L.-A. Ragnarsson, R. Ritzenthaler, S. Mukhopadhyay, H. Arimura, P.J. Roussel, E. Bury, N. Horiguchi, D. Linten, G. Groeseneken, A. Thean, in *IEEE International Reliability Physics Symposium Proceedings*, 4B.2.1 (2016)
20. G. Jiao, M. Toledano-Luque, K.-J. Nam, N. Toshiro, S.-H. Lee, J.-S. Kim, T. Kauerauf, E. Chung, D. Bae, G. Bae, D.-W. Kim, K. Hwang, in *IEEE International Electron Devices Meeting Technical Digest*, 31.2.1 (2016)
21. N. Parihar, R.G. Southwick, U. Sharma, M. Wang, J. H. Stathis, S. Mahapatra, in *IEEE International Reliability Physics Symposium Proceedings*, 2D.4.1 (2017)
22. N. Parihar, R. Southwick, M. Wang, J.H. Stathis, S. Mahapatra, in *IEEE International Electron Devices Meeting Technical Digest*, 7.3.1 (2017)

23. N. Parihar, U. Sharma, R.G. Southwick, M. Wang, J.H. Stathis, S. Mahapatra, IEEE Trans. Electron Devices **65**, 23 (2018)
24. N. Parihar, R.G. Southwick, M. Wang, J.H. Stathis, S. Mahapatra, IEEE Trans. Electron Devices **65**, 1699 (2018)
25. N. Parihar, R.G. Southwick, M. Wang, J.H. Stathis, S. Mahapatra, IEEE Trans. Electron Devices **65**, 1707 (2018)
26. N. Parihar, R. Tiwari, S. Mahapatra, in *International Conference on Simulation of Semiconductor Processes and Devices*, 176 (2018)
27. R. Tiwari, N. Parihar, K. Thakor, H.Y. Wong, S. Motzny, M. Choi, V. Moroz, S. Mahapatra, IEEE Trans. Electron Devices **66**, 2086 (2019)
28. R. Tiwari, N. Parihar, K. Thakor, H.Y. Wong, S. Motzny, M. Choi, V. Moroz, S. Mahapatra, IEEE Trans. Electron Devices **66**, 2093 (2019)
29. N. Parihar, U. Sharma, R. G. Southwick, M. Wang, J.H. Stathis, S. Mahapatra, in *IEEE International Reliability Physics Symposium Proceedings* (2019)
30. M. Wang, J. Zhang, H. Zhou, R.G. Southwick, R. Hsin, K. Chao, X. Miao, V.S. Basker, T. Yamashita, D. Guo, G. Karve, H. Bu, in *IEEE International Reliability Physics Symposium Proceedings* (2019)
31. N. Choudhury, U. Sharma, H. Zhou, R.G. Southwick, M. Wang, S. Mahapatra, in *IEEE International Reliability Physics Symposium Proceedings* (2020)
32. H. Zhou, M. Wang, J. Zhang, K. Watanabe, C. Durfee, S. Mochizuki, R. Bao, R. Southwick, M. Bhuiyan, B. Veeraraghavan, in *IEEE International Reliability Physics Symposium Proceedings* (2020)
33. N. Choudhury, T. Samadder, R. Tiwari, H. Zhou, R.G. Southwick, M. Wang, S. Mahapatra, in *IEEE International Reliability Physics Symposium Proceedings* (2021)
34. S. Tsujikawa, T. Mine, K. Watanabe, Y. Shimamoto, R. Tsuchiya, K. Ohnishi, T. Onai, J. Yugami, S. Kimura, in *IEEE International Reliability Physics Symposium Proceedings*, 183 (2003)
35. Y. Taur, T.H. Ning, *Fundamentals of Modern VLSI Devices* (Cambridge University Press, 2009)
36. W.J. Liu, Z.Y. Liu, D. Huang, C.C. Liao, L.F. Zhang, Z.H. Gan, W. Wong, C. Shen, M.-F. Li, in *IEEE International Electron Devices Meeting Technical Digest*, 813 (2007)
37. G. Groeseneken, H.E. Maes, N. Beltran, R.F. De Keersmaecker, IEEE Trans. Electron Devices **31**, 42 (1984)
38. G. Chen, M.F. Li, C.H. Ang, J.Z. Zheng, D.L. Kwong, IEEE Electron Device Lett. **23**, 734 (2002)
39. J.H. Stathis, G. LaRosa, A. Chou, in *IEEE International Reliability Physics Symposium Proceedings*, 1 (2004)
40. J.P. Campbell, P.M. Lenahan, A.T. Krishnan, S. Krishnan, in *IEEE International Reliability Physics Symposium Proceedings*, 442 (2006)
41. A. Neugroschel, G. Bersuker, R. Choi, C. Cochrane, P. Lenahan, D. Heh, C. Young, C. Y. Kang, B.H. Lee, R. Jammy, in *International Electron Devices Meeting Technical Digest* (2006)
42. S. Mahapatra, N. Goel, S. Desai, S. Gupta, B. Jose, S. Mukhopadhyay, K. Joshi, A. Jain, A.E. Islam, M.A. Alam, IEEE Trans. Electron Devices **60**, 901 (2013)
43. S. Mukhopadhyay, K. Joshi, V. Chaudhary, N. Goel, S. De, R.K. Pandey, K.V.R.M. Murali, S. Mahapatra, in *IEEE International Reliability Physics Symposium Proceedings*, GD 3.1 (2014)
44. A. Neugroschel, Chih-Tang Sah, K.M. Han, M.S. Carroll, T. Nishida, J. T. Kavalieros, Yi Lu, IEEE Trans. Electron Devices **42**, 1657 (1995)
45. P.E. Nicollian, M. Rodder, D.T. Grider, P. Chen, R.M. Wallace, S.V. Hattangady, in *IEEE International Reliability Physics Symposium Proceedings*, 404 (1999)
46. S. Mahapatra, N. Goel, A. Chaudhary, K. Joshi, S. Mukhopadhyay, in *Fundamentals of Bias Temperature Instability in MOS Transistors* (Springer India, 2015), pp. 43–92
47. T. Grasser, W. Gos, V. Sverdlov, B. Kaczer, in *IEEE International Reliability Physics Symposium Proceedings*, 268 (2007)
48. H. Reisinger, O. Blank, W. Heinrigs, A. Muhlhoff, W. Gustin, C. Schlunder, in *IEEE International Reliability Physics Symposium Proceedings*, 448 (2006)

49. K. Sakuma, D. Matsushita, K. Muraoka, Y. Mitani, in *IEEE International Reliability Physics Symposium Proceedings*, 454 (2006)
50. J.H. Lee, W.H. Wu, A.E. Islam, M.A. Alam, A.S. Oates, in *IEEE International Reliability Physics Symposium Proceedings*, 745 (2008)
51. V. Huard, in *IEEE International Reliability Physics Symposium Proceedings*, 33 (2010)
52. B. Kaczer, T. Grasser, J. Martin-Martinez, E. Simoen, M. Aoulaiche, P.J. Roussel, G. Groeseneken, in *IEEE International Reliability Physics Symposium Proceedings*, 55 (2009)
53. K.K. Hung, P.K. Ko, C. Hu, Y.C. Cheng, IEEE Trans. Electron Devices **37**, 654 (1990)
54. G. Kapila, N. Goyal, V.D. Maheta, C. Olsen, K. Ahmed, S. Mahapatra, in *IEEE International Electron Devices Meeting Technical Digest* (2008)
55. J.W. McPherson, H.C. Mogul, J. Appl. Phys. **84**, 1513 (1998)
56. M. A. Alam, J. Bude, A. Ghetti, in *IEEE International Reliability Physics Symposium Proceedings*, 21 (2000)
57. E. Wu, J. Sune, C. LaRow, R. Dufresne, in *2012 International Electron Devices Meeting Technical Digest*, 28.5.1 (2012)
58. M.A. Alam, IEEE Trans. Electron Devices **49**, 226 (2002)
59. J. Yang, M. Masuduzzaman, K. Joshi, S. Mukhopadhyay, J. Kang, S. Mahapatra, M. A. Alam, in *IEEE International Reliability Physics Symposium Proceedings*, 5D.4.1 (2012)
60. S. Mahapatra, P. Bharath Kumar, M.A. Alam, IEEE Trans. Electron Devices **51**, 1371 (2004)
61. C.L. Chen, Y.M. Lin, C. J. Wang, K. Wu, in *IEEE International Reliability Physics Symposium Proceedings*, 704 (2005)
62. S. Mahapatra, D. Saha, D. Varghese, P. Bharath Kumar, IEEE Trans. Electron Devices **53**, 1583 (2006)
63. K.F. Schuegraf, C. Hu, IEEE Trans. Electron Devices **41**, 761 (1994)
64. K. Okada, H. Kubo, A. Ishinaga, K. Yoneda, in *Symposium on VLSI Technology Digest of Technical Papers*, 158 (1998)
65. T. Nigam, R. Degraeve, G. Groeseneken, M.M. Heyns, H.E. Maes, in *IEEE International Reliability Physics Symposium Proceedings*, 381 (1999)
66. A. Ghetti, J. Bude, G. Weber, in *Symposium on VLSI Technology Digest of Technical Papers*, 218 (2000)
67. E.Y. Wu, J. Sune, W. Lai, IEEE Trans. Electron Devices **49**, 2141 (2002)
68. P.E. Nicollian, A.T. Krishnan, C. Bowen, S. Chakravarthi, C.A. Chancellor, R.B. Khamankar, in *IEEE International Electron Devices Meeting Technical Digest*, 392 (2005)
69. W.L. Chang, J.H. Stathis, E. Cartier, in *IEEE International Reliability Physics Symposium Proceedings*, 787 (2010)

Chapter 4
BTI Analysis Tool (BAT) Model Framework—Generation of Interface Traps

Souvik Mahapatra, Narendra Parihar, Subhadeep Mukhopadhyay, and Nilesh Goel

4.1 Introduction

As discussed in the earlier chapters, Negative Bias Temperature Instability (NBTI) became an important reliability concern with the migration from Silicon Dioxide (SiO_2) to Silicon Oxynitride (SiON) gate insulator MOSFETs [1–7]. It continues to remain as a concern for dual layer (SiO_2 or SiON Interlayer (IL) and Hafnium Dioxide (HfO_2) High-K layer) High-K Metal Gate (HKMG) gate insulator-based bulk [8–13] and Fully Depleted Silicon On Insulator (FDSOI) [14, 15] planar MOSFETs, bulk and SOI FinFETs [15–28], as well as Gate All Around Stacked Nanosheet FETs [29–32], with either Silicon (Si) or Silicon Germanium (SiGe) channel. The key features of NBTI are summarized hereinafter (reproduced from Chap. 3, Sect. 3.1).

As described in Chap. 1, Sect. 1.3, NBTI results in gradual buildup of positive charges in a p-MOSFET gate insulator and causes threshold voltage (ΔV_T) in time under the application of a negative gate bias (V_G). ΔV_T accelerates at higher magnitude of V_G during stress ($V_G = V_{GSTR}$) and higher temperature (T), governed by the Voltage Acceleration Factor (VAF) and Arrhenius T activation energy (E_A) respectively. The parametric shift accrued during stress partially recovers after stress when the magnitude of V_G is reduced or removed ($V_G = V_{GREC}$ or 0 V), and therefore, AC stress results in lower ΔV_T than DC. The AC to DC ratio depends on the Pulse Duty Cycle (PDC) and pulse low bias (V_{GLOW}); however, it may or may not depend on the frequency (f) of the gate pulse (depends on AC stress mode). On the other hand, NBTI recovery necessitates the use of ultra-fast methods for artifact free measurements, which is discussed in Chap. 1, Sect. 1.2.

S. Mahapatra (✉) · N. Parihar · S. Mukhopadhyay · N. Goel
Department of Electrical Engineering, Indian Institute of Technology Bombay, Powai, Mumbai 400076, India
e-mail: souvik@ee.iitb.ac.in

S. Mahapatra (ed.), *Recent Advances in PMOS Negative Bias Temperature Instability*,
https://doi.org/10.1007/978-981-16-6120-4_4

As described in Chap. 2, the time kinetics of measured ΔV_T and related parameters, such as the power law slope (n) at longer stress time (t_{STR}), VAF, E_A, T dependence of VAF during stress and Fraction Remaining (FR) during recovery after stress (FR is defined as ΔV_T at $t = t_{REC}$ after stress to that at $t = t_{STR}$ at the end of stress) depend on different transistor processes. In modern HKMG gate insulator-based p-MOSFETs, some of the key processes that influence NBTI are Nitrogen content (N%) in the gate insulator and Germanium content (Ge%) in the channel. The magnitude of ΔV_T increases, while the associated n, VAF, E_A, T dependence of VAF and FR reduce with higher N%. On the other hand, the magnitude of ΔV_T reduces, while the associated n, VAF, E_A, T dependence of VAF and FR increase with higher Ge%. Moreover, NBTI reduces with fin length and fin width scaling in FinFETs, sheet length scaling in GAA-SNS FETs and larger spacing (SA) between the Shallow Trench Isolation (STI) and device active in FDSOI MOSFETs; however, it increases with sheet width scaling in GAA-SNS FETs.

Any practical and technologically relevant modeling framework should be able to explain the experimental features listed in Chap. 3, Sect. 3.1. The BTI Analysis Tool (BAT) framework described in this book models NBTI parametric drift using uncorrelated contributions from generated interface traps (density ΔN_{IT}) and bulk gate insulator traps (density ΔN_{OT}), and hole trapping in preexisting bulk gate insulator traps (density ΔN_{HT}). Several independent experimental evidences regarding the impact of these underlying subcomponents are demonstrated in Chap. 3. In this chapter, the Reaction Diffusion (RD) model is explained to calculate the time kinetics of ΔN_{IT}. Other subcomponents are modeled in Chaps. 5 and 6.

4.2 BTI Analysis Tool (BAT) Framework

Figure 4.1 illustrates the BAT framework used throughout this book and the underlying subcomponents of NBTI degradation [12]. The time kinetics of measured ΔV_T is due to uncorrelated contributions from generated interface (ΔV_{IT}) and bulk gate

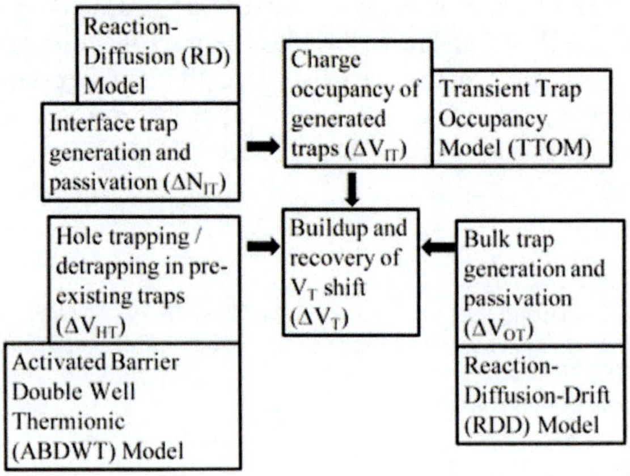

Fig. 4.1 Schematic of the BTI Analysis Tool (BAT) framework used in this book to model measured ΔV_T kinetics during and after DC and AC NBTI stress

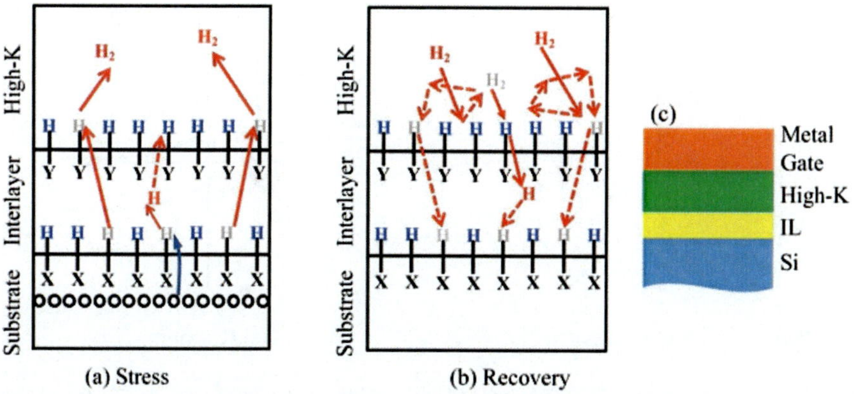

Fig. 4.2 Schematic of the Reaction Diffusion (RD) model with defect-assisted dimerization and reverse process to calculate the time kinetics of trap generation and passivation respectively (a) during and (b) after NBTI stress, example is shown for (c) HKMG gate insulator. The RD model equations are described in Sect. 4.3

insulator (ΔV_{OT}) traps, and hole trapping in preexisting bulk gate insulator defects (ΔV_{HT}). Note that the electrically active gate insulator defects are denoted as traps, and these terms would be interchangeably used in this book.

Figure 4.2 illustrates the double interface Reaction Diffusion (RD) model [12, 33, 34], which is used to calculate the generation and passivation of interface traps in this book. The RD model calculates the depassivation and re-passivation of Hydrogen (H) passivated defects at the channel/interlayer (IL) interface and inside the bulk of the HKMG gate stack. For simplicity, all bulk defects are lumped into a suitable "second interface" which is assigned to the IL/High-K interface for HKMG gate insulator, as illustrated in this example. The "second interface" can be defined at the center of the gate insulator for a single-layer SiO_2 or SiON gate stack. The RD model is discussed in Sect. 4.3.

Figure 4.3 illustrates the energy band diagram of a dual layer HKMG stack (a) during and (b) after stress. RD model calculates trap kinetics at the channel/IL and IL/High-K interfaces. The traps are presumed donor type, and therefore, only the ones energetically located above the Fermi level of the substrate would be positively charged and contribute (ΔV_{IT}) to ΔV_T during stress (any generated traps located below the Fermi level is not shown). During recovery, some of these traps would go below the Fermi level and would capture electrons to neutralize. Therefore, they would not contribute to ΔV_T, although they can physically exist (and at a later time can get re-passivated). Trap occupancy is calculated by the Transient Trap Occupancy Model (TTOM) [12], which is discussed in Chap. 5.

Trapping (and detrapping) of holes into (and out of) preexisting gate insulator defects give rise to hole trapping contribution (ΔV_{HT}) to ΔV_T as shown in Fig. 4.1. This is calculated by the Activated Barrier Double Well Thermionic (ABDWT) model [35], which is discussed in Chap. 5. Contribution due to generated bulk gate insulator traps (ΔV_{OT}) also contributes, Fig. 4.1. This is calculated by the Reaction Diffusion Drift (RDD) model [36] and is discussed in Chap. 6.

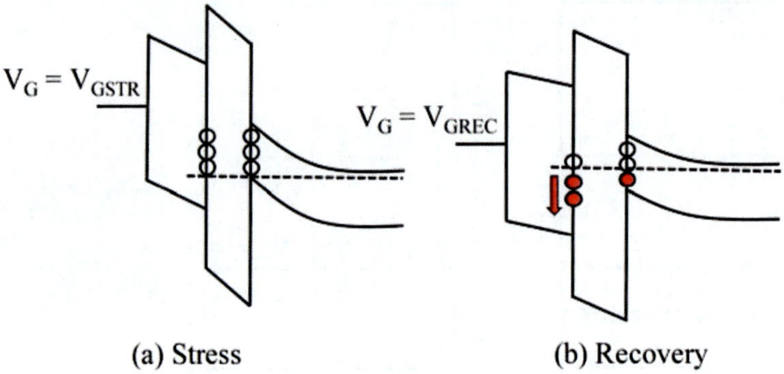

Fig. 4.3 Transient Trap Occupancy Model (TTOM) for the calculation of interface trap occupancy (a) during and (b) after stress; the example is shown for a HKMG gate insulator. The TTOM equations are described in Chap. 5

Note that the bulk trap generation is possibly due to breaking of Si-O-Si bonds (Si: Silicon, O: Oxygen) or other (different) H passivated defects to create Si-Si dimer or other forms of Oxygen vacancy (O_V) [37] (note that the chemical nature of defects is discussed in Sect. 4.6). The bulk trap generation is related to hot holes generated by the Anode Hole Injection (AHI) mechanism [38], which is discussed in Chap. 6. However, the interface trap generation is triggered by the tunneling of inversion layer (cold) holes, which is discussed in this chapter.

4.3 Reaction Diffusion (RD) Model

Figure 4.4 illustrates different versions of the RD model proposed in the literature. The time kinetics of interface traps has been modeled first using the conventional RD framework [39, 40], which suggests the depassivation (during stress) and repassivation (after stress) of H passivated defects at the channel/gate insulator interface. The released H atoms diffuse into the oxide (and beyond) during stress, as illustrated in Fig. 4.4 (a), and diffuse back toward the interface after stress. The framework is reaction limited at short time and atomic H diffusion limited at long time during stress [40]. Note, this basic version of the RD model could explain the power law time dependence of measured NBTI kinetics during stress with a time slope $n \sim 1/4$, which has been reported in older publications [1, 2, 5, 7]. It could also explain the frequency independence of NBTI during AC stress [41], reported in early experiments [42] (see Chap. 1, Sect. 1.3 and Chap. 14 for a discussion on the f dependence or independence of NBTI).

However, all the direct methods to measure interface trap generation are intrinsically slow and suffer from recovery-related artifacts (lower magnitude and higher time slope n) as the stress is interrupted for measurement. As shown in Chap. 3, Sect. 3.2, measured time kinetics of ΔN_{IT} after delay correction results in $n \sim 1/6$, which is universally observed across different stress conditions.

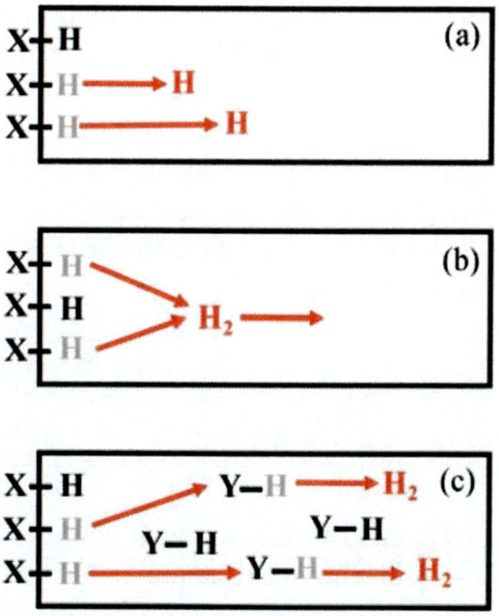

Fig. 4.4 Different versions of RD model: (a) basic with only H diffusion, (b) dimerization of H into H_2 and diffusion of H_2, and (c) defect-assisted dimerization of H into H_2 and diffusion of H_2 (also shown in Fig. 4.2 where the bulk defects are lumped into an effective second interface for simplicity)

Therefore, the RD model has been suitably modified by adding dimerization of atomic H into molecular H_2 as illustrated in Fig. 4.4 (b) [3, 33, 34, 43]. This model is reaction limited at shorter time, governed by the conversion of H to H_2 at moderate time and H_2 diffusion limited at longer time during stress; the opposite processes occur during recovery after stress. The model can explain the measured and delay corrected $n \sim 1/6$ time slope at long-time stress and frequency independence for AC stress [33]. However, it is shown later that the dimerization of two H atoms into H_2 is stochastically less probable in small area devices [44], and moreover, the reverse dissociation of H_2 into H requires a very large T activation energy of $E_A = 4.5$ eV [45]. Therefore, the defect-assisted dimerization model is preferred, Fig. 4.4 (c), which is discussed next. The defect-assisted version is consistent between the deterministic and stochastic implementations, as demonstrated in [46]. The interested reader may refer to [34] for further details on various versions of the RD model.

4.3.1 *RD Model with Defect-Assisted Dimerization*

Figure 4.2 illustrates the RD model with defect-assisted (a) dimerization of H into H_2 during stress and (b) reverse conversion of H_2 to H during recovery after stress for a (c) HKMG gate insulator stack. During stress, inversion layer holes break the H passivated defects (X-H) at the channel/IL interface, and the detailed mechanism is discussed later in this section. The released H atoms from broken X-H bonds diffuse into the gate insulator and react with other H passivated defects (Y-H bonds, lumped at the IL/High-K interface for simplicity) to produce H_2 molecules (defect-assisted dimerization). The generated H_2 molecules diffuse into the gate insulator bulk and backend. During recovery, H_2 molecules diffuse back and passivate the bulk insulator defects, and the generated H atoms subsequently diffuse and passivate the defects at the channel/gate insulator interface. The chemical nature of defect precursors is discussed in Sect. 4.6, and due to uncertainties, they are denoted as X-H and Y-H bonds in this book. The model remains valid for single-layer SiO_2 or SiON gate insulators, where the bulk defects can be lumped into an imaginary interface at the center of the gate insulator stack.

The RD model with defect-assisted dimerization is explained by the following chemical reactions:

$$X - H + (\text{hole}) \leftrightarrow X - + H \tag{4.1}$$

$$Y - H + H \leftrightarrow Y - + H_2 \tag{4.2}$$

The charged state (occupancy) of X—at the channel/gate insulator interface and of Y—inside the gate insulator bulk is determined by whether these defects, presumed donor like, are energetically located above (positively charged) or below (neutral) the Fermi level (of the substrate) during stress and post-stress phases, see Fig. 4.3. As mentioned before, this aspect is handled using the TTOM framework and is discussed in Chap. 5. Total ΔV_{IT} is calculated by summing the contributions from ΔV_{IT1} at the channel/IL and ΔV_{IT2} at the IL/High-K interfaces, by using appropriate capacitance ratios, and counting only traps that are energetically above the Fermi level during or after stress from the TTOM framework.

Note that during stress, the released H atoms have to find H passivated defects in the gate insulator bulk to initiate the depassivation reactions, and during recovery, the returning H_2 molecules have to find the un-passivated defects to initiate the re-passivation reactions. However, the diffusivity of H atoms is much larger than that of H_2 molecules [3]. Therefore, the H atoms can quickly find the required precursors during stress, Fig. 4.2 (a), but the H_2 molecules would need to hop till they can find the un-passivated defects during recovery, Fig. 4.2 (b), before the respective forward and reverse reactions can proceed [33]. The H_2 molecules would hop more when the difference between the stress and recovery time is large (relatively shorter stress time and longer recovery time), due to limited availability of the un-passivated defects that can take part in the reverse reaction. Furthermore, a fraction of the H atoms and/or H_2 molecules can get temporarily locked out of the diffusion domain, due to trapping/bonding or otherwise, and become unavailable for the reaction and/or diffusion processes [39, 47]. The H_2 hopping and lock-in processes have been simulated and their roles in slowing down the recovery kinetics have been verified in the stochastic simulation domain [46]. However, in the continuum (deterministic) simulation domain, these effects are handled by slowing down the diffusivity of H_2 molecules with the passage of time *only* during recovery after stress [33, 48].

The forward and reverse reactions at the channel/IL (first) and IL/High-K (second) interfaces, as per the chemical reactions shown in Eq. 4.1 and Eq. 4.2 respectively, are given by the following equations [12]:

$$\frac{dN_{\mathrm{IT}(1)}}{\mathrm{d}t} = K_{\mathrm{F1}}\left(N_{\mathrm{O}(1)} - N_{IT(1)}\right) - K_{R1}N_{\mathrm{IT}(1)}N_{\mathrm{H}(1)} \tag{4.3}$$

$$\frac{\mathrm{d}N_{\mathrm{IT}(2)}}{\mathrm{d}t} = K_{\mathrm{F2}}\left(N_{\mathrm{O}(2)} - N_{\mathrm{IT}(2)}\right)N_{\mathrm{H}(2)} - K_{R2}N_{\mathrm{IT}(2)}N_{\mathrm{H2}(2)} \tag{4.4}$$

where N_0, N_{IT}, N_{H} and N_{H2}, respectively, are the H passivated defect, trap (after H depassivation), atomic and molecular Hydrogen densities at the first (1) and second (2) interfaces, while K_{F1}, K_{F2} and K_{R1}, K_{R2} are the corresponding forward and reverse reaction rates. The flux balance is done by the following equations:

$$\frac{\delta}{2}\frac{\mathrm{d}N_{\mathrm{H}(1)}}{\mathrm{d}t} = D_{\mathrm{H}}\frac{\mathrm{d}N_{\mathrm{H}(1)}}{\mathrm{d}x} + \frac{\mathrm{d}N_{\mathrm{IT}(1)}}{\mathrm{d}t} \tag{4.5}$$

$$\frac{\delta}{2}\frac{\mathrm{d}N_{\mathrm{H2}(2)}}{\mathrm{d}t} = D_{\mathrm{H2}}\frac{\mathrm{d}N_{\mathrm{H2}(2)}}{\mathrm{d}x} \tag{4.6}$$

where δ is the interfacial layer thickness (=1.5 Å), and D_{H} and D_{H2} are the diffusivities of atomic and molecular Hydrogen respectively. The diffusion of H and H_2 species is governed by the following equations:

$$\frac{\mathrm{d}N_{\mathrm{H}}}{\mathrm{d}t} = D_{\mathrm{H}}\frac{\mathrm{d}^2N_{\mathrm{H}}}{\mathrm{d}x^2} \tag{4.7}$$

$$\frac{\mathrm{d}N_{\mathrm{H2}}}{\mathrm{d}t} = D_{\mathrm{H2}}\frac{\mathrm{d}^2N_{\mathrm{H2}}}{\mathrm{d}x^2} \tag{4.8}$$

The hopping and lock-in related slowing down of H_2 diffusion during recovery is handled by the following equation:

$$D_{\mathrm{H2}}(t) = \frac{D_{\mathrm{H2\_STRESS}}}{\left(1 + A * \left(\frac{\mathrm{t}}{t_{\mathrm{STR}}}\right)\right)} \tag{4.9}$$

where t is the recovery time, t_{STR} is stress time and $D_{\mathrm{H2\_STRESS}}$ is the diffusivity value used during stress and A is the diffusivity reduction parameter.

The first interface forward reaction rate (K_{F1}) depends on V_{GSTR}, T and transistor process and materials, and is explained below. All other forward (second interface) and reverse (first and second interfaces) reaction rates and diffusivities of H and H_2 are only Arrhenius T activated and are process independent. The diffusivity reduction parameter only depends on the device architecture ($A = 7$ is used for planar MOSFETs and $A = 35$ for FinFETs and GAA-SNS FETs). The process-independent RD model parameters is listed in Table 4.1, and same values are used to analyze different devices throughout this book.

Table 4.1 Process (or technology)-independent RD model parameters used throughout this book. These parameters are Arrhenius T activated: $X = X_0 \exp(-E_A/kT)$, where X_0 is the pre-factor and E_A is the T activation energy of the parameter of interest (X). The diffusivity pre-factors are mentioned for IL//High-K and beyond

Parameter	Unit	Pre-factor	E_A (eV)
K_{F2}	cm^3/s	5750	0.235
K_{R1}	cm^3/s	5×10^{-6}	0.12
K_{R2}	cm^3/s	7.5×10^{-4}	0.2
D_H	cm^2/s	$2 \times 10^{-2}//4 \times 10^{-5}$	0.2
D_{H2}	cm^2/s	$9.5 \times 10^{-11}//9.5 \times 10^{-8}$	0.5

4.3.2 Physical Mechanism of Interfacial Defect Dissociation

Figure 4.5 illustrates the H passivated defect dissociation process at the first interface and lists the equations governing K_{F1} [12, 33, 34]. During stress, the inversion layer holes aided by oxide electric field (E_{OX}) tunnel to the interfacial X-H bonds. These bonds are already polarized (by factor p) in the presence of E_{OX}, and upon hole capture they become weak and are subsequently dissociated by thermal activation. The forward reaction rate K_{F1} depends on the pre-factor K_{F10} (which is proportional to hole density, p_H, tunneling coefficient, T_H, and capture cross section, σ), field acceleration (Γ_E) and T activation of bond dissociation (E_{AKF1}). The field acceleration factor (Γ_E) is a sum of the T-independent (Γ_0) and T-dependent (α/kT) terms, where α is the polarization coefficient of the X-H bond.

The process-dependent RD model parameters are K_{F10}, Γ_0, α and E_{AKF1}, among which, the parameters K_{F10} and Γ_0 depend on the effective mass (m_T) and barrier (φ_B) of the hole tunneling process (and can be determined using bandstructure calculations). Different processes such as Ge% in the channel, N% in the gate insulator stack (near the channel/IL interface) and mechanical strain in the channel can impact the bandstructure and hence m_T and φ_B. These in turn impact the parameters K_{F10} and Γ_0. It is important to note that the pre-factor K_{F10} would also depend on the capture cross section and the quality of the gate insulator. Therefore, when the capture cross section and quality of the gate insulator stay similar, bandstructure calculations can

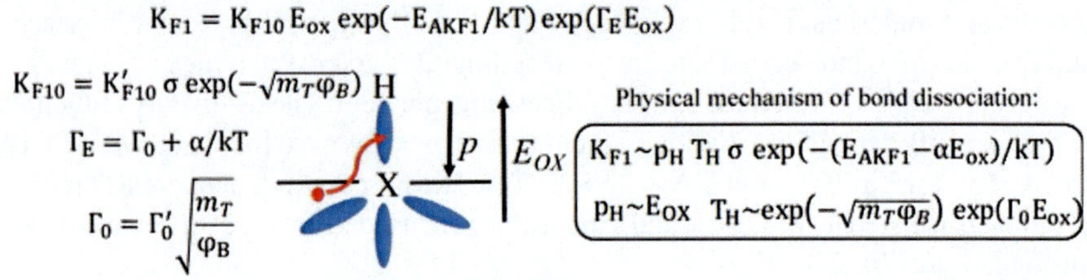

Fig. 4.5 Schematic of the inversion layer hole and oxide electric field induced dissociation of H passivated defects at the channel/gate insulator interface [34]

be used to determine the relative K_{F10} changes across different processes. The other two (α and E_{AKF1}) are pure fitting parameters.

The following process-dependent trends have been observed for the RD model parameters (discussed in detail in Sect. 4.5 and also in later chapters):

- Higher Ge% in the channel increases the valence band offset and hence φ_B with no significant impact on m_T, obtained from bandstructure calculations using the tight binding method [26, 49]. Therefore, both K_{F10} and Γ_0 reduces, although the overall Γ_E (and hence VAF) often increases due to the increase in α at higher Ge%. This remains valid for both (100) and (110) interfaces and also across different devices (bulk and FDSOI planar MOSFETs and FinFET), provided the N% is kept identical between different Ge% devices. Moreover, E_{AKF1} increases at higher Ge%, but the reason for this is not yet identified. The impact of Ge% changes on NBTI is analyzed and modeled in Chap. 8, Chap. 9 and Chap. 11, respectively, for bulk and FDSOI planar MOSFETs and FinFETs.
- Higher N% in the gate insulator near the channel/IL interface reduces the valence band offset (note that SiON has smaller bandgap than SiO_2) and hence reduces φ_B. Lower φ_B results in higher Γ_0, unless m_T also reduces at higher N%. Analysis of different devices indicates reduction in Γ_0 for the Si (100) surface but slight increase in Γ_0 for the SiGe (100) and (110) surfaces at higher N%. The relative changes in φ_B (always reduces) and m_T (reduces for the Si (100) surface but likely remains unchanged for the SiGe (100) and (110) surfaces) impact the overall change in Γ_0 at higher N%. Since α does not change, the variation in Γ_E with N% depends only on that of Γ_0. However, K_{F10} would always increase at higher N%, due to reduction in φ_B and also in m_T when applicable. The T activation E_{AKF1} reduces at higher N% and has been verified using atomistic calculations [50]. Note that the Equivalent Oxide Thickness (EOT) of the gate insulator reduces at higher N%, due to higher dielectric constant of the IL (ε_{IL}) and/or High-K (ε_{HK}), depending on the N profile in the gate stack. Therefore, a device having higher N% in the gate stack shows higher E_{OX} when compared to a lower N% device at iso-V_{GSTR}. Higher E_{OX} results in higher ΔV_{IT}; however, the resultant increase would be lower than expected, since higher ΔV_{IT} in turn would also reduce the effective NBTI stress at fixed V_{GSTR}, due to reduction in the channel electric field and inversion hole density. Therefore, even if Γ_E stays constant (due to the balancing of φ_B and m_T), the stress reduction effect results in lower VAF in higher N% devices (as the magnitude of ΔV_{IT} is higher due to higher K_{F10}). The impact of N% changes on NBTI is analyzed and modeled in Chap. 7, Chap. 9 and Chap. 11, for bulk and FDSOI planar MOSFETs and FinFETs respectively.
- Higher uniaxial compressive stress increases φ_B but does not significantly impact m_T for the (100) surface (relevant for planar, FDSOI, GAA-SNS FET top sheet), while it increases m_T but does not significantly impact φ_B for the (110) surface (relevant for FinFET sidewalls and GAA-SNS FET sheet sides), as obtained from bandstructure calculations. Therefore, both K_{F10} and Γ_0 are appropriately changed, if strain is changed due to changes in the layout or device dimensions. However, no noticeable strain impact is noted for α and E_{AKF1}. The impact of

mechanical strain on NBTI is analyzed and modeled for layout changes in FDSOI MOSFETs in Chap. 9, and for dimension changes in FinFETs and GAA-SNS FETs in Chaps. 12 and 13.

4.4 Experimental Validation of RD model

The RD model is validated using DCIV measured and delay corrected ΔN_{IT} time kinetics from planar MOSFETs and FinFETs, see Chap. 3, Sect. 3.2 for measurement and other details. Table 4.2 lists the four process-dependent RD model parameters for the Gate First (GF) HKMG planar p-MOSFETs (D1 and D2, different N% in the gate insulator) and Replacement Metal Gate (RMG) HKMG p-FinFETs (D3 and D4, different Ge% in the channel and N% in the gate insulator) used in this work. The process-independent parameters are listed in Table 4.1.

The X-H defect density at the channel/IL interface ($N_{0(1)}$) would be different for different devices, and it depends on channel orientation, gate stack thermal budget and Ge% in the channel. Note that $N_{0(1)}$ (in /cm^{2-}) values of 5×10^{12} and 7×10^{12} are, respectively, used for thermal and low T Chemical Oxide IL in (100) surface, and 1×10^{13} is used for Chemical Oxide IL in (110) surface for Si channel throughout the book. Moreover, the values are suitably reduced for SiGe channel depending on Ge%. On the other hand, the Y-H defect density at the second interface ($N_{0(2)}$) is taken as 5×10^{13} /cm^2 for all cases.

Figure 4.6 shows the time evolution of measured and modeled ΔN_{IT} in D1 (left panels) and D2 (right panels) GF planar devices at different V_{GSTR} and T during DC stress. As DCIV is a slow method, the measured data can be obtained only at longer stress time ($t_{STR} > 1$ s), while RD model simulation is shown from short to long time. Note that measured ΔN_{IT} increases with more negative V_{GSTR} and larger T as expected, increases with higher N% in the gate insulator (*e.g.*, for D2 compared to

Table 4.2 Process-dependent RD model parameters for different devices having Silicon (Si) and Silicon Germanium (SiGe) channels. The K_{F10} parameter is not listed for the D3 and D4 FinFETs to maintain confidentiality. The parameters for SiGe devices are strongly dependent on the details of the IL formation process (see Chap. 11 for further details)

Device	Unit	D1	D2	D3	D4
Type	–	Planar	Planar	FinFET	FinFET
Channel	–	Si	Si	Si	SiGe
IL type	–	Thermal	Thermal	Chem-Ox	Chem-Ox
Nitrogen	–	Low	High	Low	Medium
K_{F10}	cm/Vs	0.22	0.05	–	–
E_{AKF1}	eV	0.40	0.18	0.29	0.80
Γ_0	cm/MV	0.38	0.10	0.29	0.43
α	qÅ	1.2	1.2	1.8	2.3

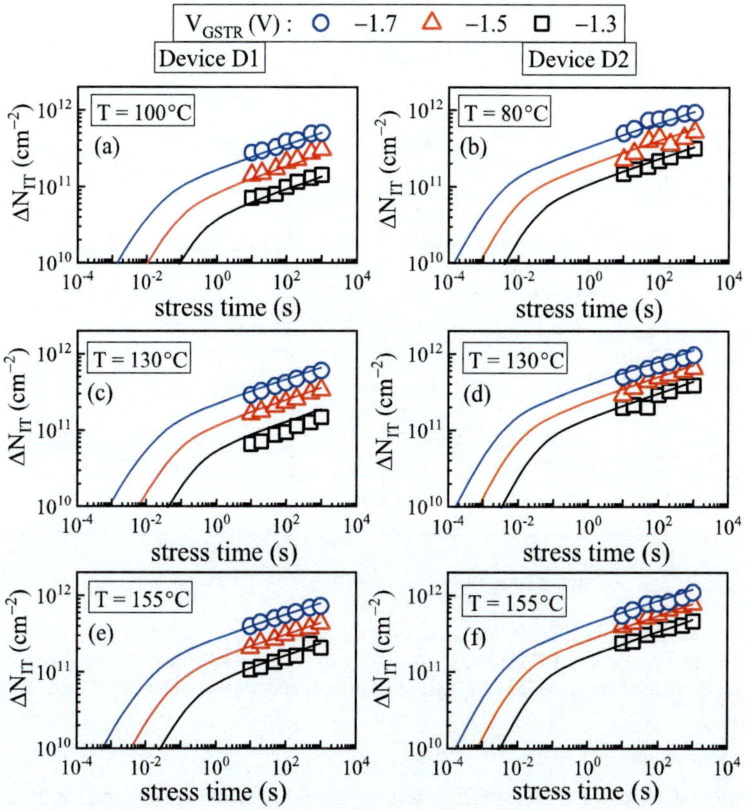

Fig. 4.6 Time evolution of DCIV measured (and delay corrected) and RD model simulated ΔN_{IT} at different V_{GSTR} and T ($V_{GSTR} \times T$ matrix) for DC stress in GF HKMG Si channel p-MOSFETs with low N% (left panels) and high N% (right panels) in the gate insulator stack. Symbols: experiment, lines: model calculation. Data from [12]

D1, see Table 4.2), and shows power law time dependence with long-time slope of $n \sim 1/6$. It is important to remark that this characteristic time slope is ubiquitously observed across stress conditions and devices.

The simulated ΔN_{IT} time kinetics evolves rapidly at the initiation of stress and asymptotically settles into a power law dependence with identical $n \sim 1/6$ slope at different V_{GSTR}, T and for both devices. As mentioned before, simulated ΔN_{IT} time kinetics using RD model with defect-assisted dimerization is governed by the first interface reaction at shorter time, by defect-assisted dimerization at the second interface at intermediate time, while the long-time part is governed by molecular H_2 diffusion in the gate oxide and beyond. Note that the long-time slope of $n \sim 1/6$ is a *parameteragnostic feature* of the RD model, driven purely by molecular H_2 diffusion [34].

Figure 4.7 shows the time evolution of measured and modeled ΔN_{IT} in RMG HKMG D3 and D4 FinFETs under DC stress at different V_{GSTR} and T (left panels)

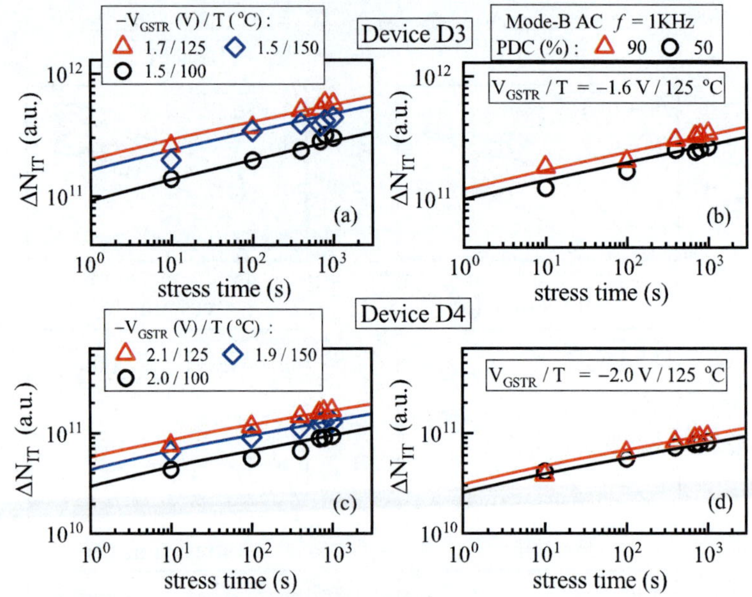

Fig. 4.7 Time evolution of DCIV measured (and delay corrected) and RD model simulated ΔN_{IT} at different (a, c) V_{GSTR} and T for DC stress and (b, d) PDC for AC stress in RMG HKMG (a, b) Si and (c, d) SiGe channel p-FinFETs. Symbols: experiment, lines: model calculation. Data from [23, 24]

and under AC stress at different PDC (right panels). The measured and modeled time kinetics of ΔN_{IT} show power law dependence with slope $n \sim 1/6$ across devices (only the long-time data can be obtained for measurement and are plotted for the simulation), for different V_{GSTR} and T during DC stress and different PDC for AC stress. Note that the ΔN_{IT} magnitude reduces at higher Ge%. Identical model parameters are used to explain the DC and AC stress for a particular device.

The measured and modeled ΔN_{IT} at fixed t_{STR} of 1Ks as a function of V_{GSTR} at different T are shown for DC stress in devices D1 and D2 in Fig. 4.8 and in devices D3 and D4 in Fig. 4.9, and also for AC stress in device D4 in Fig. 4.9, see Table 4.2 for device details. The magnitude of ΔN_{IT} increases with V_{GSTR} and T as expected. The VAF reduces at higher N% but increases at higher Ge% at a given T. Note that the VAF for a particular device reduces at higher T, and the T dependence of VAF is larger (*i.e.*, larger VAF reduction at higher T) for SiGe compared to Si devices, while no significant impact is observed for changes in N%. The reduction in VAF at higher T is due to the bond polarization effect, although the stress reduction effect (*i.e.*, reduction in the effective stress at longer time due to higher degradation-related electrostatic effect) also contributes. The model can explain the T and process dependence of measured VAF. Note that the ΔN_{IT} kinetics during stress depends on E_{OX} and not V_{GSTR}. Therefore, the process dependencies of Γ_E (see Fig. 4.5) and E_{OX} in the IL are responsible for the process dependence of VAF when ΔN_{IT} is plotted as a function of V_{GSTR}. This is discussed in the following section.

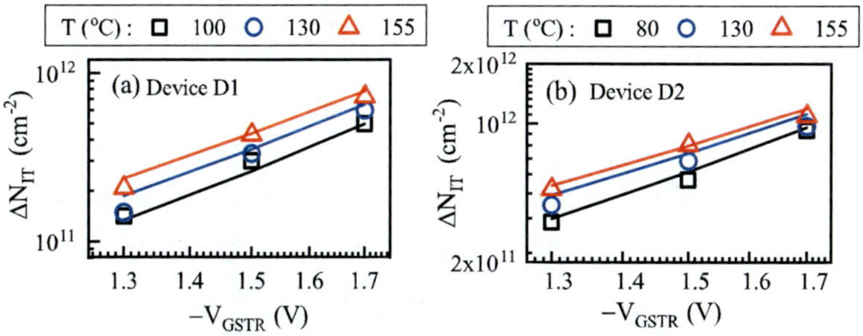

Fig. 4.8 DCIV measured (and delay corrected) and RD model simulated ΔN_{IT} at fixed t_{STR} of 1Ks as a function of V_{GSTR} at different T for DC stress in (a) D1 and (b) D2 devices listed in Table 4.2. Symbols: experiment, lines: model calculation. Data from [12]

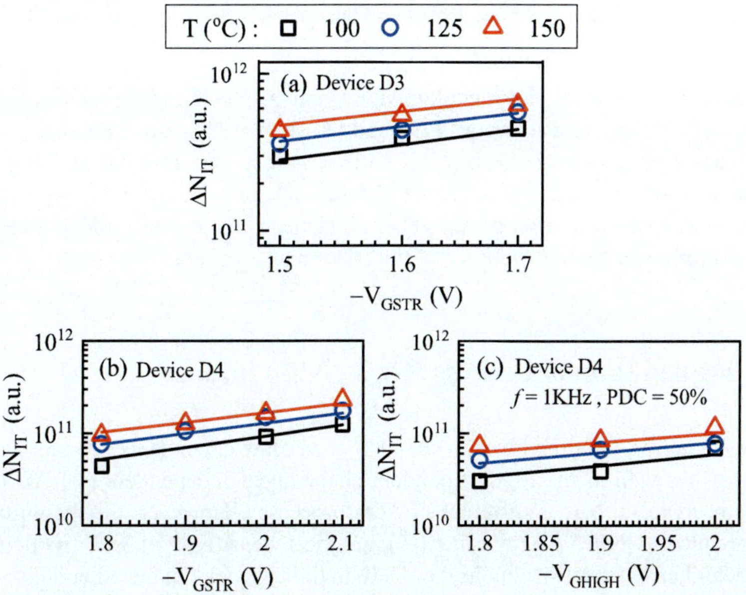

Fig. 4.9 DCIV measured (and delay corrected) and RD model simulated ΔN_{IT} at fixed t_{STR} of 1Ks as a function of V_{GSTR} at different T, for (a, b) DC stress in D3 and D4 devices and (c) AC stress in D4 device listed in Table 4.2. Symbols: experiment, lines: model calculation. Data from [23, 24]

Figure 4.10 shows the measured and modeled ΔN_{IT} at fixed t_{STR} of 1Ks as a function of (a) PDC and (b) frequency of the AC pulse in different devices. Note that identical AC to DC ratio (all data are normalized to DC stress), PDC-dependent shape and f independence are observed for all devices. One interesting aspect to note is the absence of a large jump or "kink" in the PDC dependence of ΔN_{IT} near DC, which is unlike that of the ultra-fast measured PDC dependence of ΔV_T shown in Chap. 1, Sect. 1.4 (also see Chap.14). This aspect is related to occupancy of

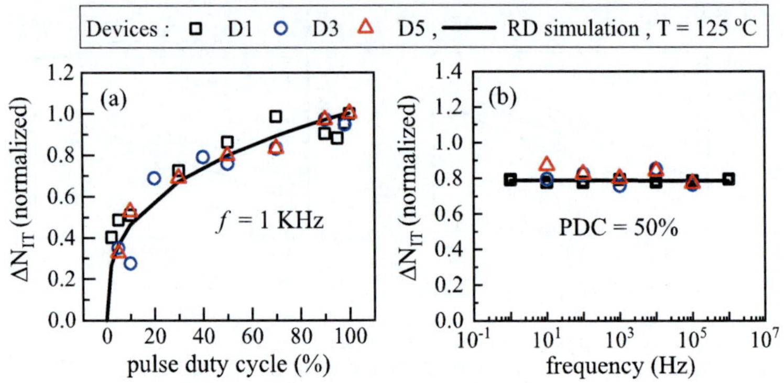

Fig. 4.10 DCIV measured (delay corrected) and RD model simulated ΔN_{IT} at fixed t_{STR} of 1Ks as a function of (a) PDC and (b) frequency. All AC data are normalized to the DC data of the particular device under consideration. Symbols: experiment, lines: model calculation. Data from [12, 24]

generated interface traps and is explained in Chap. 5. The RD model can explain the measured AC to DC ratio at various PDC and f as shown. The remarkable universality of the ΔN_{IT} time kinetics during DC and AC stress (see Fig. 4.6 and Fig. 4.7), as well as the PDC and f dependence during AC stress (see Fig. 4.10), suggests universality of the underlying trap generation mechanism. The f independence is another *parameteragnostic feature* of the RD model.

4.5 Explanation of Process (Ge%, N%) Impact

As shown above, the time kinetics (n ~ 1/6 power law dependence) during DC and AC stress, as well as the PDC-dependent shape and f independence of AC to DC ratio are universal across different devices/processes. However, the magnitude of ΔN_{IT} reduces, while VAF (at a fixed T) and the T sensitivity of VAF (reduction of VAF at higher T) increase with higher Ge% in the channel. On the other hand, ΔN_{IT} increases while VAF slightly reduces with higher N% in the gate stack, and there is no noticeable impact on the T sensitivity of VAF.

As also shown above, except the parameters governing the forward reaction of X-H bond dissociation at the channel/IL interface listed in Table 4.2, all other RD model parameters are process independent as shown in Table 4.1. The E_{OX} and T during stress and the parameters K_{F10}, E_{AKF1} and Γ_E control the bond dissociation rate $_{F1}$, where Γ_E is due to Γ_0 and α, see Fig. 4..4.5. Note, E_{OX} is determined by the thickness (T_{IL} and T_{HK}) and dielectric constant (ε_{IL} and ε_{HK}) of the IL and High-K layers in HKMG gate insulators (or T_{OX} and ε_{OX} for a single-layer gate insulator). K_{F10} and Γ_0 depend on m_T and φ_B, and these can be obtained from bandstructure calculations using the tight binding approach [49] as discussed below.

Figure 4.11 shows the simulated bandstructure, *i.e.*, the light hole (LH) and the

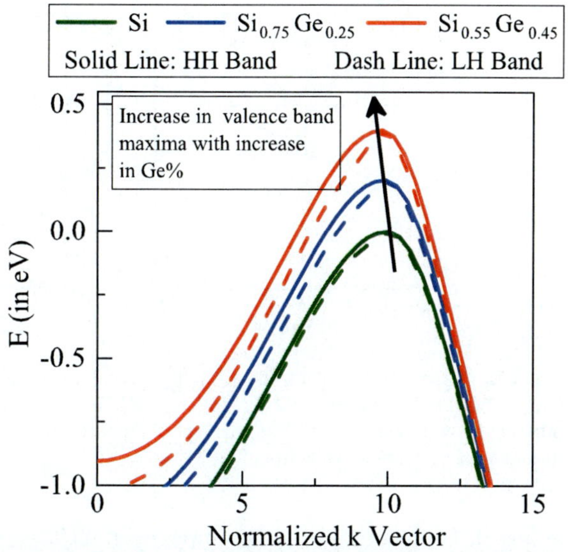

Fig. 4.11 Tight binding model [49] calculated E-k diagrams for valence bands for different Ge% in the channel. Calculations are done for the (110) surface relevant for FinFET sidewalls

heavy hole (HH) sub-bands of the valence band for different Ge% in the channel. Both the LH and HH bands are lifted up at higher Ge%, which indicate increase in φ_B. However, there is negligible change in the curvature of the bands and hence m_T remains unchanged. Higher φ_B reduces both K_{F10} (via the tunneling coefficient T_H) and Γ_0, see Fig. 4.5. The reduction in K_{F10} would result in reduction in ΔN_{IT} at higher Ge%. It is important to remark that above discussion on the impact of φ_B on K_{F10} is valid only if E_{AKF1} remains constant across different Ge%. Since E_{AKF1} is higher at higher Ge%, the relative K_{F10} value is also higher, see Table 2.2. However, the product of K_{F10} and exp $(-E_{AKF1}/kT)$ is lower at higher Ge%, and hence explains the reduction of ΔN_{IT} at higher Ge% shown in Figs. 4.7 and 4.9. Moreover, the above discussion on φ_B on Γ_0 is valid only if N% is kept same across different devices (more on this in Chaps. 8, 9 and 11).

For a particular type of channel (Si or SiGe), higher N% increases K_{F10} due to the reduction in φ_B and m_T when applicable (see Fig. 4.5), while E_{AKF1} reduces, see Table 4.2. The product of $_{F10}$ and exp $(-E_{AKF1}/kT)$ is higher at higher N% and hence explains the increase of ΔN_{IT} at higher N% shown in Figs. 4.6 and 4.8. As mentioned before, the impact of N% on Γ_0 depends on the relative changes in φ_B (always reduces at higher N%) and m_T (reduces for (100) but likely increases for (110) surface at higher N%), more on this in Chap.7, 9 and 11.

Figure 4.12 shows the Arrhenius T dependence of Γ_E, calculated using the T-dependent VAF data for (a) GF devices having different N% and (b) RMG devices having different Ge% and N%. The intercept Γ_0 reduces at higher N% but the slope (~polarization term α) does not change for GF Si channel devices (D2 versus D1). As mentioned before, the barrier φ_B reduces as the bandgap of IL reduces at higher N% near the channel/IL interface. This would imply increase in Γ_0, but the opposite is observed. This is possible if m_T also reduces at higher N%. Reduction in φ_B and

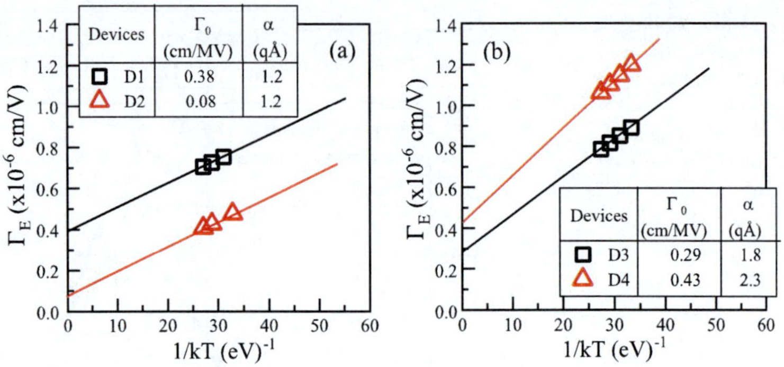

Fig. 4.12 Measured T dependence of the field acceleration factor (Γ_E), using data from (a) GF HKMG MOSFETs at different N% and (b) RMG HKMG FinFETs at different Ge% (the N% is different between D3 and D4). The intercept (Γ_0) and slope (~α) are shown. Symbols: experiment, lines: model calculation. Data from [12, 23]

m_T result in higher K_{F10} for D2 compared to D1 device (as also mentioned before, due to difference in E_{AKF1}, the term K_{F10}*exp($-E_{AKF1}/kT$) is higher at higher N%). Moreover, note that ε_{IL} increases at higher N% in the IL, resulting in lower E_{OX} at a particular V_{GSTR} and can further reduce the VAF. E_{AKF1} reduces with increase in N%, and this is addressed by atomistic calculations [50].

Interestingly, the intercept Γ_0 increases at higher Ge% (D4 versus D3), which is not expected if only changes (increase) in φ_B is considered. Due to differences in N% for the Si and SiGe devices, the m_T is different (higher for SiGe in this case), resulting in higher Γ_0. Note that Γ_0 indeed reduce at higher Ge% if N% is kept low for all devices, see Chaps. 8, 9 and 11. However, the slope increases at higher Ge% and indicates increase in the polarization factor α. Moreover, E_{AKF1} also increases with Ge%. First principles calculation is needed to explain the impact of Ge% on E_{AKF1} and α, which is beyond the scope of this analysis. Note that the term K_{F10}*exp($-E_{AKF1}/kT$) is lower at higher Ge% and explains the reduction of ΔN_{IT}. Note that besides surface orientation, Ge% and N%, the parameters for the SiGe devices are strongly dependent on the details of the IL formation process during gate stack formation.

4.6 Discussion on RD model

The chemical nature of H passivated gate insulator defects, validity of the inversion hole-assisted X-H defect dissociation mechanism at the channel/IL (gate insulator) interface, and the parameters used in RD model are debated in the literature [51] and hence are discussed in this section.

4.6.1 Nature of H Passivated Defects (Defect Precursors)

The H passivated defects at the channel/gate insulator interface are usually presumed to be Si-H bonds, which has been identified as P_b centers in Si (111)/SiO_2 interface and as P_{b0} and P_{b1} centers in Si (100)/SiO_2 interface by using the Electron Spin Resonance (ESR) and Electron Paramagnetic Resonance (EPR) studies [45, 52–54]. However, it has been suggested that not all electrically active defects are paramagnetic and detectable by ESR or EPR; rather the electrically active defects can indeed have much larger density than the spin active defects [53, 55]. Similar conclusions were also drawn from Spin-Dependent Recombination (SDR) studies [56]. Moreover, other (not P_b like) defects were suggested to be more important in technologically relevant gate insulators containing Nitrogen (which is always the case for modern gate insulators) [51–60]. Therefore, due to these uncertainties, the H passivated channel/gate insulator interfacial defects are denoted as X-H bonds in this book, Eq. 4.1.

Moreover, the H passivated bulk insulator defects for defect-assisted dimerization (and reverse reaction) can be H passivated Si-H, N-H, O-H, Oxygen vacancy (O_V-H) defects [61], different H passivated E' centers [62], and/or other complex H-related defects (Hydrogen bridge, Hydroxyl E' centers) [37]. Therefore, they are collectively denoted as Y-H bonds for simplicity, Eq. 4.2, to distinguish them from the interfacial defects (although it should be noted that the difference between interface and bulk becomes blurry in ultrathin HKMG stacks).

Note that the RD model with defect dissociation (at some rate) and subsequent H/H_2 diffusion is agnostic to the nature of H passivated defects at the channel/gate insulator interface and inside the gate insulator bulk. As long as there are H passivated defects, the RD model is applicable and would provide $n \sim 1/6$ power law time dependence during DC and AC stress and f independence during AC stress as discussed earlier in Sect. 4.4.

4.6.2 Dissociation of H Passivated Defects

Although a generic X-H defect is used in Sect. 4.3 due to the uncertainties listed in Sect. 4.6.1, the original reports suggested that inversion layer holes during NBTI stress tunnel to the interfacial Si-H defect precursors, get captured to make them weak, and the weak bonds can subsequently get broken by thermal activation [34, 43]. However, the charge neutrality level (0/ +) of Si-H bond was shown at ~4 eV below the Si valence band [63], and therefore, it is suggested that Si-H bonds cannot capture holes via tunneling [51].

As discussed above, there exists a strong possibility that the dominant electrically active defects are something other than the usual paramagnetic Si-H bonds. Moreover, the Si-H charge neutrality level is determined using Density Functional Theory (DFT) calculations in bulk amorphous Si (a-Si) in [63] and not in SiO_2 (or more appropriately, SiON). Moreover, while DFT calculations presumably work well for thicker and bulk material systems, it faces challenges when a thin amorphous layer is sandwiched between two interfaces, *e.g.*, a thin SiO_2 (or SiON) layer between Si/SiO_2 and SiO_2/poly-Si interfaces for conventional single-layer gate insulator, and even thinner SiO_2 (or SiON) IL between Si/IL and IL/High-K interfaces for HKMG gate insulators. DFT also faces challenges in the presence of different species, *e.g.*, the presence of Hf in IL due to penetration from HfO_2 High-K and also the presence of N in IL due to penetration from spacer or gate material [9], and in the presence of defects (*e.g.*, O_v- [61]). Therefore, DFT simulations are challenging in realistic gate insulator stacks.

The thermal dissociation of Si-H bonds requires very high energy ($E_A = 2.6$ eV) [45, 64] and therefore is presumed to be impossible under normal NBTI experimental conditions [65]. However, the chemical reaction Si-H + H → Si + H_2 is suggested to have very low barrier [45], and hence, it is presumed that the release of H atoms bonded with channel acceptors initiate the H dissociation from interfacial Si-H bonds [65]. However, note that the NBTI degradation magnitude is similar between planar (high channel doping) and FinFET (negligible channel doping) devices for similar V_{GSTR} and T [16, 17], which is inconsistent with the concept of released H from the channel acceptors being responsible for bond dissociation (this theory would imply negligible NBTI in FinFETs). Another recent report has suggested the release of H atoms from the gate and subsequent diffusion toward the channel initiate the Si-H bond dissociation near the channel/gate insulator interface (the Gate Side Hydrogen Release model) [66]. However, this framework is rate limited by H release from the gate and therefore cannot explain the impact of higher Ge% in the channel (reduction in NBTI magnitude and increase in VAF at higher Ge%) as discussed before and later in Chaps. 8, 9 and 11.

On the other hand, the mechanism discussed in Sect. 4.3 can explain measured data under wide range of experimental conditions and channel/gate insulator process changes as discussed in Sect. 4.4 and also in later chapters of this book.

4.6.3 Model Parameters

The Si-H bond dissociation energy has been found to be ~2.6 eV using Electron Paramagnetic Resonance (EPR) studies [45], which is much larger than the E_{AKF1} values listed in Table 4.2. Note that the bond dissociation studied in [45] is under vacuum thermal anneal in Si (111)/SiO_2 (50 nm thickness) samples and is unlikely to represent NBTI defects in Si (100) or Si (110) interfaces with thinner gate insulators as mentioned above. Moreover, as discussed above, the electrically active defects exposed to NBTI in modern devices are not necessarily the standard P_b centers that were studied using EPR in [45].

The Arrhenius T activation values used in Table 4.1 are similar for the forward and reverse reactions governed by Eq. 4.4 (corresponding to the chemical reaction of Eq. 4.2). However, they are respectively found to be exothermic and have very large thermal barrier from EPR studies in Si (111)/SiO_2 system [45, 67]. Therefore, the choice of these RD model parameter values seems questionable [51]. It is important to note that Eq. 4.4 (or Eq. 4.2) and the corresponding parameter values are for H passivate defects inside the SiO_2 (or SiON) bulk or at the IL/High-K interface, and therefore, identical values as [45, 67] are not expected.

Finally, the H_2 diffusivity values listed in Table 4.1 are presumably different as compared to those in other reports [51]. It is noteworthy that modern gate stacks containing Nitrogen can have significantly lower H_2 diffusion and hence, reduced diffusivity values are expected [3].

It is noteworthy that although the objections raised in [51] seem plausible at the face value, none of them can be justified under a rigorous inspection. On the other hand, the RD model with fixed and adjustable parameters listed in Table 4.1 and Table 4.2, respectively, can quantitatively model the DCIV measured ΔN_{IT} time kinetics during DC and AC stress under different experimental conditions, and on various types of devices. The RD model can also explain a variety of other process dependence as discussed in later chapters.

4.7 Summary

According to the BAT-NBTI framework, device parametric drift is due to uncorrelated sum of different underlying processes, *e.g.*, $\Delta V_{\mathrm{T}} = \Delta V_{\mathrm{IT}} + \Delta V_{\mathrm{HT}} + \Delta V_{\mathrm{OT}}$, where ΔV_{IT} is calculated by the TTOM-enabled RD model. RD model with defect-assisted dimerization remains consistent between the deterministic and stochastic implementations and is used throughout this book. RD model calculates the time kinetics of $\Delta \mathrm{N}_{\mathrm{IT}}$ during and after stress, while TTOM computes their contribution (ΔV_{IT}) to overall ΔV_{T}. The ΔN_{IT} kinetics simulated by the RD model shows power law time dependence at longer time with exponent n ~ 1/6 for both DC and AC stress and f independence for AC stress. The model uses an inversion layer hole-assisted defect dissociation mechanism at the channel/gate insulator interface, with four parameters to quantify process changes. The model is validated using DCIV measured ΔN_{IT} time kinetics during DC and AC stress, in planar and FinFET devices having different processes (Ge% in the channel and N% in the gate stack), and for different experimental conditions. Two of the process-dependent parameters can be obtained by bandstructure calculations, and so the other two are truly adjustable across process changes. The validity of the physical mechanisms governing RD model and the model parameter values are also discussed.

Other components of the BAT framework for NBTI, *i.e.*, TTOM, hole trapping and bulk trap generation, are described in Chaps. 5 and 6.

Acknowledgements All experimental data presented in this chapter are re-plotted from previously published reports. One of the authors (Souvik Mahapatra) acknowledges Muhammad Ashraful Alam and Ahmed Ehteshamul Islam for discussion on defect kinetics modeling. The authors acknowledge applied materials for providing planar MOSFETs and IBM for providing measurement facilities and FinFETs. One of the authors (Narendra Parihar) acknowledges Richard Southwick, Miaomiao Wang and James Stathis for help with measurement facilities. Karansingh Thakor is acknowledged for help with manuscript preparation.

References

1. N. Kimizuka, K. Yamaguchi, K. Imai, T. Iizuka, C.T. Liu, R.C. Keller, T. Horiuchi, in *Symposium on VLSI Technology Digest of Technical Papers* **92** (2000)
2. V. Huard, M. Denais, F. Perrier, N. Revil, C. Parthasarathy, A. Bravaix, E. Vincent, Microelectron. Reliab. **45**, 83 (2005)
3. A. T. Krishnan, C. Chancellor, S. Chakravarthi, P. E. Nicollian, V. Reddy, A. Varghese, R. B. Khamankar, and S. Krishnan, in *IEEE International Electron Devices Meeting Technical Digest*, 688 (2005)
4. D. Varghese, D. Saha, S. Mahapatra, K. Ahmed, F. Nouri, M. Alam, in IEEE International Electron Devices Meeting Technical Digest, 684 (2005)
5. Y. Mitani, T. Yamaguchi, H. Satake, A. Toriumi, in*IEEE International Reliability Physics Symposium Proceedings*, 226 (2007)
6. S. Mahapatra, K. Ahmed, D. Varghese, A.E. Islam, G. Gupta, L. Madhav, D. Saha, M.A. Alam, in *IEEE International Reliability Physics Symposium Proceedings*, 1 (2007)
7. Y. Mitani, H. Satake, A. Toriumi, IEEE Trans. Device Mater. Reliab. **8**, 6 (2008)
8. S. Pae, M. Agostinelli, M. Brazier, R. Chau, G. Dewey, T. Ghani, M. Hattendorf, J. Hicks, J. Kavalieros, K. Kuhn, M. Kuhn, J. Maiz, M. Metz, K. Mistry, C. Prasad, S. Ramey, A. Roskowski, J. Sandford, C. Thomas, J. Thomas, C. Wiegand, J. Wiedemer, in *IEEE International Reliability Physics Symposium Proceedings*, 352 (2008)
9. K. Joshi, S. Hung, S. Mukhopadhyay, V. Chaudhary, N. Nanaware, B. Rajamohanan, T. Sato, M. Bevan, A. Wei, A. Noori, B. McDougal, C. Ni, G. Saheli, C. Lazik, P. Liu, D. Chu, L. Date, S. Datta, A. Brand, J. Swenberg, S. Mahapatra, in *IEEE International Reliability Physics Symposium Proceedings*, 4C.2.1 (2013)
10. J. Franco, B. Kaczer, P.J. Roussel, J. Mitard, M. Cho, L. Witters, T. Grasser, G. Groeseneken, IEEE Trans. Electron Devices **60**, 396 (2013)
11. P. Srinivasan, J. Fronheiser, K. Akarvardar, A. Kerber, L.F. Edge, R.G. Southwick, E. Cartier, H. Kothari, in *IEEE International Reliability Physics Symposium Proceedings*, 6A.3.1 (2014)
12. N. Parihar, N. Goel, S. Mukhopadhyay, S. Mahapatra, IEEE Trans. Electron Devices **65**, 392 (2018)
13. N. Parihar, S. Mahapatra, in *IEEE International Reliability Physics Symposium Proceedings*, TX.5.1 (2018)
14. V. Huard, C. Ndiaye, M. Arabi, N. Parihar, X. Federspiel, S. Mhira, S. Mahapatra, A. Bravaix, in *IEEE International Reliability Physics Symposium Proceedings*, TX.4.1 (2018)
15. N. Parihar, R. Tiwari, C. Ndiaye, M. Arabi, S. Mhira, H. Wong, S. Motzny, V. Moroz, V. Huard, S. Mahapatra, in International Conference on Simulation of Semiconductor Processes and Devices, 167 (2018)
16. S. Ramey, A. Ashutosh, C. Auth, J. Clifford, M. Hattendorf, J. Hicks, R. James, A. Rahman, V. Sharma, A.S. Amour, C. Wiegand, in *IEEE International Reliability Physics Symposium Proceedings*, 4C.5.1 (2013)
17. K.T. Lee, W. Kang, E-A Chung, G. Kim, H. Shim, H. Lee, H. Kim, M. Choe, N.-I. Lee, A. Patel, J. Park, J. Park, in *IEEE International Reliability Physics Symposium Proceedings*, 2D.1.1 (2013)
18. J. Franco, B. Kaczer, A. Chasin, H. Mertens, L.-A. Ragnarsson, R. Ritzenthaler, S. Mukhopadhyay, H. Arimura, P. J. Roussel, E. Bury, N. Horiguchi, D. Linten, G. Groeseneken, A. Thean, in *IEEE International Reliability Physics Symposium Proceedings*, 4B.2.1 (2016)

19. G. Jiao, M. Toledano-Luque, K.-J. Nam, N. Toshiro, S.-H. Lee, J.-S. Kim, T. Kauerauf, E. Chung, D. Bae, G. Bae, D.-W. Kim, K. Hwang, in *IEEE International Electron Devices Meeting Technical Digest*, 31.2.1 (2016).
20. N. Parihar, R.G. Southwick, U. Sharma, M. Wang, J.H. Stathis, S. Mahapatra, in *IEEE International Reliability Physics Symposium Proceedings*, 2D.4.1 (2017)
21. N. Parihar, R. Southwick, M. Wang, J.H. Stathis, S. Mahapatra, in *IEEE International Electron Devices Meeting Technical Digest*, 7.3.1 (2017).
22. N. Parihar, U. Sharma, R.G. Southwick, M. Wang, J.H. Stathis, S. Mahapatra, IEEE Trans. Electron Devices **65**, 23 (2018)
23. N. Parihar, R.G. Southwick, M. Wang, J.H. Stathis, S. Mahapatra, IEEE Trans. Electron Devices **65**, 1699 (2018)
24. N. Parihar, R.G. Southwick, M. Wang, J.H. Stathis, S. Mahapatra, IEEE Trans. Electron Devices **65**, 1707 (2018)
25. N. Parihar, R. Tiwari, S. Mahapatra, in International Conference on Simulation of Semiconductor Processes and Devices, **176** (2018)
26. R. Tiwari, N. Parihar, K. Thakor, H. Y. Wong, S. Motzny, M. Choi, V. Moroz, S. Mahapatra, IEEE Trans. Electron Devices **66**, 2086 (2019)
27. R. Tiwari, N. Parihar, K. Thakor, H. Y. Wong, S. Motzny, M. Choi, V. Moroz, S. Mahapatra, IEEE Trans. Electron Devices **66**, 2093 (2019)
28. N. Parihar, U. Sharma, R.G. Southwick, M. Wang, J. H. Stathis, S. Mahapatra, in *IEEE International Reliability Physics Symposium Proceedings* (2019)
29. M. Wang, J. Zhang, H. Zhou, R. G. Southwick, R. Hsin, K. Chao, X. Miao, V.S. Basker, T. Yamashita, D. Guo, G. Karve, H. Bu, in *IEEE International Reliability Physics Symposium Proceedings* (2019)
30. N. Choudhury, U. Sharma, H. Zhou, R.G. Southwick, M. Wang, S. Mahapatra, in *IEEE International Reliability Physics Symposium Proceedings* (2020)
31. H. Zhou, M. Wang, J. Zhang, K. Watanabe, C. Durfee, S. Mochizuki, R. Bao, R. Southwick, M. Bhuiyan, B. Veeraraghavan, in *IEEE International Reliability Physics Symposium Proceedings* (2020)
32. N. Choudhury, T. Samadder, R. Tiwari, H. Zhou, R.G. Southwick, M. Wang, S. Mahapatra, in *IEEE International Reliability Physics Symposium Proceedings* (2021)
33. S. Mahapatra, N. Goel, S. Desai, S. Gupta, B. Jose, S. Mukhopadhyay, K. Joshi, A. Jain, A.E. Islam, M.A. Alam, IEEE Trans. Electron Devices **60**, 901 (2013)
34. A.E. Islam, N. Goel, S. Mahapatra, M.A. Alam, in *Fundamentals of Bias Temperature Instability in MOS Transistors* (Springer India, 2015), pp. 181–207
35. N. Choudhury, N. Parihar, N. Goel, A. Thirunavukkarasu, S. Mahapatra, IEEE J. Electron Devices Soc. **8**, 1281 (2020)
36. T. Samadder, N. Choudhury, S. Kumar, D. Kochar, N. Parihar, S. Mahapatra, IEEE Trans. Electron Devices **68**, 485 (2021)
37. S. Ling, A.M. El-Sayed, F. Lopez-Gejo, M.B. Watkins, V.V. Afanas'ev, A.L. Shluger, Microelectron. Eng. **109**, 310 (2013)
38. K.F. Schuegraf, C. Hu, IEEE Trans. Electron Devices **41**, 761 (1994)
39. K.O. Jeppson, C.M. Svensson, J. Appl. Phys. **48**, 2004 (1977)
40. M.A. Alam, S. Mahapatra, Microelectron. Reliab. **45**, 71 (2005)
41. M.A. Alam, in *IEEE International Electron Devices Meeting Technical Digest*, 14.1.1 (2003)
42. G. Chen, M.F. Li, C.H. Ang, J.Z. Zheng, D.L. Kwong, IEEE Electron Device Lett. **23**, 734 (2002)
43. A.E. Islam, H. Kufluoglu, D. Varghese, S. Mahapatra, M.A. Alam, IEEE Trans. Electron Devices **54**, 2143 (2007)
44. F. Schanovsky, T. Grasser, in *IEEE International Reliability Physics Symposium Proceedings*, XT.10.1 (2012).
45. K.L. Brower, S.M. Myers, Appl. Phys. Lett. **57**, 162 (1990)
46. S. Kumar, R. Anandkrishnan, N. Parihar, S. Mahapatra, IEEE Trans. Electron Devices **67**, 4741 (2020).

47. S. Rangan, N. Mielke, E.C.C. Yeh, in *IEEE International Electron Devices Meeting Technical Digest*, 14.3.1 (2003)
48. N. Goel, S. Mahapatra, in *Fundamentals of Bias Temperature Instability in MOS Transistors* (Springer India, 2015), pp. 209–263
49. https://nanohub.org/resources/bandstrlab
50. S.S. Tan, T.P. Chen, C.H. Ang, L. Chan, IEEE Electron Device Lett. **25**, 504 (2004)
51. J.H. Stathis, S. Mahapatra, T. Grasser, Microelectron. Reliab. **81**, 244 (2018)
52. P.M. Lenahan, P.V. Dressendorfer, J. Appl. Phys. **55**, 3495 (1984)
53. E. Cartier, J.H. Stathis, D.A. Buchanan, Appl. Phys. Lett. **63**, 1510 (1993)
54. J.H. Stathis, E. Cartier, Phys. Rev. Lett. **72**, 2745 (1994)
55. E. Cartier, J.H. Stathis, Microelectron. Eng. **28**, 3 (1995)
56. J.H. Stathis, D.J. DiMaria, Appl. Phys. Lett. **61**, 2887 (1992)
57. J.P. Campbell, P.M. Lenahan, A.T. Krishnan, S. Krishnan, in *IEEE International Reliability Physics Symposium Proceedings*, **442** (2006)
58. S. Fujieda, Y. Miura, M. Saitoh, E. Hasegawa, S. Koyama, K. Ando, Appl. Phys. Lett. **82**, 3677 (2003)
59. J.P. Campbell, P.M. Lenahan, A.T. Krishnan, S. Krishnan, in *2007 IEEE International Reliability Physics Symposium Proceedings, 503* (2007)
60. J.T. Ryan, P.M. Lenahan, A.T. Krishnan, S. Krishnan, J.P. Campbell, in *IEEE International Reliability Physics Symposium Proceedings*, **988** (2009).
61. S. Mukhopadhyay, K. Joshi, V. Chaudhary, N. Goel, S. De, R.K. Pandey, K.V.R.M. Murali, S. Mahapatra, in *IEEE International Reliability Physics Symposium Proceedings*, GD 3.1 (2014)
62. C.J. Nicklaw, Z. Lu, D.M. Fleetwood, R.D. Schrimpf and S.T. Pantelides, IEEE Trans. on Nucl. Sci. **49**, 2667 (2002)
63. D.P. DiVincenzo, J. Bernholc, M.H. Brodsky, Phys. Rev. B **28**, 3246 (1983)
64. R. Khatri, P. Asoka-Kumar, B. Nielsen, L.O. Roellig, K.G. Lynn, Appl. Phys. Lett. **65**, 330 (1994)
65. S.T. Pantelides, L. Tsetseris, S.N. Rashkeev, X.J. Zhou, D.M. Fleetwood, R.D. Schrimpf, Microelectron. Reliab. **47**, 903 (2007)
66. T. Grasser, M. Waltl, Y. Wimmer, W. Goes, R. Kosik, G. Rzepa, H. Reisinger, G. Pobegen, A. El-Sayed, A. Shluger, B. Kaczer, in *IEEE International Electron Devices Meeting Technical Digest*, 20.1.1 (2015).
67. K.L. Brower, Phys. Rev. B **38**, 9657 (1988)

Chapter 5
BTI Analysis Tool (BAT) Model Framework—Interface Trap Occupancy and Hole Trapping

Souvik Mahapatra, Narendra Parihar, Nilotpal Choudhury, and Nilesh Goel

5.1 Introduction

As shown in the previous chapters, Negative Bias Temperature Instability (NBTI) became an important reliability concern during the migration from Silicon Dioxide (SiO_2) to Silicon Oxynitride (SiON) gate insulator MOSFETs [1–6]. It continues to remain as a concern in dual layer (SiO_2 or SiON Interlayer (IL) and Hafnium Dioxide (HfO_2) high-K layer) High-K Metal Gate (HKMG) insulator-based bulk [7–12] and Fully Depleted Silicon On Insulator (FDSOI) [13, 14] planar MOSFETs, bulk and SOI FinFETs [14–27], as well as Gate All Around Stacked Nanosheet FETs [28–31], with either Silicon (Si) or Silicon Germanium (SiGe) channel. The key features of NBTI are summarized as follows (reproduced from Chap. 3, Sect. 3.1):

As shown in Chap. 1, Sect. 1.3, NBTI results in the buildup of positive charges in the gate insulator of a p-MOSFET during stress under negative gate bias ($V_G = V_{GSTR}$), resulting in device parametric drift, *e.g.*, threshold voltage shift (ΔV_T), in time. The positive charges and resulting ΔV_T reduce when the V_G is reduced or removed after stress ($V_G = V_{GREC}$ or 0 V). Hence, AC stress results in lower ΔV_T compared to DC stress. Note that ΔV_T during DC and AC stress gets accelerated at more negative V_{GSTR} (V_{GHIGH} for AC pulse) and at elevated temperature (T), and these dependencies are, respectively, governed by the voltage acceleration factor (VAF) and Arrhenius T activation energy (E_A). The ratio of AC to DC ΔV_T shows a typical "S"-shaped characteristic as a function of Pulse Duty Cycle (PDC), with a large kink or jump near DC. The AC-to-DC ratio also depends on the pulse low bias (V_{GLOW}) and may or may not depend on the frequency (f) of the gate pulse. Note that the PDC and V_{GLOW} dependence and f (in) dependence are also governed by the AC stress mode (Mode-A or Mode-B).

S. Mahapatra (✉) · N. Parihar · N. Choudhury · N. Goel
Department of Electrical Engineering, Indian Institute of Technology Bombay, Powai, Mumbai 400076, India
e-mail: souvik@ee.iitb.ac.in

S. Mahapatra (ed.), *Recent Advances in PMOS Negative Bias Temperature Instability*,
https://doi.org/10.1007/978-981-16-6120-4_5

As described in Chap. 2, the ΔV_T magnitude and its time kinetics during and after stress, *i.e.*, the power-law time slope (n) at long stress time (t_{STR}) during stress and the Fraction Remaining (FR) during recovery after stress (FR is defined as ΔV_T at $t = t_{REC}$ after stress to that at $t = t_{STR}$ at the end of stress), VAF, E_A and the T dependence of VAF depend on the nitrogen content (N%) in the gate insulator and germanium content (Ge%) in the channel. The following features are seen (reproduced from Chap. 4, Sect. 4.1): The ΔV_T magnitude increases, but n, VAF, E_A, T dependence of VAF and FR reduce with higher N%. However, the ΔV_T magnitude reduces, but n, VAF, E_A, T dependence of VAF and FR increase with higher Ge%. Moreover, ΔV_T reduces with fin length and width scaling in FinFETs, sheet length scaling in GAA-SNS FETs and larger spacing (SA) between the Shallow Trench Isolation (STI) and device active in FDSOI MOSFETs, but increases with sheet width scaling in GAA-SNS FETs. Any relevant NBTI model should be able to explain the above features, summarized in Chap. 3, Sect. 3.1. The BTI Analysis Tool (BAT) framework described in Chap. 4 through Chap. 6 is used to explain all the above-listed features of NBTI in this book.

Figure 5.1 illustrates the BAT framework for NBTI. As mentioned in Chap. 4, Sect. 4.2, the positive gate insulator charges and the resulting ΔV_T are due to uncorrelated contributions from the generated interface traps (ΔV_{IT}), hole trapping in

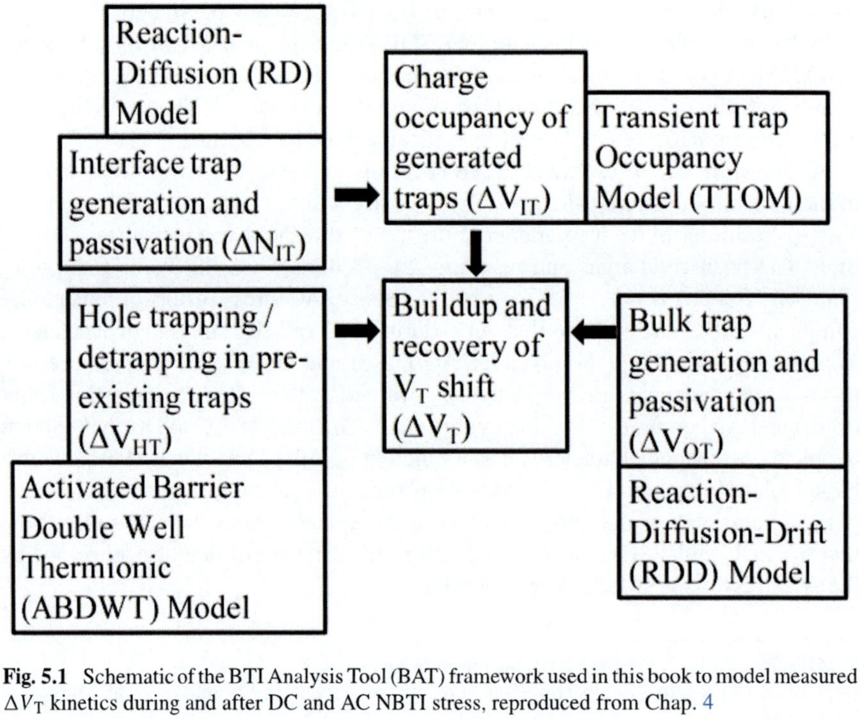

Fig. 5.1 Schematic of the BTI Analysis Tool (BAT) framework used in this book to model measured ΔV_T kinetics during and after DC and AC NBTI stress, reproduced from Chap. 4

preexisting process-related gate insulator traps (ΔV_{HT}) and generated bulk gate insulator traps (ΔV_{OT}). The time kinetics of interface trap density (ΔN_{IT}) is calculated using the Reaction–Diffusion (RD) model [11, 32, 33] and their contribution (ΔV_{IT}) to overall ΔV_T by the Transient Trap Occupancy Model (TTOM) [11]. ΔV_{HT} and ΔV_{OT} contributions are calculated by the Activated Barrier Double Well Thermionic (ABDWT) model [34] and the Reaction–Diffusion Drift (RDD) model [35], respectively. The RD model was described in detail and independently validated in Chap. 4. In this chapter, the TTOM augmentation of the RD modelas well as the ABDWT model are described and validated. The RDD model is described and validated in Chap. 6.

5.2 Interface Trap Generation and Reaction–Diffusion (RD) Model

In Chap. 3, Sect. 3.2, experimental methods such as Charge Pumping (CP) and gated diode or Direct Current I-V (DCIV) are used to independently measure the time kinetics of ΔN_{IT} during DC and AC NBTI stress. Since these are slow measurement methods and implemented in the Measure–Stress–Measure (MSM) mode, ΔN_{IT} can only be obtained at longer stress time (t_{STR} ~ 1s and higher), and the as-measured time kinetics requires measurement delay correction. Note that the delay-corrected ΔN_{IT} kinetics exhibits power-law time dependence with a universal n ~ 1/6 time slope across V_{GSTR}, T, PDC and f during DC and AC stress in different devices. Although ΔN_{IT} increases with PDC, it does not show the large kink or jump near DC (unlike the PDC dependence of ΔV_T, see Chap. 1, Sect. 1.3). Note that the ΔN_{IT} magnitude does not depend on V_{GLOW} and shows f independence. Furthermore, measured ΔN_{IT} reduces in devices having SiGe channel (*i.e.*, with higher Ge% in the channel) when compared to the conventional Si channel and increases in devices having higher N% in the gate insulator.

Figure 5.2 illustrates the RD model with defect-assisted dimerization of hydrogen atoms (H) into molecules (H_2) during stress (and reverse process after stress) as described in Chap. 4, Sect. 4.3. Although this example is illustrated for a HKMG insulator having dual layer stack, the model is also applicable for single-layer SiO_2 and SiON gate insulators. During stress, Fig. 5.2 (a), RD model calculates the dissociation of H passivated defects at the channel/IL interface, diffusion of released H and further reaction at the IL/high-K interface, formation of H_2 (defect-assisted dimerization of H to H_2) and subsequent diffusion of H_2 into the gate stack and back end. The reverse diffusion–reaction processes happen after the stoppage of stress, as shown in Fig. 5.2 (b). Note that the stochastic hopping and lock-in-related effects are included in the deterministic framework by slowing down the H_2 diffusion only in the recovery phase after stress; see [32, 36] for details.

The H passivated defect precursors are denoted as X–H and Y–H due to several uncertainties discussed in Chap. 4, Sect. 4.6. Furthermore, although the Y–H bonds

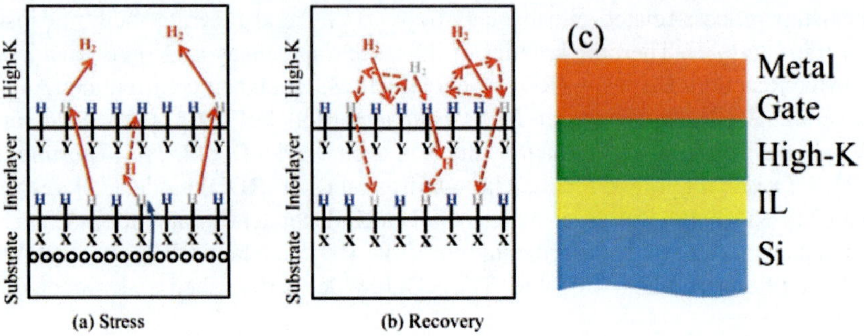

Fig. 5.2 Schematic of the Reaction–Diffusion (RD) model with defect-assisted dimerization and reverse process to calculate the time kinetics of trap generation and passivation, respectively, (a) during and (b) after NBTI stress; example is shown for (c) HKMGgate insulator. The RD model equations are described in Chap. 4, Sect. 4.3. Reproduced from Chap. 4

are actually scattered throughout the gate insulator bulk, they are all lumped at the IL/high-K interface for simplicity (for a single-layer gate insulator, they can all be lumped at an "imaginary interface" at the center of the gate dielectric). Except the dissociation of defects at the channel/IL interface, all other reaction and diffusion processes are only Arrhenius T activated (no bias dependence), and the associated parameters (see Chap. 4, Table 4.1) remain fixed across all devices analyzed in this book. Note that the electrically active gate insulator defects are also denoted as traps, and these terms would be interchangeably used throughout this book.

Figure 5.3 illustrates the channel/IL interfacial defect dissociation mechanism, also discussed in Chap. 4, Sect. 4.3. The oxide electric field (E_{OX}) during stress aids in the tunneling of inversion layer holes into interfacial X–H bonds, which remain polarized (term p) due to E_{OX}. The bonds get weak due to hole capture and subsequently dissociate by thermal excitation. The forward reaction rate K_{F1} depends on the pre-factor K_{F10} (which in turn depends on hole density, p_H, tunneling coefficient, T_H, and capture cross section, σ), field acceleration (Γ_E) and T activation of bond dissociation (E_{AKF1}). The field acceleration factor (Γ_E) is a sum of the T-independent (Γ_0) and T-dependent (α/kT) terms, with α being the polarization

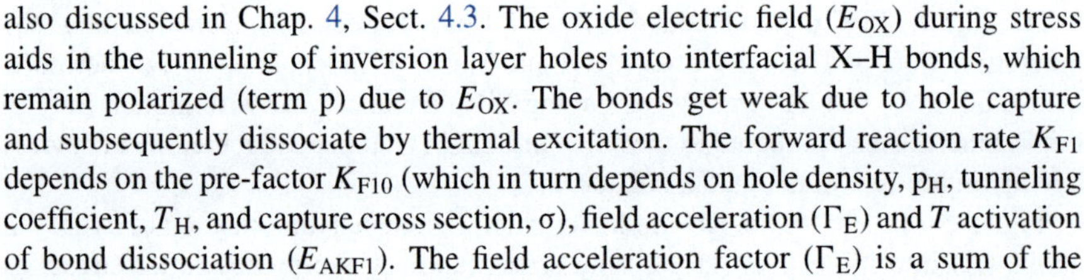

Fig. 5.3 Schematic of the inversion layer hole and oxide electric field-induced dissociation of H passivated defects at the channel/gate insulator interface, details in Chap. 4, Sect. 4.3

coefficient of the X–H bond. The dependence of K_{FI} on Ge% and N% can explain the process dependence of DCIV measured ΔN_{IT} as discussed in Chap. 4, Sect. 4.5. In addition, changes in the mechanical strain on the channel of a device also impact ΔN_{IT} via K_{FI}, which is discussed later (see Chap. 9 and Chap. 13).

Note that the RD model needs to be augmented by TTOM to convert ΔN_{IT} to ΔV_{IT}, and contributions due to ΔV_{HT} and ΔV_{OT} also need to be considered to calculate overall ΔV_{T}.

5.3 Occupancy of Interface Traps

Figure 5.4 illustrates the energy band diagram of a HKMGinsulator stack, showing RD model calculated ΔN_{IT} at the channel/IL and IL/high-K interfaces (a) during and (b) after stress. Generated defects are presumed donor like, and the ones located above the Fermi level (of the substrate) result in positive gate insulator charges. Note that any generated defects below the Fermi level are not considered and not shown during stress.

It is important to remark that the energy locations of the traps shown in Fig. 5.4 are only for illustrative purpose, since due to the uncertainties in the chemical nature of defects discussed in Chap. 4, Sect. 4.6, their exact energy location is not known. However, it is presumed that the energy location is such that a fraction (f_{FAST}) of these traps would go below the Fermi level when the magnitude of V_{G} is reduced from V_{GSTR} to V_{GREC} after stress. These traps would neutralize or "recover" via electron capture, before they eventually get physically re-passivated at a later time by the H back diffusion process (only the first event is relevant from an electrostatic perspective). The remaining fraction ($f_{\mathrm{SLOW}} = 1 - f_{\mathrm{FAST}}$) would neutralize or "recover" via trap passivation.

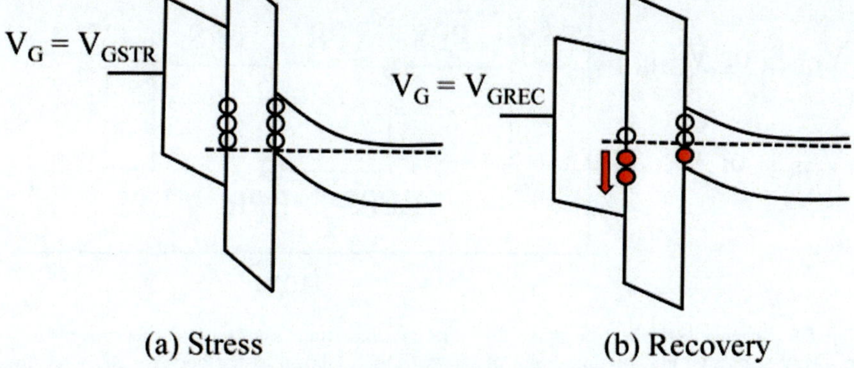

Fig. 5.4 Transient Trap Occupancy Model (TTOM) for the calculation of interface trap occupancy (a) during and (b) after stress; the example is shown for a HKMG gate insulator

5.3.1 *Transient Trap Occupancy Model (TTOM)*

Figure 5.5 illustrates the gate pulse sequence to explain TTOM. The first half cycle is the stress (pulse on) phase, and $\Delta V_{IT} = q^*\Delta N_{IT1}/C_{OX1} + q^*\Delta N_{IT2}/C_{OX2}$ in this case, where q is the electronic charge, C_{OX1} and C_{OX2} are the respective capacitances, and ΔN_{IT1} and ΔN_{IT2} are calculated by the RD model (channel/IL and IL/high-K interfaces are denoted as the first and second interfaces, respectively, and marked as (1) and (2)). All traps are presumed to be above the Fermi level in this case as a starting reference. The TTOM equations (explained below) calculate ΔV_{IT} from the second half cycle (pulse off) and for the successive on/off cycles. If the Current Segment (CS) is stress as shown in Fig. 5.5 (a), PR and PS, respectively, denote the Previous Recovery and Previous Stress segments, and EPR and EPS are the time corresponding to the end of those segments. If CS is recovery as shown in Fig. 5.5 (b), PS, PR and PPS denote the Previous Stress, Recovery and Previous–to-Previous Stress segments, respectively, and EPS, EPR and EPPS are the time corresponding to the end of those segments.

In this framework, any segment (either stress or recovery) can have any value of V_G (*i.e.*, any value of V_{GSTR} or V_{GREC}), which would determine the trap fraction below (f_{FAST}) or above (f_{SLOW}) the Fermi level. Each segment can also have different T, although the T impact comes via the RD model and not TTOM, since the TTOM parameters are T independent. Stretched exponentials are used to model the electron capture and emission kinetics.

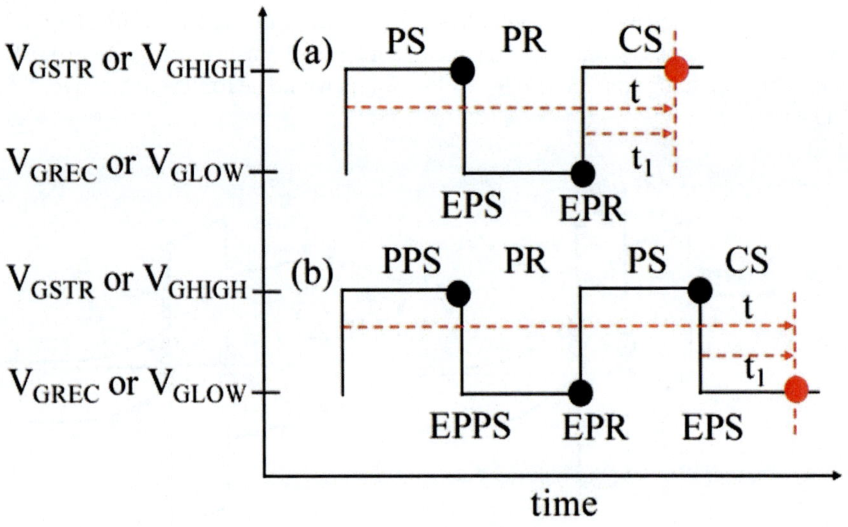

Fig. 5.5 Schematic of the gate pulse to model interface trap occupancy in successive stress–recovery segments. For the time point of interest in CS (shown as red dot), t is the total time since the beginning of stress, while t_1 is the time since the beginning of CS

Table 5.1 Process (or technology)-independent fixed TTOM parameters used in this book

Parameter	Unit	Value (#: at $V_{GREC} = 0$ V)	
τ_{EE}	s	1.0×10^{-2}	
β_{EE}	–	0.4	
β_{EC} (#)	–	0.15	
Device:	–	D1	D2
Type	–	Planar	Planar
Channel	–	Si	Si
Nitrogen	–	Low	High
f_{FAST} (#)	–	0.78	0.60
τ_{EC} (#)	s	2×10^{-2}	8×10^{-5}

The adjustable, process-dependent parameters are listed for the D1 and D2 devices shown in Chap. 4, Table 4.2, having different N% in the gate stack. Some of the parameters depend on V_{GREC} (marked as (#)); see Fig. 5.6

The RD model which calculated ΔN_{IT} time evolution for subsequent stress–recovery cycles is used in TTOM to obtain ΔV_{IT} for a particular stress (Eq. 5.1) or recovery (Eq. 5.2) segment [11]. The different terms of the TTOM equations are explained below. The TTOM parameters are listed in Table 5.1 for the devices used in this chapter (D1 and D2 devices of Chap. 4, Table 4.2) for model validation.

- For stress CS, see Fig. 5.5 (a), ΔV_{IT} depends on its value at EPR, generated traps that are above the Fermi level during CS and electron emission from the fraction of previously generated traps during PS that went below the Fermi level and captured electrons during PR, and once again got back above the Fermi level during CS; see Eq. (5.1).
- For recovery CS, see Fig. 5.5 (b), ΔV_{IT} depends on the fraction of traps that were generated and remained above the Fermi level during PS and also stay above the Fermi level during CS and recover only by the H/H_2-induced passivation process, fast electron capture in traps that were generated and remained above the Fermi level during PS and go below the Fermi level during CS, as well as fast electron capture in traps that were generated during PPS and do not undergo any electron capture during PR but now go below the Fermi level during CS; see Eq. (5.2).

The stretched exponential parameters τ_{EC}, β_{EC} and τ_{EE}, β_{EE} are for electron capture and emission, respectively. Only f_{FAST} and τ_{EC} are process dependent; all other parameters are process independent, and identical values are used across different devices throughout the book. Note that f_{FAST} is a free parameter due to the uncertainty of trap energy location (*i.e.*, the nature of defects). For example, it is higher for the Gate First (GF) HKMG planar devices with thinner IL [11] than the Replacement Metal Gate (RMG) FinFETs with slightly thicker IL [22], and these devices are analyzed in more detail in Chap. 7 and Chap. 11, respectively. Moreover, τ_{EC} is lower in the planar devices than FinFETs.

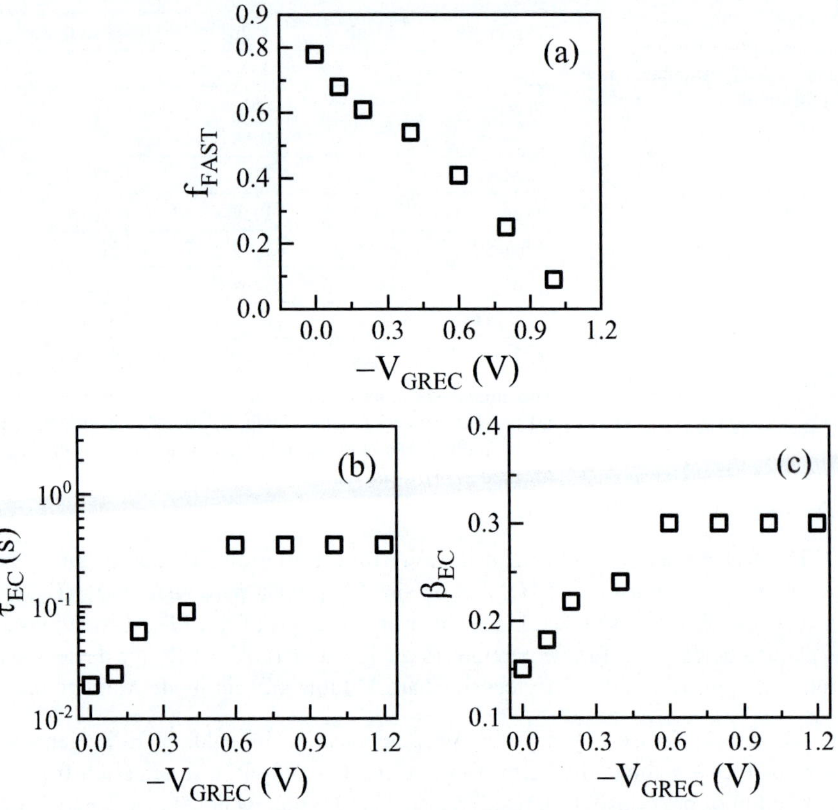

Fig. 5.6 Dependence of TTOM parameters (f_{FAST}, τ_{EC} and β_{EC}) on V_{GREC}, shown for the D1 device of Table 5.1. Data from [11]

Note that although the empirical TTOM equations are used throughout this book, in [31, 34], the ABDWT model (described in Sect. 5.4) has been used to verify the electron capture time constant used in the empirical TTOM framework. Also note that the electron capture process is not considered to model the interruption of stress for measurement, for the typical delay (~10μs) of the ultra-fast Measure–Stress–Measure (MSM) method (see Chap. 1, Sect. 1.2) used throughout the book (in all chapters except Chap. 8). Therefore, TTOM is not used to model DC stress data from MSM method, and ΔV_{IT} is obtained using RD model itself. This approximation holds good if the measure (sense) V_{G} stays above V_{T0} (prestress threshold voltage) so that the electron capture process is not significant. However, if TTOM is invoked, the K_{F10} parameter of the RD model has to be readjusted to account for the difference (albeit small) in ΔV_{IT} for such case. For large measurement delay (see Chap. 7, Sect. 7.6, and Chap. 8), recovery phases are inserted for measurement during stress, and hence, the TTOM process would impact the ΔV_{IT} subcomponent of obtained (simulated to mimic actual measurement) ΔV_{T} during stress.

$$\Delta V_{IT}(t) = \Delta V_{IT}(EPR) + \frac{q f_{SLOW,CS}}{C_{OX}} \left[\Delta N_{IT}(t) - \Delta N_{IT}(EPR)\right] + \frac{q f_{FAST,PR}}{C_{OX}} \left[f_{SLOW,PS} \left\{ \Delta N_{IT}(EPR) - \Delta N_{IT}(EPPS) e^{-\left(\frac{t_{PR}}{\tau_{EC}}\right)^{\beta_{EC}}} \right\} \right] \left[1 - e^{-\left(\frac{t_1}{\tau_{EE}}\right)^{\beta_{EE}}} \right] \tag{5.1}$$

$$\Delta V_{IT}(t) = \frac{q f_{SLOW,CS} f_{SLOW,PS} \Delta N_{IT}(t)}{C_{OX}} + \frac{q f_{FAST,CS}}{C_{OX}} \left[f_{SLOW,PS} \{ \Delta N_{IT}(EPS) - \Delta N_{IT}(EPR) \} \right] e^{-\left(\frac{t_1}{\tau_{EC}}\right)^{\beta_{EC}}} + \frac{q f_{FAST,CS}}{C_{OX}} \left[f_{SLOW,PPS} \left\{ \Delta N_{IT}(EPR) - \Delta N_{IT}(EPPS) e^{-\left(\frac{t_{PR}}{\tau_{EC}}\right)^{\beta_{EC}}} \right\} \right] \left[1 - e^{-\left(\frac{t_{PS}}{\tau_{EE}}\right)^{\beta_{EE}}} \right] \left[e^{-\left(\frac{t_1}{\tau_{EC}}\right)^{\beta_{EC}}} \right] \tag{5.2}$$

As shown in Table 5.1, some of the parameters depend on V_{GREC} (or V_{GLOW} for AC pulse). Figure 5.6 shows the V_{GREC} dependence of (a) f_{FAST}, (b) τ_{EC} and (c) β_{EC} for the D1 device of Table 5.1. Note that f_{FAST} reduces but τ_{EC} and β_{EC} increase as the magnitude of V_{GREC} (or V_{GLOW}) is made high. The reduction in f_{FAST} implies lower fraction of traps that go below the Fermi level and undergo electron capture at higher magnitude of V_{GREC}. The increase in τ_{EC} and β_{EC} implies slowing down of the electron capture process. These dependencies are modeled by suitable empirical equations. As mentioned before, only f_{FAST} and τ_{EC} are process dependent, and similar functional dependencies are observed for other devices. Identical V_{GREC} dependence of β_{EC} as shown in Fig. 5.6 (c) is used for all devices in this book. The parameters τ_{EE} and β_{EE} do not depend on V_{GREC}.

Figure 5.7 shows the time kinetics of ΔV_{IT} calculated using the TTOM-enabled RD model during (a) stress and (b) recovery for the D1 device of Table 5.1. Note that the fixed and adjustable RD model parameters are listed, respectively, in Table 4.1 and Table 4.2 of Chap. 4, and the fixed and adjustable TTOM parameters are listed in Table 5.1 and Fig. 5.6. All traps generated during stress at $V_G = V_{GSTR}$ are

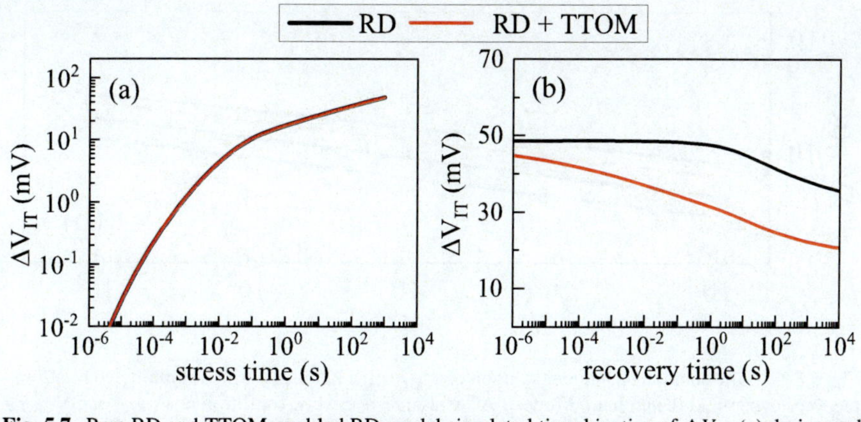

Fig. 5.7 Pure RD and TTOM-enabled RD model simulated time kinetics of ΔV_{IT} (a) during and (b) after DC stress for the D1 device of Table 5.1

presumed to be energetically located above the Fermi level (as a starting reference), and therefore, the RD and TTOM-enabled RD models show same results. However, since a fraction of the generated traps goes below the Fermi level and captures electrons during recovery as the magnitude of V_G is reduced to V_{GREC}, see Fig. 5.4, the RD and TTOM-enabled RD models show significantly different results. The ΔV_{IT} recovery calculated using the TTOM-enabled RD model is due to two components, $\Delta V_{IT,FAST}$ due to electron capture in the f_{FAST} fraction of traps (faster time constant) and $\Delta V_{IT,SLOW}$ for the remaining traps via trap passivation (slower time constant). If TTOM is ignored, ΔV_{IT} recovery is only due to the slower trap passivation process (RD only).

Figure 5.8 shows the time kinetics of ΔV_{IT} calculated using the TTOM-enabled RD model during (a) multiple DC stress–recovery cycles at different V_{GREC} and (b) higher f AC stress at different V_{GLOW} for the D1 device of Table 5.1. Note that, due

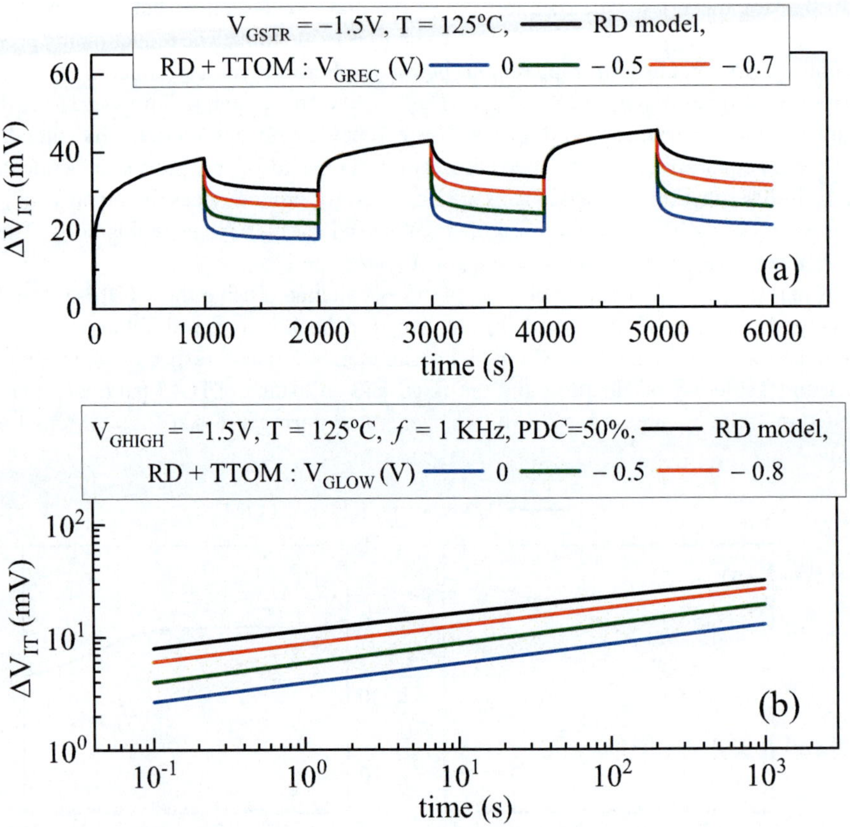

Fig. 5.8 TTOM-enabled RD model solution of ΔV_{IT} time kinetics during (a) multiple DC stress–recovery cycles and (b) higher f Mode-B AC stress, respectively, for different V_{GREC} and V_{GLOW} values, for the D1 device of Table 5.1. The pure RD model solution is also plotted, which does not change with V_{GREC} or V_{GLOW} values (not explicitly shown)

to large number of on/off cycles, only data from certain time points are plotted in Fig. 5.8 (b). The RD model solution for ΔV_{IT} ($= q^*\Delta N_{IT}/C_{OX}$) is also shown. The V_G value at a particular stress or recovery segment of Fig. 5.8 determines the f_{FAST}, τ_{EC} and β_{EC} values for that segment for TTOM calculations. The TTOM-enabled RD solution for ΔV_{IT} is lower compared to the pure RD solution for ΔV_{IT} from the first recovery half cycle after DC stress as shown in Fig. 5.8 (a). Therefore, the ΔV_{IT} time evolution during AC stress obtained from the TTOM-enabled RD model is lower compared to the pure RD solution as shown in Fig. 5.8 (b); note that only data corresponding to longer t_{STR} is plotted, which is after many cycles. The difference between the TTOM-enabled RD and pure RD solutions becomes larger for lower magnitude of V_{GREC} (or lower V_{GLOW} for AC stress). However, the time slope n is same for all V_{GLOW} values in Fig. 5.8 (b).

5.3.2 *Validation of TTOM-Enabled RD model*

The RD model has been directly validated against DCIV measured and delay-corrected ΔN_{IT} under different experimental conditions, which is discussed in Chap. 4, Sect. 4.4. However, the TTOM-enabled RD model needs to be validated against ultra-fast measured ΔV_T. As illustrated in Fig. 5.1, overall ΔV_T also gets impacted by ΔV_{HT} and ΔV_{OT} subcomponents in addition to ΔV_{IT}. However, it will be shown in subsequent chapters that ΔV_{HT} contribution is negligible under Mode-B AC stress (refer to Chap. 1, Sect. 1.2, for the details of AC stress gate pulse). Moreover, ΔV_{OT} is not significant unless the magnitude of pulse ($V_{GHIGH} = V_{GSTR}$ during AC stress) and T are high [11, 21, 23, 27]. Note that the ΔV_{IT} dominance of ΔV_T is clearly evident from the $n \sim 1/6$ power-law time dependence observed for different Mode-B AC stress conditions (changes in V_{GSTR}, T, PDC, f and V_{GLOW}), as shown in Chap. 1, Fig. 1.14. Therefore, in such situation (Mode-B AC stress under moderate V_{GSTR} and T), the ultra-fast measured ΔV_T can be assigned entirely to ΔV_{IT} and can be used to validate the TTOM-enabled RD model.

Figure 5.9 (a) shows the time evolution of measured and modeled ΔV_T during Mode-B AC stress at different V_{GLOW} for the D1 device of Table 5.1. The ultra-fast full sweep MSM method (see Chap. 1, Sect. 1.2) is used for measurements. As discussed above, the ΔV_T is due to ΔV_{IT} in this case, and the model lines are obtained using TTOM-enabled RD simulations. The RD model parameters are listed in Chap. 4, Tables 4.1 and 4.2, except K_{F10}, which is higher due to the larger energy bandgap scanned by V_T compared to DCIV measurements (see Chap. 7 for further details regarding model parameters for this device).

It is important to remark that the DCIV measurements and RD modeling of the ΔN_{IT} time kinetics are independent of V_{GLOW} (not explicitly shown). Although the time kinetics of ultra-fast measured ΔV_T ($= \Delta V_{IT}$) shows power-law dependence with $n \sim 1/6$, its magnitude reduces at lower magnitude of V_{GLOW} due to electron capture in the generated traps, which can be modeled by the TTOM-enabled RD framework.

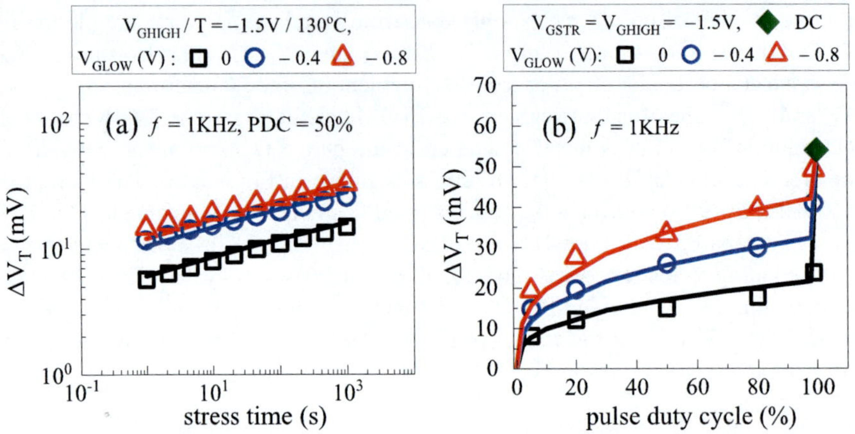

Fig. 5.9 Validation of TTOM-enabled RD model using ultra-fast measured (a) ΔV_T time kinetics and (b) ΔV_T at fixed t_{STR} as a function of PDC, for Mode-B AC stress at different V_{GLOW} for the D1 device of Table 5.1. Symbols: experiment, lines: model calculation. Data from [11]. See Chap. 7 for additional information on model parameters

As a further proof, Fig. 5.9 (b) plots the measured and modeled ΔV_T ($= \Delta V_{IT}$) at fixed time ($t_{STR} = 1$Ks) during Mode-B AC stress as a function of PDC for different V_{GLOW}, for the same D1 device. Note that the "S"-shaped PDC dependence is observed with a kink or jump near DC, and the magnitude of this jump depends on the V_{GLOW} value of the AC pulse (also shown in Chap. 1, Fig. 1.15). This is due to the electron capture process, as this jump is absent in the PDC dependence of DCIV measured ΔN_{IT}, as shown in Chap. 4, Fig. 4.10. The TTOM-enabled RD framework can model the PDC dependence of ΔV_T ($= \Delta V_{IT}$) at different V_{GLOW}. Note that the small difference between the ultra-fast measured ΔV_T and the TTOM-enabled RD model simulated ΔV_{IT} at DC stress (100% PDC) is due to non-negligible contribution from ΔV_{HT} (note that the ΔV_{OT} contribution is not significant due to use of moderate V_{GSTR} and T in this case). The kinetics of ΔV_{HT} is discussed next.

5.4 Hole Trapping in Preexisting Defects

The simplest approach is to use empirical stretched exponential equations for hole trapping and detrapping; refer to [11–14, 19–24, 29] for details.

The earlier physical modeling methods used hole tunneling from the inversion layer and subsequent trapping in the gate insulator traps based on the conventional Shockley–Read–Hall (SRH) theory [37–39]. However, hole trapping is followed by structural relaxation in the vicinity of the traps and phonons are involved [40]. The hole trapping and related structural relaxation were simulated by the Double Well Thermionic (DWT) [41], Harry Diamond Laboratories (HDL) [42], Nonradiative

Multi-Phonon (NMP) [43] and extended NMP (e-NMP) [44] models; the interested reader can refer to the original references for details. The DWT model has been modified into ABDWT model to make it compatible with data measured over extended *T* range [32, 34, 45].

Note that the e-NMP model has large number of adjustable parameters and hence is of limited practical use (as of now, the model is not used to explain data from different gate insulator processes). The ABDWT model is better suited than the NMP model to explain the hole trapping–detrapping time kinetics over extended *T* range during NBTI stress–recovery studies and for Random Telegraph Noise (RTN) and Time Dependent Defect Spectroscopy (TDDS) experiments in small area devices, as shown in [46, 47]. Hence, the ABDWT model is preferred and used throughout this book.

5.4.1 *Activated Barrier Double Well Thermionic (ABDWT) Model*

Figure 5.10 illustrates the ABDWT model schematic [34]. The reservoir state E_1 (valence band for holes) is separated from the trap level E_2 by the energy barrier E_B. The barrier is normally distributed with a mean (E_{BM}) and spread (E_{BS}), and these

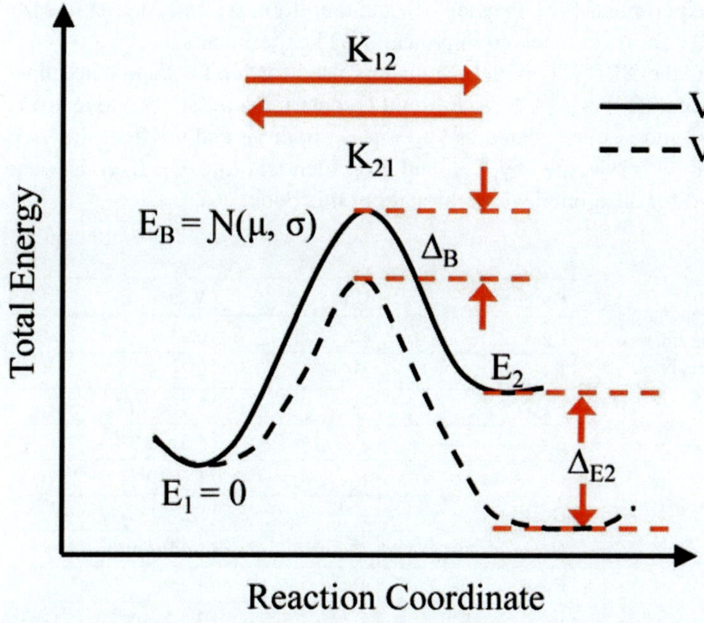

Fig. 5.10 Schematic of the ABDWT model, showing levels E_1, E_2 and barrier E_B along the reaction coordinate

are thermally activated with energies E_{AM} and E_{AS}, respectively. The barrier E_B and level E_2 go down by Δ_B and Δ_{E2}, respectively, under the application of V_{GSTR}, which trigger the forward thermionic, over the barrier, reaction K_{12}, and populate the state E_2. The barrier E_B and level E_2 go back up after stress (depend on the magnitude of V_{GREC}) and trigger the reverse reaction K_{21} to depopulate the state E_2.

The state E_2 is considered charged, and its occupancy governs the magnitude of ΔV_{HT}, after multiplying with q/C_{OX} and the density of preexisting defects (N_{0HT}); note that trapping is lumped at the channel/IL (or channel/gate insulator) interface for simplicity. The ABDWT model equations are as follows [34]:

$$K_{12} = \nu e^{-(E_B - E_1 - \Delta_B)/kT} \tag{5.3}$$

$$K_{21} = \nu e^{-(E_B - E_2 - \Delta_B + \Delta_{E2})/kT} \tag{5.4}$$

$$\Delta_B = \gamma_B E_{OX} + \gamma'_B E_{OX}^2 \tag{5.5}$$

$$\Delta_{E2} = \gamma_{E2} E_{OX} + \gamma'_{E2} E_{OX}^2 \tag{5.6}$$

where ν is the attempt to escape frequency, E_1 is chosen as reference (= 0 eV) and the terms γ_B, γ'_B and γ_{E2}, γ'_{E2} are related to the E_{OX} dependence of Δ_B and Δ_{E2}, respectively. Note that γ'_B and γ'_{E2} are only used to explain the behavior of switching traps in TDDS experiments [46]; they are ~ 0, and therefore, Δ_B and Δ_{E2} are linearly dependent on E_{OX} for traps involved in typical NBTI experiments.

Table 5.2 lists the ABDWT model parameters used for NBTI. N_{0HT} determines the maximum (saturated) ΔV_{HT}, E_{BM}, E_{BS} and E_2 determine the stress and recovery time kinetics, γ_B and γ_{E2} are related as $\gamma_{E2} = m\gamma_B$, both γ_B and m affect the VAF, and the T activation is governed by E_{AM} and E_{AS}. Identical E_{BS}, E_2, E_{AM}, E_{AS} and ν values are used for different devices throughout this book.

Table 5.2 Process (or technology)-independent ABDWT model parameters used throughout this book

Parameter	Unit	Value
E_{BS}	eV	0.24
E_2	eV	0.21
E_{AM}	eV	8.5×10^{-3}
E_{AS}	eV	5.5×10^{-3}
ν	1/s	1×10^{13}
Device:	–	D2
N_{0HT}	1/cm^2	4.64×10^{12}
E_{BM}	eV	1.25
γ_B	C.cm	3.6×10^{-9}
m	–	3.5

The process-dependent parameters are listed for the D2 device of Table 5.1. It will be shown in later chapters that N_{0HT} varies with process changes; the saturated ΔV_{HT} magnitude increases with higher N% but reduces with higher Ge%. However, note that the other parameters may change (*e.g.*, with changes in Ge% and N%; see Chap. 7, Chap. 9 and Chap. 11) or may not change (*e.g.*, with changes in the device dimensions, but no changes in the channel material or gate insulator stack; see Chap. 12).

It is important to note that the ΔV_{HT} contribution depends on the quality of the gate insulator stack. A poor quality gate insulator would naturally have higher preexisting trap density and as a consequence higher ΔV_{HT}. However, this subcomponent is always small for all production quality gate insulator stacks. Furthermore, it is negligible for SiGe channel devices, and for AC stress in all devices at high f, refer to later chapters for further details.

5.4.2 Validation of TTOM-Enabled RD and ABDWT Models

All data shown in this section are from the D2 device of Table 5.1. As discussed in Chap. 7, GF HKMG MOSFET with high N% in the gate stack has negligible ΔV_{OT} contribution for stress using moderate V_{GSTR} and T [11]. Therefore, the time kinetics of measured ΔV_T is governed by the kinetics of the ΔV_{IT} and ΔV_{HT} subcomponents. Measurements are done using the ultra-fast, full sweep MSM method (see Chap. 1, Sect. 1.2). The parameters are listed in Chap. 4, Tables 4.1 and 4.2, for the RD model, in Table 5.1 for TTOM and in Table 5.2 for ABDWT model. Note that the K_{F10} parameter of the RD model is higher (than in Table 4.2) due to larger energy bandgap scanned by V_T compared to DCIV measurements (see Chap. 7).

Figure 5.11 shows the time evolution of measured and modeled ΔV_T (a, c) during and (b, d) after DC stress. The measured and modeled ΔV_T kinetics together with the underlying ΔV_{IT} and ΔV_{HT} subcomponents are shown in Fig. 5.11 (a) and (b) at fixed V_{GSTR} and T, while the modeling of overall ΔV_T at multiple V_{GSTR} and T is shown in Fig. 5.11 (c) and (d). Note that the adjustable model parameters governing the ΔV_{IT} subcomponent are independently verified by DCIV measurements, and identical Γ_0, E_{AKF1} and α values are used in this case (see Chap. 4, Sect. 4.4). However, the K_{F10} would be different between delay-corrected DCIV and I-V measurements as mentioned before.

Note that the ΔV_{IT} subcomponent evolves gradually in time at the initiation of stress and asymptotically approaches the $n \sim 1/6$ power-law time dependence for $t_{STR} > 1$s. It recovers over several decades in time after stress, which is governed by electron capture (~ TTOM) and trap passivation processes as discussed before. On the other hand, the ΔV_{HT} subcomponent evolves rapidly in time at the initiation of stress; however, it saturates at longer t_{STR}, which is consistent with the results shown in Chap. 3, Sect. 3.3. ΔV_{HT} recovers quickly after the end of stress. The model can explain measured stress–recovery kinetics at multiple V_{GSTR} and T.

Figure 5.12 shows the measured and modeled ΔV_T at fixed t_{STR} and the underlying

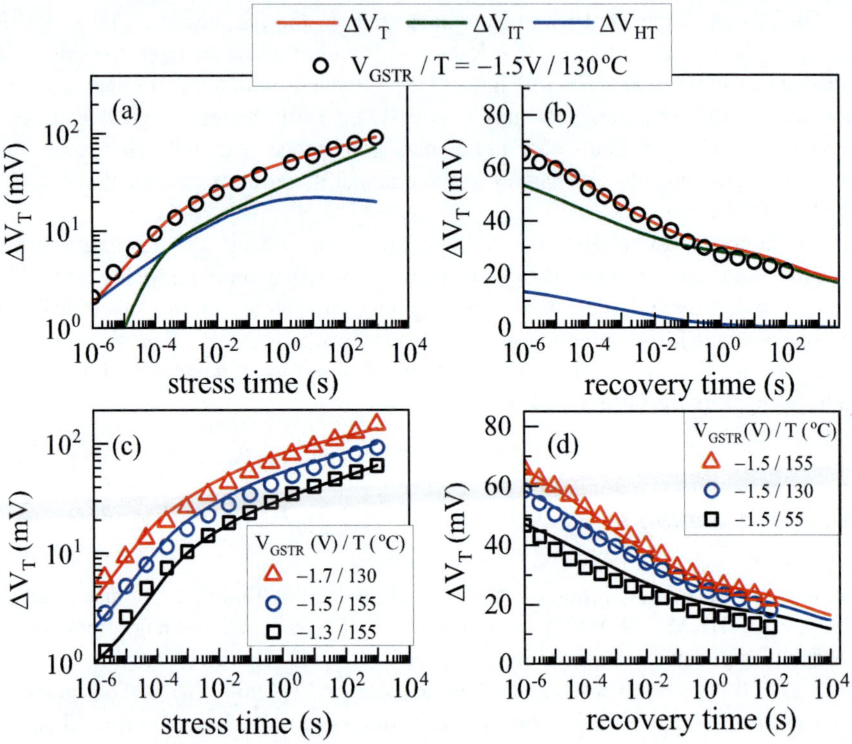

Fig. 5.11 Validation of TTOM-enabled RD and ABDWT models using ultra-fast measured ΔV_{T} time kinetics (a, c) during and (b, d) after stress. Data are shown for the D2 device of Table 5.1 (a, b) with underlying subcomponents at fixed V_{GSTR}, T, and (c, d) at multiple V_{GSTR} and T. Symbols: experiment, lines: model calculation. Data from [11]. See Chap. 7 for additional information on model parameters

ΔV_{IT} and ΔV_{HT} subcomponents as a function of (a) V_{GSTR} at fixed T and (b) stress T at fixed V_{GSTR}. Note that ΔV_{IT} dominates overall ΔV_{T} for the conditions used in this experiment. Figure 5.12 also shows the measured and modeled ΔV_{T} at fixed t_{STR} during stress as a function of (c) V_{GSTR} at different T and (d) T at different V_{GSTR}. The VAF at multiple T and therefore E_{A} at multiple V_{GSTR} can be modeled. Note that the VAF reduces slightly at higher T (E_{A} reduces slightly at higher V_{GSTR}) in this device.

The relative contribution of ΔV_{HT} increases with higher N% in the gate stack (especially in the IL) and reduces with higher Ge% in the channel, more on this in Chap. 7, Chap. 9 and Chap. 11. Hence, a comparison between the GF HKMG planar devices D1 and D2, or the RMG HKMG FinFETs D3 and D4 of Chap. 4, Table 4.2, would further verify the differences in the time and T dependencies between the ΔV_{IT} and ΔV_{HT} subcomponents. However, the contribution from the ΔV_{OT} subcomponent is not negligible in other devices and needs to be considered before

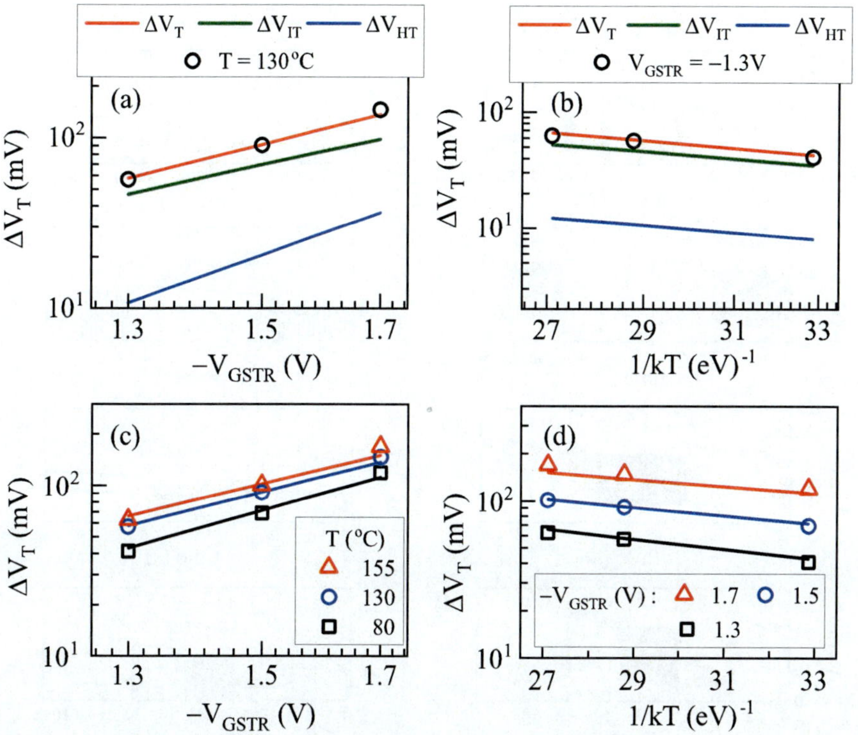

Fig. 5.12 Validation of TTOM-enabled RD and ABDWT models using ultra-fast measured ΔV_T at fixed t_{STR} during stress as a function of (a, c) V_{GSTR} and (b, d) T. Data are shown for the D2 device of Table 5.1, with underlying subcomponents at fixed (a) T and (b) V_{GSTR}, and (c, d) at multiple T and V_{GSTR}. Symbols: experiment, lines: model calculation. Data from [11]. See Chap. 7 for additional information on model parameters

any such process dependencies are analyzed. The ΔV_{OT} component is modeled and validated in Chap. 6.

5.5 Multi-segment Arbitrary DC–AC Stress

All data shown in this section are from the D2 device of Table 5.1. Measurements are done using the ultra-fast, full sweep MSM method (see Chap. 1, Sect. 1.2). The parameters are listed in Chap. 4, Tables 4.1 and 4.2, for the RD model, in Table 5.1 for TTOM and in Table 5.2 for ABDWT model. The ΔV_{OT} subcomponent is negligible for this device, as analyzed before in Sect. 5.4. As mentioned earlier, the K_{F10} parameter of the RD model is higher (than in Table 4.2) due to larger energy bandgap scanned by V_T compared to DCIV measurements (see Chap. 7).

Figure 5.13 shows the time evolution of measured and modeled ΔV_T (left panels)

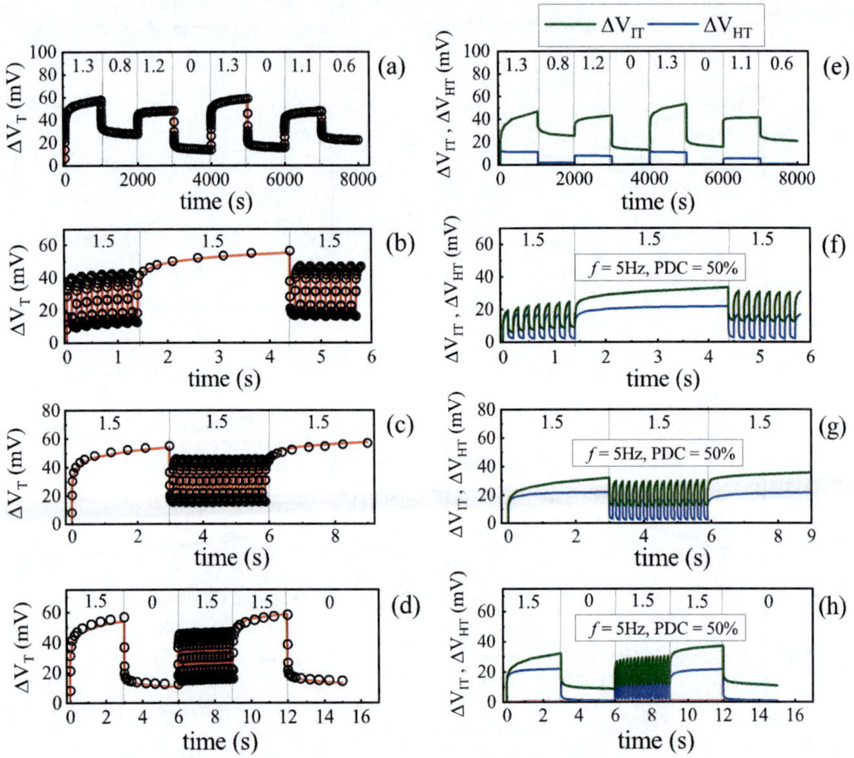

Fig. 5.13 Modeling of measured ΔV_T kinetics in D2 device under arbitrary gate excitation (see text). Symbols: experiment, lines: model calculation. The subcomponents are shown in the right panels. All biases are negative, in units of volts. Data from [11]

along with the underlying ΔV_{IT} and ΔV_{HT} subcomponents (right panels) under diverse experimental conditions with consecutive segments of: (a, e) DC stress at different V_{GSTR} (or V_{GREC}), as well as (b, f) AC, DC and AC stress, (c, g) DC, AC and DC stress, and (d, h) DC stress and recovery, AC stress and DC stress and recovery, at fixed V_{GSTR} (V_{GHIGH} for AC) for different DC and AC segments and fixed PDC and f for AC segments.

Figure 5.14 shows the time evolution of measured and modeled ΔV_T (left panels) along with the underlying ΔV_{IT} and ΔV_{HT} subcomponents (right panels) for mixed DC–AC segments with different (a, c) V_{GSTR} and (b, d) time between various segments.

Figure 5.15 shows the time evolution of measured and modeled ΔV_T (left panels) along with the underlying ΔV_{IT} and ΔV_{HT} subcomponents (right panels) for consecutive AC segments having arbitrary (a, e) frequency, (b, f) V_{GHIGH}, (c, g) V_{GHIGH} as well as f and (d, h) PDC between various segments.

The AC pulses (low f is used to demonstrate the cycle-by-cycle modeling capability) used in Fig. 5.13 through Fig. 5.15 are with $V_{GLOW} = 0$ V.

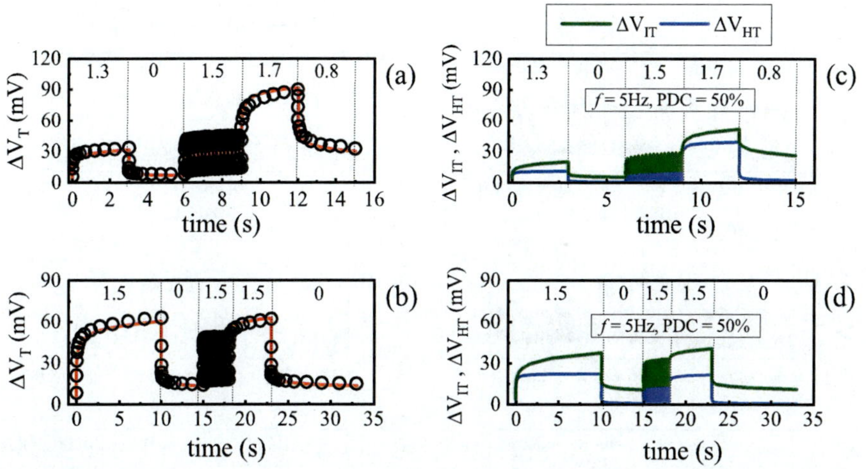

Fig. 5.14 Modeling of measured ΔV_T kinetics in D2 devices under arbitrary gate excitation (see text). Symbols: experiment, lines: model calculation. The subcomponents are shown in the right panels. All biases are negative, in units of volts. Data from [11]

5.6 Summary

In general, uncorrelated contributions from ΔV_{IT}, ΔV_{HT} and ΔV_{OT} subcomponents govern the stress and recovery time kinetics of measured ΔV_T at different V_{GSTR}, V_{GREC} and T. RD model calculates the generation and passivation of ΔN_{IT} during and after stress, and is independently validated by DCIV measured data in Chap. 4. The dependence of ΔV_T on V_{GLOW} during AC stress, while ΔN_{IT} shows no such dependence, implies the importance of trap occupancy, which is calculated using TTOM. The ΔV_{HT} kinetics is calculated using the ABDWT model. TTOM-enabled RD as well as ABDWT plus TTOM-enabled RD models are validated using specific stress conditions and/or devices, to have ΔV_T contribution dominated by only ΔV_{IT}, as well as both ΔV_{IT} and ΔV_{HT}, respectively. The RDD model for bulk traps is explained and validated in Chap. 6.

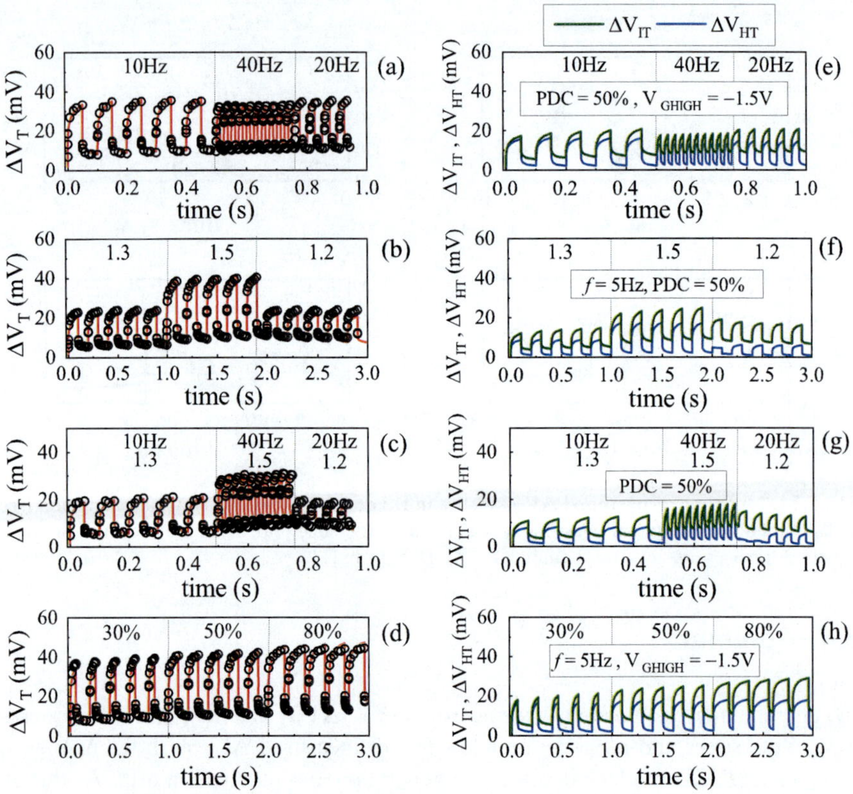

Fig. 5.15 Modeling of measured ΔV_T kinetics in D2 devices under arbitrary gate excitation (see text). Symbols: experiment, lines: model calculation. The subcomponents are shown in the right panels. All biases are negative, in units of volts. Data from [11]

Acknowledgements All experimental data presented in this chapter are re-plotted from previously published reports. The authors acknowledge Applied Materials for providing planar MOSFETs. Karansingh Thakor is acknowledged for help with manuscript preparation.

References

1. N. Kimizuka, K. Yamaguchi, K. Imai, T. Iizuka, C. T. Liu, R. C. Keller, T. Horiuchi, in Symposium on VLSI Technology Digest of Technical Papers (2000), p. 92
2. V. Huard, M. Denais, F. Perrier, N. Revil, C. Parthasarathy, A. Bravaix, E. Vincent, Microelectron. Reliab. **45**, 83 (2005)
3. A.T. Krishnan, C. Chancellor, S. Chakravarthi, P.E. Nicollian, V. Reddy, A. Varghese, R.B. Khamankar, S. Krishnan, in IEEE International Electron Devices Meeting Technical Digest (2005), p. 688

4. D. Varghese, D. Saha, S. Mahapatra, K. Ahmed, F. Nouri, M. Alam, in IEEE International Electron Devices Meeting Technical Digest (2005), p. 684
5. S. Mahapatra, K. Ahmed, D. Varghese, A.E. Islam, G. Gupta, L. Madhav, D. Saha, M. A. Alam, in IEEE International Reliability Physics Symposium Proceedings (2007), p. 1
6. Y. Mitani, H. Satake, A. Toriumi, IEEE Trans. Device Mater. Reliab. **8**, 6 (2008)
7. S. Pae, M. Agostinelli, M. Brazier, R. Chau, G. Dewey, T. Ghani, M. Hattendorf, J. Hicks, J. Kavalieros, K. Kuhn, M. Kuhn, J. Maiz, M. Metz, K. Mistry, C. Prasad, S. Ramey, A. Roskowski, J. Sandford, C. Thomas, J. Thomas, C. Wiegand, J. Wiedemer, in IEEE International Reliability Physics Symposium Proceedings (2008), p. 352
8. K. Joshi, S. Hung, S. Mukhopadhyay, V. Chaudhary, N. Nanaware, B. Rajamohanan, T. Sato, M. Bevan, A. Wei, A. Noori, B. McDougal, C. Ni, G. Saheli, C. Lazik, P. Liu, D. Chu, L. Date, S. Datta, A. Brand, J. Swenberg, S. Mahapatra, in IEEE International Reliability Physics Symposium Proceedings (2013), p. 4C.2.1
9. J. Franco, B. Kaczer, P. J. Roussel, J. Mitard, M. Cho, L. Witters, T. Grasser, G. Groeseneken, IEEE Trans. Electron Devices. **60**, 396 (2013)
10. P. Srinivasan, J. Fronheiser, K. Akarvardar, A. Kerber, L.F. Edge, R.G. Southwick, E. Cartier, H. Kothari, in IEEE Int. Reliab. Phys. Symp. Proc. **6A.3.1** (2014)
11. N. Parihar, N. Goel, S. Mukhopadhyay, S. Mahapatra, IEEE Trans. Electron Devices. **65**, 392 (2018)
12. N. Parihar, S. Mahapatra, in IEEE International Reliability Physics Symposium Proceedings (2018), p. TX.5.1
13. V. Huard, C. Ndiaye, M. Arabi, N. Parihar, X. Federspiel, S. Mhira, S. Mahapatra, A. Bravaix, in IEEE International Reliability Physics Symposium Proceedings (2018), p. TX.4.1
14. N. Parihar, R. Tiwari, C. Ndiaye, M. Arabi, S. Mhira, H. Wong, S. Motzny, V. Moroz, V. Huard, S. Mahapatra, in International Conference on Simulation of Semiconductor Processes and Devices (2018), p. 167
15. S. Ramey, A. Ashutosh, C. Auth, J. Clifford, M. Hattendorf, J. Hicks, R. James, A. Rahman, V. Sharma, A. St. Amour, C. Wiegand, in IEEE International Reliability Physics Symposium Proceedings (2013), p. 4C.5.1
16. K.T. Lee, W. Kang, E.-A. Chung, G. Kim, H. Shim, H. Lee, H. Kim, M. Choe, N.-I. Lee, A. Patel, J. Park, J. Park, in IEEE International Reliability Physics Symposium Proceedings (2013), p. 2D.1.1
17. J. Franco, B. Kaczer, A. Chasin, H. Mertens, L.-A. Ragnarsson, R. Ritzenthaler, S. Mukhopadhyay, H. Arimura, P. J. Roussel, E. Bury, N. Horiguchi, D. Linten, G. Groeseneken, A. Thean, in IEEE International Reliability Physics Symposium Proceedings (2016), p. 4B.2.1
18. G. Jiao, M. Toledano-Luque, K.-J. Nam, N. Toshiro, S.-H. Lee, J.-S. Kim, T. Kauerauf, E. Chung, D. Bae, G. Bae, D.-W. Kim, K. Hwang, in IEEE International Electron Devices Meeting Technical Digest (2016), p. 31.2.1
19. N. Parihar, R.G. Southwick, U. Sharma, M. Wang, J.H. Stathis, S. Mahapatra, in IEEE International Reliability Physics Symposium Proceedings (2017), p. 2D.4.1
20. N. Parihar, R. Southwick, M. Wang, J.H. Stathis, S. Mahapatra, in IEEE International Electron Devices Meeting Technical Digest (2017), p. 7.3.1
21. N. Parihar, U. Sharma, R.G. Southwick, M. Wang, J.H. Stathis, S. Mahapatra, IEEE Trans. Electron Devices. **65**, 23 (2018)
22. N. Parihar, R.G. Southwick, M. Wang, J.H. Stathis, S. Mahapatra, IEEE Trans. Electron Devices. **65**, 1699 (2018)
23. N. Parihar, R.G. Southwick, M. Wang, J.H. Stathis, S. Mahapatra, IEEE Trans. Electron Devices. **65**, 1707 (2018)
24. N. Parihar, R. Tiwari, S. Mahapatra, in International Conference on Simulation of Semiconductor Processes and Devices (2018), p. 176
25. R. Tiwari, N. Parihar, K. Thakor, H.Y. Wong, S. Motzny, M. Choi, V. Moroz, S. Mahapatra, IEEE Trans. Electron Devices. **66**, 2086 (2019)
26. R. Tiwari, N. Parihar, K. Thakor, H.Y. Wong, S. Motzny, M. Choi, V. Moroz, S. Mahapatra, IEEE Trans. Electron Devices. **66**, 2093 (2019)

27. N. Parihar, U. Sharma, R.G. Southwick, M. Wang, J.H. Stathis, S. Mahapatra, in IEEE International Reliability Physics Symposium Proceedings (2019)
28. M. Wang, J. Zhang, H. Zhou, R.G. Southwick, R. Hsin, K. Chao, X. Miao, V.S. Basker, T. Yamashita, D. Guo, G. Karve, H. Bu, in IEEE International Reliability Physics Symposium Proceedings (2019)
29. N. Choudhury, U. Sharma, H. Zhou, R.G. Southwick, M. Wang, S. Mahapatra, in IEEE International Reliability Physics Symposium Proceedings (2020)
30. H. Zhou, M. Wang, J. Zhang, K. Watanabe, C. Durfee, S. Mochizuki, R. Bao, R. Southwick, M. Bhuiyan, B. Veeraraghavan, in IEEE International Reliability Physics Symposium Proceedings (2020)
31. N. Choudhury, T. Samadder, R. Tiwari, H. Zhou, R.G. Southwick, M. Wang, S. Mahapatra, in IEEE International Reliability Physics Symposium Proceedings (2021)
32. S. Mahapatra, N. Goel, S. Desai, S. Gupta, B. Jose, S. Mukhopadhyay, K. Joshi, A. Jain, A.E. Islam, M.A. Alam, IEEE Trans. Electron Devices. **60**, 901 (2013)
33. A.E. Islam, N. Goel, S. Mahapatra, M.A. Alam, in Fundamentals of Bias Temperature Instability in MOS Transistors (Springer India, 2015), pp. 181–207
34. N. Choudhury, N. Parihar, N. Goel, A. Thirunavukkarasu, S. Mahapatra, IEEE J. Electron Devices Soc. **8**, 1281 (2020)
35. T. Samadder, N. Choudhury, S. Kumar, D. Kochar, N. Parihar, S. Mahapatra, IEEE Trans. Electron Devices. **68**, 485 (2021)
36. S. Kumar, R. Anandkrishnan, N. Parihar, S. Mahapatra, IEEE Trans. Electron Devices. **67**, 4741 (2020)
37. S. Christensson, I. Lundström, C. Svensson, Solid-State Electron. **11**, 797 (1968)
38. T.L. Tewksbury, Hae-Seung L, IEEE J Solid-State Circ. **29**, 239 (1994)
39. N. Zanolla, D. Siprak, P. Baumgartner, E. Sangiorgi, C. Fiegna, in International Conference on Ultimate Integration of Silicon (2008), p. 137
40. T. Grasser, Microelectron. Reliab. **52**, 39 (2012)
41. D. Ielmini, M. Manigrasso, F. Gattel, G. Valentini, in IEEE International Reliability Physics Symposium Proceedings (2009), p. 26
42. A.J. Lelis, T.R. Oldham, IEEE Trans. Nucl. Sci. **41**, 1835 (1994)
43. G. Rzepa, J. Franco, B. O'Sullivan, A. Subirats, M. Simicic, G. Hellings, P. Weckx, M. Jech, T. Knobloch, M. Waltl, P.J. Roussel, D. Linten, B. Kaczer, T. Grasser, Microelectron. Reliab. **85**, 49 (2018)
44. W. Goes, Y. Wimmer, A.-M. El-Sayed, G. Rzepa, M. Jech, A.L. Shluger, T. Grasser, Microelectron. Reliab. **87**, 286 (2018)
45. S. Desai, S. Mukhopadhyay, N. Goel, N. Nanaware, B. Jose, K. Joshi, S. Mahapatra, in IEEE International Reliability Physics Symposium Proceedings (2013), p. XT.2.1
46. S. Bhagdikar, S. Mahapatra, in International Conference on Simulation of Semiconductor Processes and Devices (2019)
47. S. Bhagdikar, S. Mahapatra, in International Conference on Simulation of Semiconductor Processes and Devices (2020), p. 117

Chapter 6
BTI Analysis Tool (BAT) Model Framework—Generation of Bulk Traps

Souvik Mahapatra, Narendra Parihar, Tarun Samadder, Nilotpal Choudhury, and Akshay Raj

6.1 Introduction

As shown in the previous chapters, Negative Bias Temperature Instability (NBTI) became an important reliability concern with the migration from Silicon Dioxide (SiO_2) to Silicon Oxynitride (SiON) gate insulator MOSFETs [1–6]. It continues to remain as a concern in dual layer (SiO_2 or SiON Interlayer (IL) and Hafnium Dioxide (HfO_2) High-K layer) High-K Metal Gate (HKMG) gate insulator based bulk [7–12] and Fully Depleted Silicon On Insulator (FDSOI) [13, 14] planar MOSFETs, bulk and SOI FinFETs [14–27], as well as Gate All Around Stacked Nanosheet FETs [28–31], with either Silicon (Si) or Silicon Germanium (SiGe) channel. The key features of NBTI are listed hereinafter (reproduced from Chap. 3, Sect. 3.1).

As described in Chap. 1, Sect. 1.3, NBTI results in the buildup of positive gate insulator charges during stress under negative gate bias ($V_G = V_{GSTR}$), which results in device parametric drift, e.g., threshold voltage shift (ΔV_T), over time. The positive charges and resulting ΔV_T reduce when the V_G is reduced or removed after stress ($V_G = V_{GREC}$ or 0 V). Hence, AC stress results in lower ΔV_T compared to DC stress. Note, ΔV_T during DC and AC stress gets accelerated with more negative V_{GSTR} (V_{GHIGH} for AC pulse) and at elevated temperature (T), and these dependencies are respectively governed by the Voltage Acceleration Factor (VAF) and Arrhenius T activation energy (E_A). The AC to DC ratio depends on the Pulse Duty Cycle (PDC) and pulse low value (V_{GLOW}), and may or may not depend on the frequency (f) of the gate pulse. Note that the PDC, V_{GLOW}, and f (in) dependence are also governed by the AC stress mode (Mode-A or Mode-B).

As described in Chap. 2, the ΔV_T magnitude and its time kinetics during and after stress, i.e., the power-law time slope (n) at long stress time (t_{STR}) during stress

S. Mahapatra (✉) · N. Parihar · T. Samadder · N. Choudhury · A. Raj
Department of Electrical Engineering, Indian Institute of Technology Bombay, Powai, Mumbai 400076, India
e-mail: souvik@ee.iitb.ac.in

S. Mahapatra (ed.), *Recent Advances in PMOS Negative Bias Temperature Instability*,
https://doi.org/10.1007/978-981-16-6120-4_6

and the Fraction Remaining (FR) during recovery after stress (FR is defined as ΔV_T at $t = t_{REC}$ after stress to that at $t = t_{STR}$ at the end of stress), VAF, E_A and the T dependence of VAF depend on the Nitrogen content (N%) in the gate insulator and Germanium content (Ge%) in the channel. The following signatures are observed (reproduced from Chap. 4, Sect. 4.1): The ΔV_T magnitude increases, but *n*, VAF, E_A, T dependence of VAF and FR reduce with higher N%. However, ΔV_T magnitude reduces, but *n*, VAF, E_A, T dependence of VAF and FR increase at higher Ge%. Moreover, ΔV_T reduces with fin length and width scaling in FinFETs, sheet length scaling in GAA-SNS FETs, and larger spacing (SA) between the Shallow Trench Isolation (STI) and device active in FDSOI MOSFETs, however, it increases with sheet width scaling in GAA-SNS FETs.

As mentioned in Chaps. 4 and 5, any relevant NBTI model should be able to explain the above experimental features, summarized in Chap. 3, Sect. 3.1. This is done using the BTI Analysis Tool (BAT) framework in this book.

Figure 6.1 illustrates the BAT framework. As mentioned in Chap. 4, Sect. 4.2, uncorrelated contributions from generated interface traps (ΔV_{IT}), hole trapping in pre-existing process related gate insulator traps (ΔV_{HT}), and generated bulk gate insulator traps (ΔV_{OT}) result in NBTI related positive gate insulator charges and the resulting ΔV_T. The time kinetics of interface trap density (ΔN_{IT}) is calculated using the Reaction Diffusion (RD) model [11, 32, 33], described and validated in Chap. 4,

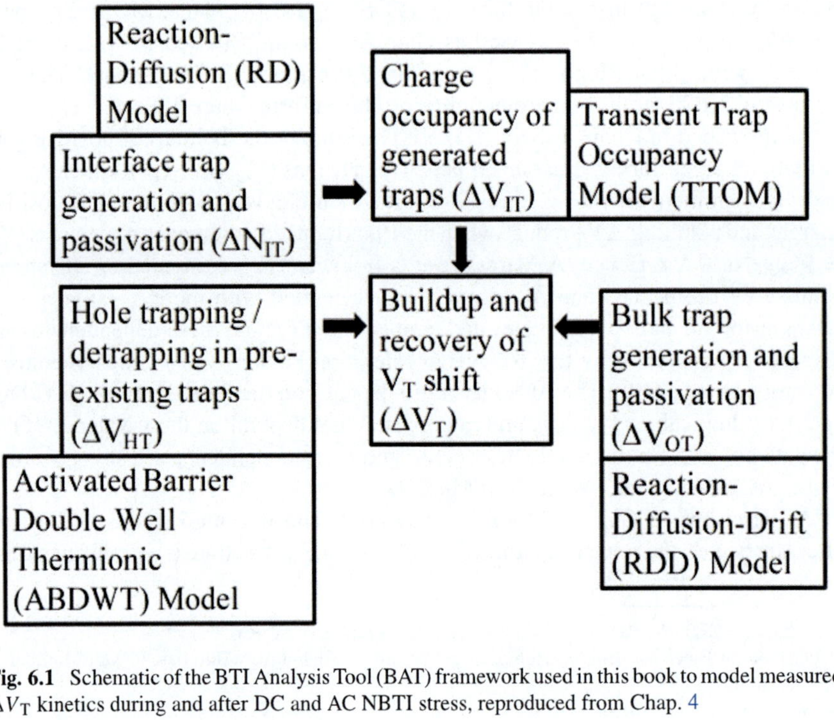

Fig. 6.1 Schematic of the BTI Analysis Tool (BAT) framework used in this book to model measured ΔV_T kinetics during and after DC and AC NBTI stress, reproduced from Chap. 4

and their contribution (ΔV_{IT}) to overall ΔV_T by the Transient Trap Occupancy Model (TTOM) [11], described and validated in Chap. 5. The contribution due to ΔV_{HT} is calculated using the Activated Barrier Double Well Thermionic (ABDWT) model [34], also described and validated in Chap. 5. In this chapter, the ΔV_{OT} contribution is calculated using the Reaction Diffusion Drift (RDD) model [35] (note, the RD model is a special case of RDD model) and validated.

6.2 Generation of Bulk Gate Insulator Traps

As mentioned in Chap.3, Sect. 3.4, NBTI stress at relatively higher V_{GSTR} and T is similar to the Time Dependent Dielectric Breakdown (TDDB) experiments, which result in the generation of bulk gate insulator traps (density ΔN_{OT}) [36–39]. Note, charges associated with these traps contribute (ΔV_{OT}) to overall ΔV_T. Trap Assisted Tunneling (TAT) via these traps gives rise to Stress Induced Leakage Current (SILC) or increase in the gate leakage current (ΔI_G) [40–48]. Typically, TDDB experiments focus on the formation of a percolation path between the channel and gate caused by these generated traps leading to a breakdown of the gate insulator. However, for NBTI, their contribution to ΔV_T before the device breaks is of interest, which is discussed in this section.

The simplest approach is to use empirical stretched exponential equations for bulk trap generation and passivation, refer to [11–14, 19–24, 29] for details. However, for a physical approach, the RDD model is used [35, 49], which is discussed next. Other models are also proposed in the literature, i.e., the Gate Side Hydrogen Release (GSHR) model [50] and the Two Well Thermionic (TWT) model [51], the interested reader can refer to the original references for details.

6.2.1 Reaction Diffusion Drift (RDD) Model

The Anode Hole Injection (AHI) process [37, 52] triggers the defect dissociation mechanism in the RDD model and is illustrated using the schematic of Fig. 6.2. As explained in Chap. 3, Sect. 3.4, under the application of stress at high V_{GSTR} and/or V_B >0 V, hot holes are injected into the gate insulator by the AHI process and dissociate the H passivated Y-H defects inside the gate insulator bulk. RDD model suggests that after dissociation, the released H atoms diffuse and further react with other Y-H defects and subsequently release H_2 molecules and/or H_2^+ ions (the ratio of H_2 to H_2^+ density depends on the energy of AHI related hot holes), and these H_2 and/or H_2^+ species diffuse and/or drift into the gate insulator and backend (an illustration similar to Chap. 4, Fig. 4.2 (a) can be considered for visualization). The reverse processes occur after stress.

However, note that both H_2 and H_2^+ would diffuse towards the IL/High-K and channel/IL interfaces during recovery under $V_{GREC} = 0$ V, as the drift of H_2^+ would

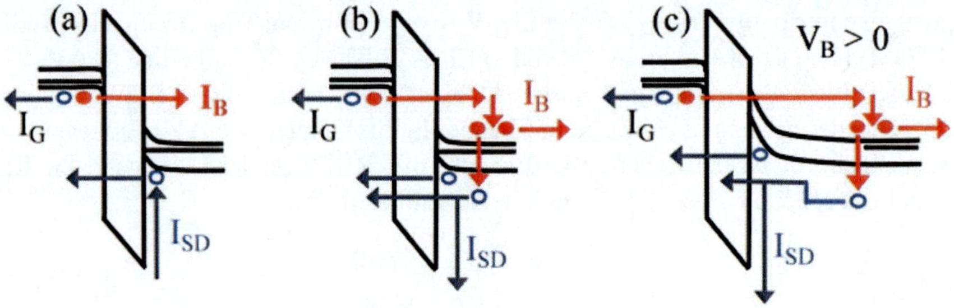

Fig. 6.2 Energy band diagram of a p-MOSFET during inversion stress under (a) moderate and (b) high V_{GSTR} but $V_B = 0$ V, and (c) moderate V_{GSTR} but $V_B > 0$V. The gate tunneling current and the underlying source drain and substrate components are shown. In (b, c) the AHI process is illustrated, which triggers bulk trap generation. Reproduced from Chap. 3

be negligible due to very low oxide electric field (E_{OX}) in this case. Furthermore, such back movement of H_2^+ would be further impeded during recovery if V_{GREC} <0 V. Note that the RDD model is similar to the RD model explained in Chap. 4, Sect. 4.3, except the initial trigger (AHI in RDD but inversion layer cold holes in RD model) and the generation and subsequent drift of ionic species (together with the diffusion of molecular species).

The following chemical equations describe the RDD model processes discussed above [35, 49]:

$$\mathrm{Y-H+(hot\,hole)\leftrightarrow Y-+H} \tag{6.1}$$

$$\mathrm{Y-H+H\leftrightarrow Y-+H_2} \tag{6.2}$$

$$\mathrm{Y-H+(hot\,hole)+H\leftrightarrow Y-+H_2^+} \tag{6.3}$$

The partial differential equations governing the chemical equations involving atoms and molecules (Eqs. 6.1 and 6.2) of the RDD model are similar to the RD model equations shown in Chap. 4, and are listed below [35]:

$$\frac{\mathrm{d}N_{OT(1)}}{\mathrm{d}t} = K_{F3}\left(N_{0(1)} - N_{OT(1)}\right) - K_{R3}N_{OT(1)}N_{H(1)} \tag{6.4}$$

$$\frac{\mathrm{d}N_{OT(2)}}{\mathrm{d}t} = K_{F4}\left(N_{0(2)} - N_{OT(2)}\right)N_{H(2)} - K_{R4}N_{OT(2)}N_{H2(2)} \tag{6.5}$$

$$\frac{\delta}{2}\frac{\mathrm{d}N_{\mathrm{H}(1)}}{\mathrm{d}t} = D_{\mathrm{H}}\frac{\mathrm{d}N_{\mathrm{H}(1)}}{\mathrm{d}x} + \frac{\mathrm{d}N_{\mathrm{OT}(1)}}{\mathrm{d}t} \tag{6.6}$$

$$\frac{\delta}{2}\frac{\mathrm{d}N_{\mathrm{H2}(2)}}{\mathrm{d}t} = D_{\mathrm{H2}}\frac{\mathrm{d}N_{\mathrm{H2}(2)}}{\mathrm{d}x} \tag{6.7}$$

$$\frac{dN_{H}}{dt} = D_{H}\frac{d^{2}N_{H}}{dx^{2}} \tag{6.8}$$

$$\frac{dN_{H2}}{dt} = D_{H2}\frac{d^{2}N_{H2}}{dx^{2}} \tag{6.9}$$

$$D_{H2}(t) = \frac{D_{H2\_STRESS}}{\left(1 + A * \left(\frac{t}{t_{STR}}\right)\right)} \tag{6.10}$$

Once again (as in RD model of Chap. 4), for simplicity, the chemical reactions are lumped into two "imaginary interfaces" (1) and (2) inside the gate insulator bulk, and N_0, N_{OT}, N_H and N_{H2} respectively are the H passivated defect, trap (after H depassivation), atomic and molecular Hydrogen densities at the first (1) and second (2) interfaces. The forward and reverse reaction rates are given by K_{F3}, K_{R3} and K_{F4}, K_{R4} respectively for Eqs. 6.1 and 6.2, and δ is the interfacial layer thickness (=1.5 Å).

K_{F3} is related to the AHI process, Fig. 6.2b, c, and is given by Eq. 6.11. Under V_G < 0 V, the electron tunneling from gate to substrate, impact ionization in the substrate, and tunneling of holes from substrate into the gate insulator individually have exponential E_{OX} dependence [37], and they are all lumped together in Eq. 6.11. The E_{OX} acceleration factor has the T independent (Γ_{0OT}) and bond polarization related (α_{OT}/kT) terms, and E_{AOT} is the Arrhenius T activation energy. It is important to note that Γ_{0OT} and E_{AOT} are related to both the AHI and bond dissociation mechanisms, while the polarization factor (α_{OT}) is related to only the bond dissociation process.

$$K_{F3} = K_{F30}e^{(\Gamma_{0OT} + \alpha_{OT}/kT)E_{OX}}e^{-(E_{AOT}/kT)} \tag{6.11}$$

Note that Eq. 6.2 and the related parameters are kept identical to that of Eq. 4.2 of Chap. 4 for the RD model. The diffusivities of H and H_2 are also identical to RD model, and the diffusivity reduction for H_2 molecules only during recovery is used to mimic the stochastic hopping and lock-in processes, as demonstrated in [49].

The ion related terms (Eq. 6.3 and drift) are unique to the RDD model and are described by the following equations, respectively governing the forward/reverse reactions, flux balance, and drift/diffusion:

$$\frac{dN_{OT(2)}}{dt} = K_{F5}\left(N_{0(2)} - N_{OT(2)}\right)N_{H(2)} - K_{R5}N_{OT(2)}N_{H2+(2)} \tag{6.12}$$

$$\frac{\delta}{2}\frac{dN_{H2+(2)}}{dt} = D_{H2+}\frac{dN_{H2+(2)}}{dx} + \mu_{H2+}E_{OX}N_{H2+(2)} \tag{6.13}$$

$$\frac{dN_{H2+}}{dt} = D_{H2+}\frac{d^{2}N_{H2+}}{dx^{2}} + \mu_{H2+}E_{OX}\frac{dN_{H2+}}{dx} \tag{6.14}$$

Table 6.1 Process (or technology) independent RDD model parameters used throughout this book. These parameters are Arrhenius T activated: $X = X_0 \exp(-E_A/kT)$, where X_0 is the pre-factor and E_A is the T activation energy. The diffusivity pre-factors are mentioned for IL // High-K and beyond

Parameter	Unit	Pre-factor	E_A (eV)
K_{F4}	cm^3/s	5750	0.235
K_{R3}	cm^3/s	5×10^{-6}	0.12
K_{R4}, K_{R5}	cm^3/s	7.5×10^{-4}	0.2
D_H	cm^2/s	2×10^{-2} // 4×10^{-5}	0.2
D_{H2}, D_{H2+}	cm^2/s	9.5×10^{-11} // 9.5×10^{-8}	0.5

where K_{F5}, K_{R5} are the forward, reverse reaction rates, N_{H2+} is the ionic Hydrogen density, D_{H2+}, μ_{H2+} are the diffusivity and mobility of ions and are related by the Einstein relation (for simplicity, identical diffusivity is assumed for H_2 and H_{2+}, see Table 4.1 of Chap. 4). Although the stochastic hopping and lock-in effects are also present for H_2^+ [49], switching it on or off has no major difference on recovery, and therefore the diffusivity slow-down process is not invoked [35]. Note, the H_2^+ related recovery is primarily controlled by asymmetry in the distance "moved" during and after stress, due to the presence and absence of drift as mentioned before. ΔV_{OT} is calculated by assigning ΔN_{OT1} and ΔN_{OT2} to their respective "imaginary" interfaces and using $\Delta V_{OT} = q * (\Delta N_{OT1}/C_{OX1} + \Delta N_{OT2}/C_{OX2})$, where the C_{OX} related to the interface of choice is used.

The trap precursor (defined at two interfaces inside the gate insulator bulk for simplicity) densities are taken as $N_{0(1)} = N_{0(2)} = 5 \times 10^{13}/cm^2$ for all the devices in this book. Except for K_{F3} and K_{F5}, all the other reaction and diffusion/drift parameters (pre-factor and Arrhenius T activation energy E_A) are kept the same for all the devices in this book. Table 6.1 lists the process independent RDD model parameters. The common parameters are kept identical between RD and RDD models (see Chap. 4, Table 4.1).

The process dependent parameters related to K_{F3} are listed later in this chapter. Note, each device has a unique pre-factor K_{F30} related to the gate insulator quality. The K_{F30} and hence ΔV_{OT} reduce at higher N%, higher Ge% (see Chap. 7 through Chaps. 9 and 11) and higher f (see Sect. 6.3 and Chap. 14). The parameters Γ_{0OT} and E_{AOT} can change across device types (e.g., between bulk and SOI FinFETs, see Sect. 6.3), but remain constant for process changes in a particular type of device (e.g., changes in the N% and Ge% in FinFETs or dimensions in GAA-SNS FETs, see Chaps. 11 and 12). The parameter α_{OT} is kept fixed for all devices used throughout this book.

The parameter K_{F5} is Arrhenius T activated, with identical E_A (same as in K_{F4}) but different pre-factor K_{F50} across different devices. Figure 6.3 shows the time evolution of ΔN_{OT} simulated (a) during and (b) after stress using the RDD model with different values of K_{F50}. The parameter K_{F5} determines the ratio of H_2 to H_2^+ density, and in turn, determines the ΔN_{OT} time kinetics during and after stress. The initial part of the stress kinetics is reaction limited, but the longer term part is diffusion/drift

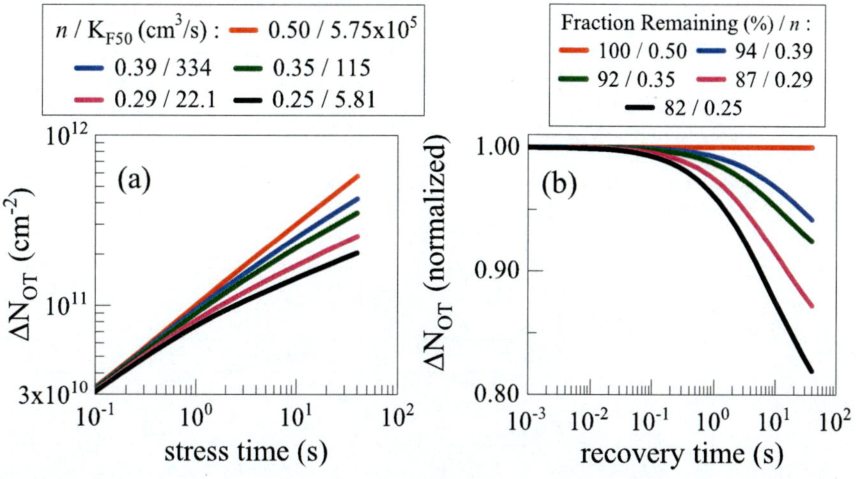

Fig. 6.3 Time evolution of RDD model simulated ΔN_{OT} (a) during and (b) after stress with different values of the parameter K_{F5}. The power-law stress slope n (in t_{STR} range of 1–1Ks) and fraction remaining (at $t_{\mathrm{REC}} = 1$Ks after $t_{\mathrm{STR}} = 1$Ks) values are listed

limited and is controlled by K_{F5}. The onset of recovery and the Fraction Remaining (FR) at a particular t_{REC} is also defined by K_{F5}.

Figure 6.3 also lists the values of n and FR. Pure H_2^+ drift (high K_{F5}) results in n ~1/2 and FR = 1 (no recovery), and pure H_2 diffusion is similar to the RD model (see Chap. 4, Sect. 4.3), and results in n ~1/6 and FR ~0.78 at $t_{\mathrm{REC}} = t_{\mathrm{STR}}$. The RDD model can explain the range of SILC slope (n ~1/4–1/2) reported in the literature [40–48] using a single adjustable parameter K_{F50}. However, for modeling of NBTI time kinetics, the K_{F50} value is chosen in the range of 23–200 cm^3/s in this book, to obtain n ~1/3 for the ΔV_{OT} subcomponent, to remain consistent with the earlier analysis using empirical model [11–14, 19–24, 29].

6.2.2 Validation of TTOM Enabled RD and RDD Models

As discussed in Chap. 11, Replacement Metal Gate (RMG) SiGe FinFET with low N% in the gate stack has negligible contribution from ΔV_{HT}. Therefore, the time kinetics of measured ΔV_{T} is governed by time kinetics of the ΔV_{IT} and ΔV_{OT} subcomponents [22]. Data are obtained by the Measure Stress Measure (MSM) One Point Drop Down (OPDD) method having 10 μs measurement delay (see Chap. 1, Sect. 1.2). The model parameters (RD, TTOM, and RDD) are listed in Table 6.2.

As described in Chap. 4, Sect. 4.3, the process dependent RD model parameters are related to the pre-factor (K_{F10}), T independent field acceleration (Γ_0), bond polarization (α), and T activation energy (E_{AKF1}) of the inversion layer hole assisted bond dissociation process, see Chap. 4, Fig. 4.5. The process dependent TTOM

Table 6.2 Process dependent RD, TTOM, and RDD parameters (as per classification done in Chaps. 4 and 5 and this section) used in this section. The parameters K_{F10} and K_{F30} are not shown to maintain confidentiality

Parameter	Unit	Value
K_{F10}	cm/Vs	–
E_{AKF1}	eV	0.67
Γ_0	cm/MV	0.05
α	qÅ	2.3
f_{FAST}	–	0.45
τ_{EC}	S	0.03
K_{F30}	1/s	–
E_{AOT}	eV	1.14
Γ_{0OT}	cm/MV	0.13
α_{0OT}	qÅ	3.6
K_{F50}	cm^3/s	23

parameters are related to fast fraction of traps that undergo electron capture (f_{FAST}) and the electron capture time constant (τ_{EC}), see Chap. 5, Sect. 5.3. The description of RDD model parameters is provided in the previous subsection. The parameters for the RDD model are already explained in the previous subsection.

Figure 6.4 shows the time evolution of measured ΔV_T (a, c) during and (b, d) after DC stress. The modeling of measured ΔV_T time kinetics and the underlying ΔV_{IT} and ΔV_{OT} subcomponents are shown in Fig. 6.4 (a, b) at fixed V_{GSTR} and T, while the modeling of overall ΔV_T at multiple V_{GSTR} and T are shown in Fig. 6.4 (c, d). Note, ΔV_{IT} evolves rapidly at the initiation of stress and shows n ~1/6 power-law time dependence at longer t_{STR}. The buildup of ΔV_{OT} is initially negligible, however, it shows n ~1/3 power-law time dependence at longer t_{STR}. The ΔV_{IT} recovery proceeds over several orders of magnitude in time and is due to the fast electron capture (ΔV_{IT_FAST}) and slow trap passivation (ΔV_{IT_SLOW}) components, see Chap. 5, Sect. 5.3. However, ΔV_{OT} shows negligible recovery after stress and is semi-permanent.

Figure 6.5 shows the measured and modeled ΔV_T at fixed $t_{STR} = 1$ Ks, with the underlying ΔV_{IT} and ΔV_{OT} subcomponents as a function of (a) V_{GSTR} at fixed T and (b) stress T at fixed V_{GSTR}. For this device, both ΔV_{IT} and ΔV_{OT} contribute to ΔV_T for the stress conditions used in this experiment. The relative contribution of ΔV_{OT} is more prominent at higher V_{GSTR} and/or T, although ΔV_{IT} dominates overall ΔV_T in the range of V_{GSTR} and T used. Figure 6.5 also shows the measured and modeled ΔV_T at fixed t_{STR} during stress as a function of (c) V_{GSTR} at different T and (d) T at different V_{GSTR}. The T dependence of VAF (note that the VAF reduces at higher T, and this feature is more prominent for SiGe devices [21, 22]) and therefore the V_{GSTR} dependence of E_A can be explained (more on this in Chap. 11).

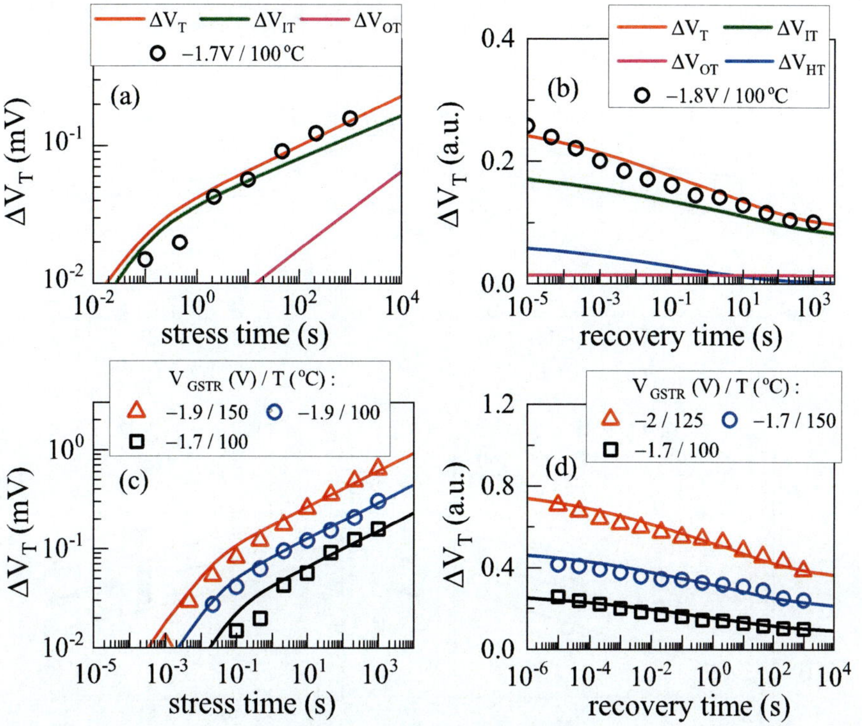

Fig. 6.4 Validation of TTOM enabled RD and RDD models using ultra-fast measured ΔV_{T} time kinetics (a, c) during and (b, d) after stress. Data are shown (a, b) with underlying subcomponents at fixed V_{GSTR}, T, and (c, d) at multiple V_{GSTR}, T. Symbols: experiment, lines: model calculation. Data from [22, 35]

6.3 Comparison of FinFET Architectures

Measured data from RMG HKMG Si channel bulk and SOI p-FinFETs are modeled in this section. The bulk FinFETs have Fin Length (FL) of 200 nm, Fin Width (FW) of 10 nm, and 24 fins, while the SOI FinFETs have FL of 20 nm, FW of 10 nm, and 24 fins. The gate insulators for both devices are based on standard Chemical Oxide based interlayer (IL), Hafnium Dioxide based High-K, and have low Nitrogen content (N%) in the IL. Both bulk and SOI devices have an equivalent oxide thickness of approximately 1.1 nm. Measurements are done using the OPDD MSM method with 10 μs measurement delay, see Chap. 1, Sect. 1.2. Note, these FinFETs are analyzed further in Chaps. 10 and 11.

As mentioned before, the measured ΔV_{T} time kinetics is modeled using uncorrelated contributions from ΔV_{IT}, ΔV_{HT}, and ΔV_{OT} subcomponents using the framework illustrated in Fig. 6.1. The model parameters are listed in Table 6.3, and their description is as follows.

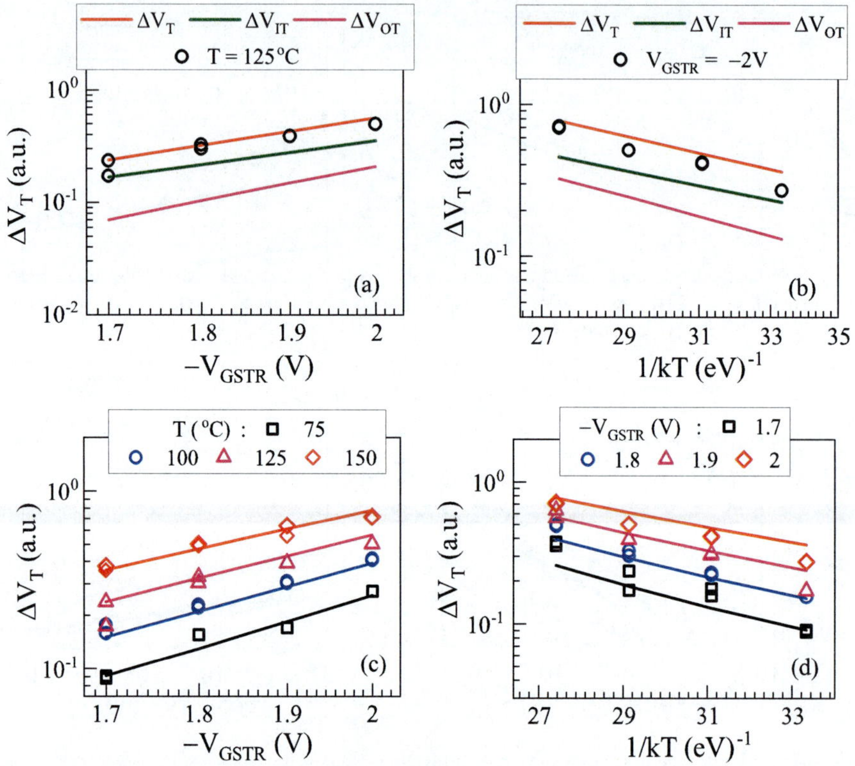

Fig. 6.5 Validation of TTOM enabled RD and RDD models using ultra-fast measured ΔV_T at fixed t_{STR} (=1 Ks) during stress as a function of (a, c) V_{GSTR} and (b, d) T. Data are shown (a, b) with underlying subcomponents at fixed (a) T and (b) V_{GSTR}, and (c, d) at multiple T and V_{GSTR}. Symbols: experiment, lines: model calculation. Data from [22, 35]

As described in Chap. 4, Sect. 4.3, the process dependent RD model parameters are related to the pre-factor (K_{F10}), T independent field acceleration (Γ_0), bond polarization (α), and T activation energy (E_{AKF1}) of the inversion layer hole assisted bond dissociation process, see Chap. 4, Fig. 4.5. The process dependent TTOM parameters are related to fast fraction of traps that undergo electron capture (f_{FAST}) and the electron capture time constant (τ_{EC}), see Chap. 5, Sect. 5.3. As described in Chap. 5, Sect. 5.4, the process dependent ABDWT model parameters are related to the density of pre-existing defects (N_{0HT}), the energy barrier (E_{BM}), and the factors associated with E_{OX} dependence of the barrier (γ_B) and trap energy level ($\gamma_{E2}= m\gamma_B$), see Chap. 5, Fig. 5.10. Finally, the process dependent RDD model parameters are related to the pre-factor (K_{F30}), T independent field acceleration (Γ_{0OT}), and T activation energy (E_{AOT}) of the Anode Hole Injection (AHI) assisted bond dissociation (note, the bond polarization factor (α_{0OT}) is not varied across devices but listed for completeness), and forward reaction rate for ions (K_{F50}), see Sect. 6.2. However, a few process dependent parameters are common between the bulk and SOI FinFETs.

Table 6.3 Process dependent RD, TTOM, ABDWT, and RDD parameters (as per classification done in Chaps. 4 and 5 and in Sect. 6.2 of this chapter) for the bulk (D1) and SOI (D2) FinFETs analyzed in this section. The parameters K_{F10}, $N_{OHT,}$ and K_{F30} are not shown to maintain confidentiality

Parameter	Unit	Bulk (D1)	SOI (D2)
K_{F10}	cm/Vs	–	–
E_{AKF1}	eV	0.29	0.3
Γ_0	cm/MV	0.29	0.19
A	qÅ	1.8	1.8
f_{FAST}	–	0.42	0.67
τ_{EC}	s	0.03	0.03
N_{OHT}	1/cm^2	–	–
E_{BM}	eV	1.3	1.3
γ_B	C.cm	4.5×10^{-9}	5.7×10^{-9}
m	–	2.4	2.4
K_{F30}	1/s	–	–
E_{AOT}	eV	1.14	0.90
Γ_{0OT}	cm/MV	0.13	0.80
α_{0OT}	qÅ	3.6	3.6
K_{F50}	cm^3/s	23	80

Other model parameters are process agnostic and are listed in the respective sections of Chaps. 4 and 5 (see Tables 4.1, 5.1, and 5.2) and Table 6.1.

Figure 6.6 shows the time evolution of measured and modeled ΔV_T together with the underlying ΔV_{IT}, $\Delta V_{HT,}$ and ΔV_{OT} subcomponents (a, b) during and (c, d) after DC stress in (a, c) D1 and (b, d) D2 devices. Note, although identical V_{GSTR} and T are used, it is difficult to make a 1:1 comparison of these devices as their FL is different because ΔV_T depends on FL, which is discussed in Chap. 2, and also in Chap. 13. Therefore, measured and modeled ΔV_T and the subcomponents for a particular device are normalized to the modeled ΔV_T at $t_{STR} = 10$Ks during stress, to compare the relative contributions of different subcomponents on overall ΔV_T. The ΔV_T recovery kinetics and the subcomponents are also normalized by the same factor for similar reasons.

Figure 6.7 shows the time evolution of measured and modeled ΔV_T (a, b) during and (c, d) after DC stress in (a, c) D1 (b, d) D2 devices at different combinations of V_{GSTR} and T.

As shown in Fig. 6.6, the relative ΔV_{IT} contribution is similar and it dominates overall ΔV_T in these devices at the V_{GSTR} and T values used in these experiments. In comparison to the bulk device, the SOI device has higher (~3X) ΔV_{OT} but lower (~2X) ΔV_{HT} relative contributions to overall ΔV_T. The relatively higher ΔV_{OT} and lower ΔV_{HT} contributions in the SOI device make the longer time slope n of overall ΔV_T higher compared to the bulk devices. Note, ΔV_{IT} and ΔV_{OT} show power-law time dependence with n ~1/6 and ~1/3 respectively, while ΔV_{HT} saturates (n ~0 in a log–log plot) at longer stress time. The ΔV_T recovery kinetics is also slower in the SOI device (FR ~55%) compared to the bulk device (FR ~50%) since ΔV_{OT}

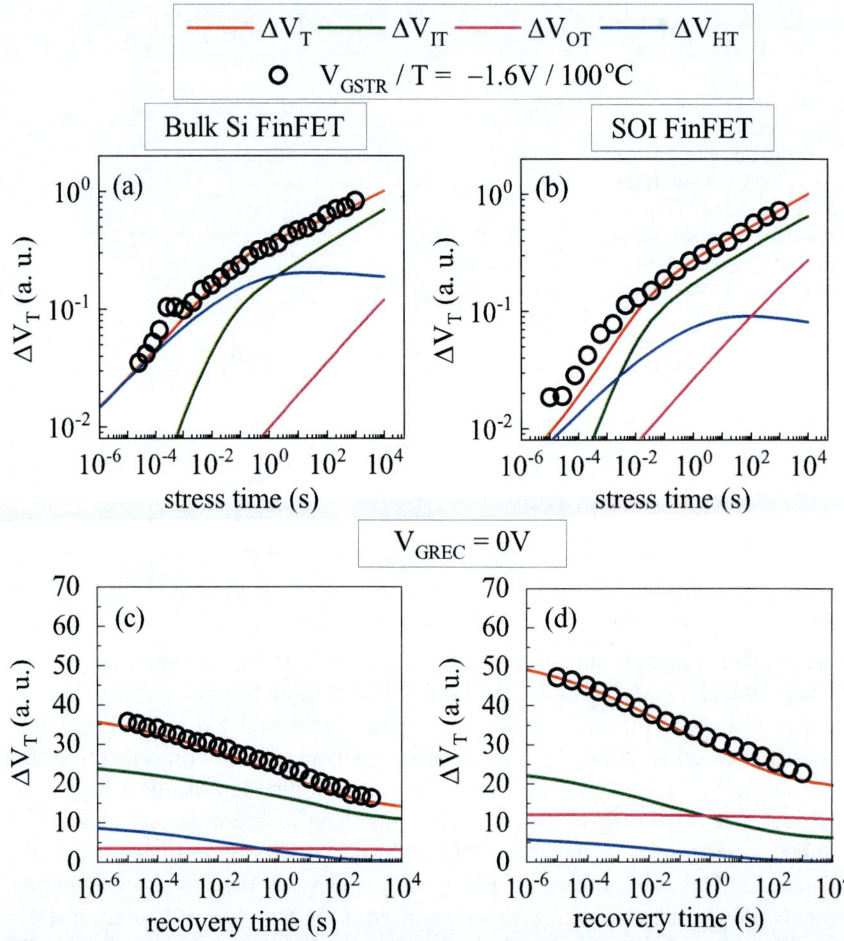

Fig. 6.6 Time evolution of measured and modeled ΔV_T and the underlying subcomponents in (a, c) D1 and (b, d) D2 devices (a, b) during and (c, d) after DC stress. Symbols: experiment, lines: model calculation. Data from [21, 22]

is semi-permanent and the contribution ΔV_{HT} (faster recovery) is relatively small. For both devices, the recovery of ΔV_{IT} extends over an extended timescale and is governed by the ΔV_{IT_FAST} and ΔV_{IT_SLOW} subcomponents as discussed in Chap. 5, Sect. 5.3. Note, these features hold across different values of V_{GSTR} and T used in Fig. 6.7 (more on this in Chaps. 10 and 11).

Figure 6.8 shows the measured and modeled fixed time ($t_{STR} = 1$Ks) ΔV_T and the underlying ΔV_{IT}, $\Delta V_{HT,}$ and ΔV_{OT} subcomponents as a function of (a, b) V_{GSTR} at fixed T and (c, d) T at fixed V_{GSTR} for (a, c) D1 and (b, d) D2 devices. Note that the overall ΔV_T is dominated by ΔV_{IT} for all values of V_{GSTR} and T used in these experiments for the bulk devices, Fig. 6.8a, c. However, ΔV_T at low to moderate

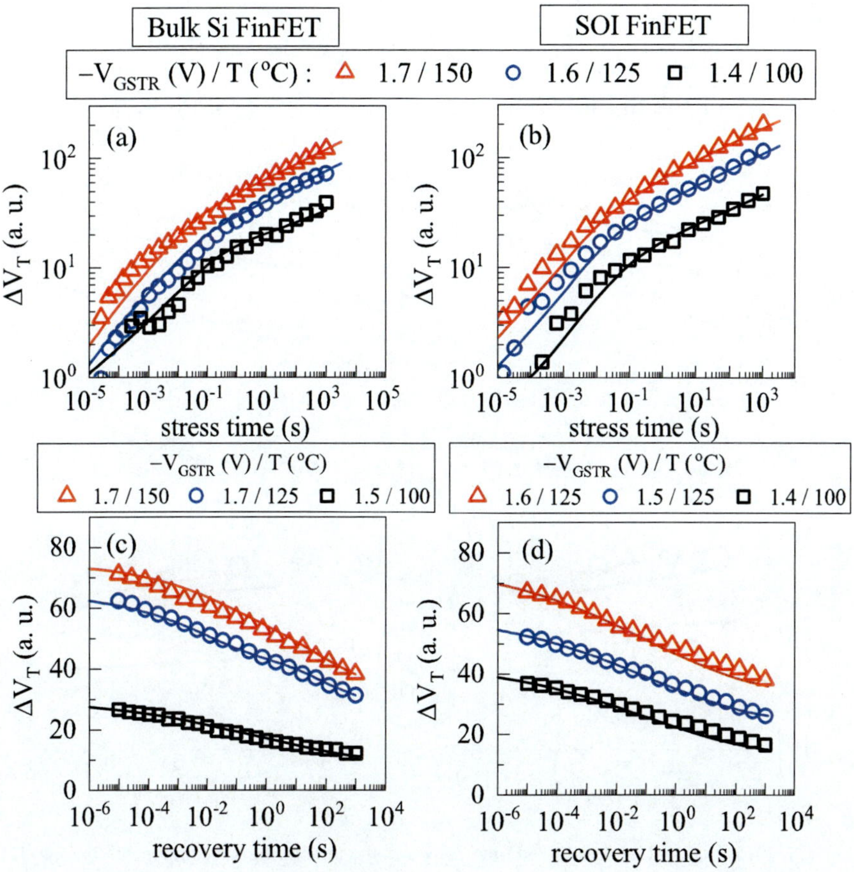

Fig. 6.7 Time evolution of measured and modeled ΔV_T in (a, c) D1 and (b, d) D2 devices (a, b) during and (c, d) after DC stress at different V_{GSTR} and T. Symbols: experiment, lines: model calculation. Data from [21, 22]

V_{GSTR} and T is dominated by ΔV_{IT} while ΔV_{IT} and ΔV_{OT} similarly contribute at very high V_{GSTR} and T in SOI devices, Fig. 6.8 (b, d). This is due to higher VAF and E_A of the ΔV_{OT} subcomponent, and also the relatively larger ΔV_{OT} contribution in SOI devices. However, it is important to note that although the contribution from ΔV_{OT} is large at high V_{GSTR} for SOI devices, its contribution drops significantly near the operating (~use) bias due to higher VAF of the ΔV_{OT} subcomponent, Fig. 6.8 (b). Hence at lower V_G that is more relevant for actual operation, the overall ΔV_T is similar and dominated by ΔV_{IT} for both bulk and SOI devices.

Figure 6.9 shows the time evolution of measured and modeled ΔV_T along with the underlying ΔV_{IT} and ΔV_{OT} subcomponents (a, b) during and (c, d) after Mode-B AC stress in (a, c) D1 and (b, d) D2 devices (refer to Chap. 1, Sect. 1.2 for the description of AC stress modes). Identical V_{GSTR} (i.e., V_{GHIGH}), T, f, and PDC values

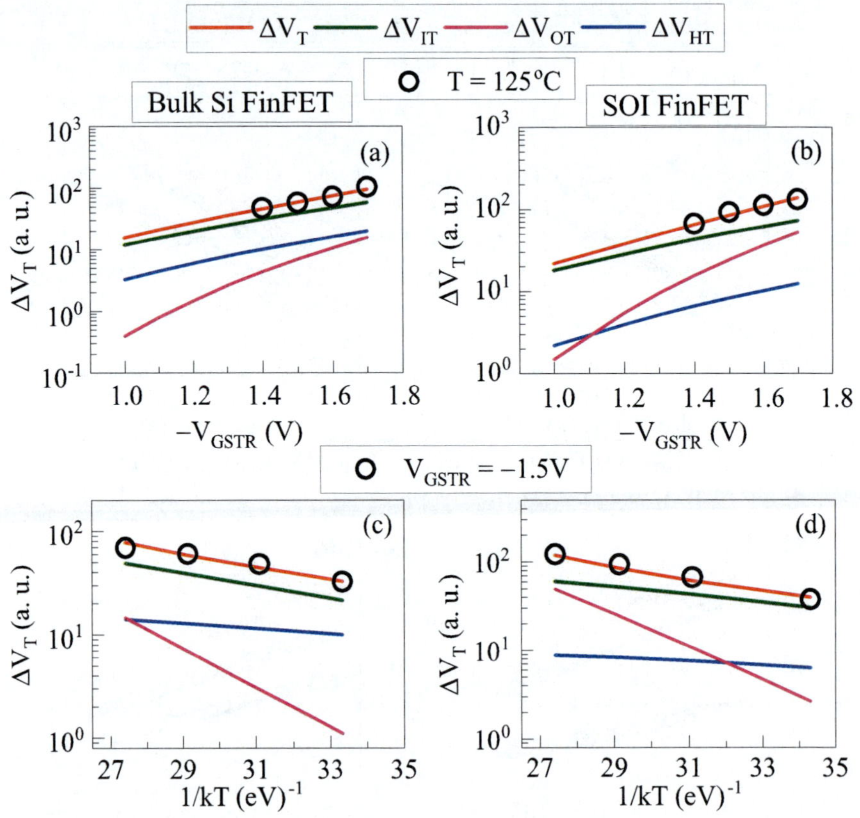

Fig. 6.8 Measured and modeled fixed time ΔV_T and the underlying subcomponents in (a, c) D1 and (b, d) D2 devices versus (a, b) V_{GSTR} at fixed T and (c, d) T at fixed V_{GSTR} during DC stress. Symbols: experiment, lines: model calculation. Data from [21, 22]

are used. Note that the RDD model parameter K_{F30} for the ΔV_{OT} subcomponent is adjusted between the DC and AC stress for both devices. This is because of the AC pulse f dependence of the ΔV_{OT} subcomponent (i.e., ΔV_{OT} reduces at higher f, which is discussed in detail in Chap. 14) [27, 53]. Although the exact reason is not well understood as of now, the f dependence of ΔV_{OT} is also consistent with the f dependence of AC TDDB experiments [54, 55]. Therefore, the pre-factor K_{F30} reduces at higher f AC stress (but it does not depend on any other condition). However, all the other model parameters (for RD, TTOM, ABDWT, and RDD) are kept same between DC and AC stress.

Note that ΔV_{IT} dominates ΔV_T for both devices. As discussed earlier in Chap. 5, Sect. 5.3, the ΔV_{HT} contribution is negligible for Mode-B AC stress at 50% PDC. The SOI device shows a somewhat higher long-time slope n due to relatively higher ΔV_{OT} contribution than bulk devices. The recovery after Mode-B AC stress shows a delayed start, which is explained in Chap. 14. Moreover, slightly higher contribution

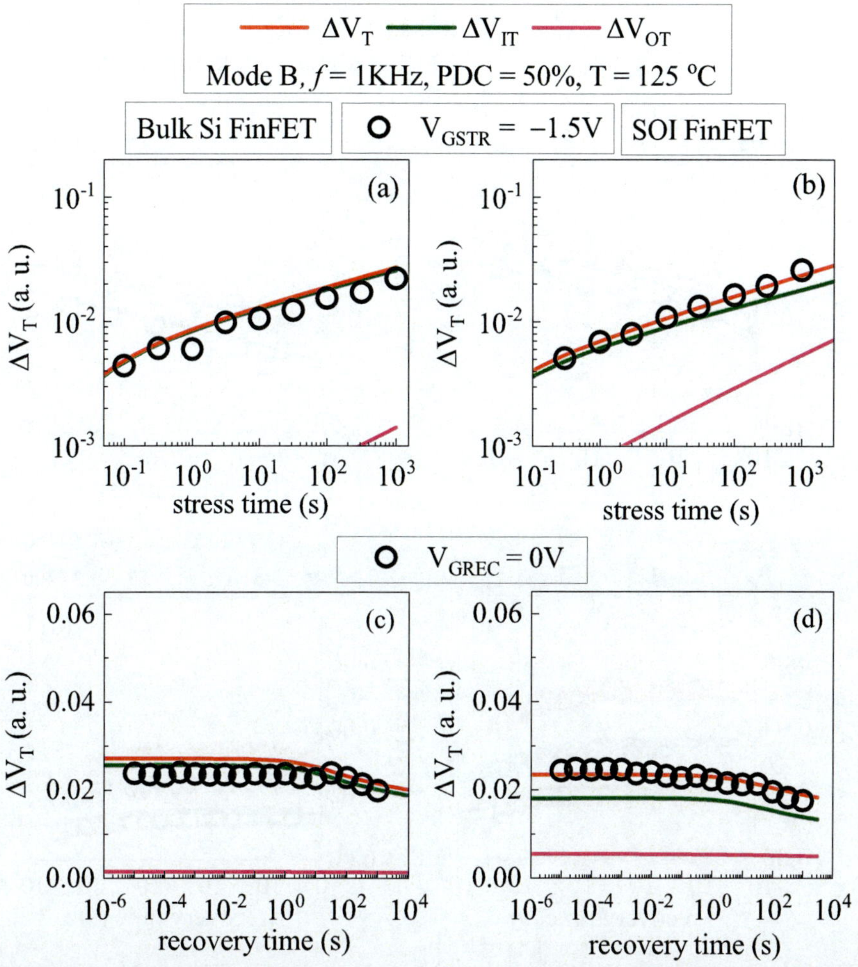

Fig. 6.9 Time evolution of measured and modeled ΔV_T in (a, c) D1 and (b, d) D2 devices (a, b) during and (c, d) after mode-B AC stress and the underlying subcomponents (ΔV_{HT} is negligible). Symbols: experiment, lines: model calculation. Data from [21, 23]

from ΔV_{OT} results in slightly lower recovery for the SOI compared to bulk devices (as ΔV_{OT} is semi-permanent).

Figure 6.10 shows the time evolution of measured and modeled ΔV_T **(a, b)** during and **(c, d)** after mode-B AC stress at different V_{GSTR} (i.e., V_{GHIGH}) and T but fixed PDC and f in **(a, c)** D1 and **(b, d)** D2 devices. Except for K_{F30}, no other parameters are readjusted to model DC and AC stress-recovery kinetics across V_{GSTR} and T for both devices (see Fig. 6.7 for DC stress-recovery modeling). The reduction of K_{F30} during AC stress only depends on f and no other experimental condition as mentioned before.

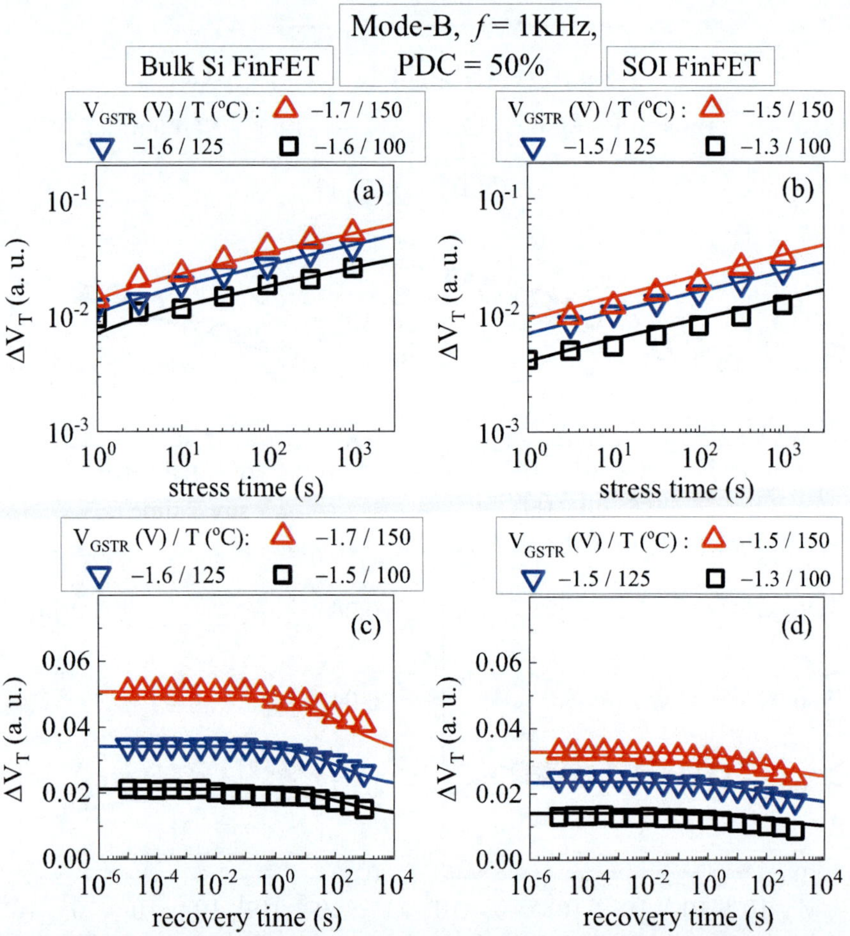

Fig. 6.10 Time evolution of measured and modeled ΔV_T in (a, c) D1 and (b, d) D2 devices (a, b) during and (c, d) after mode-B AC stress at different V_{GSTR} and T. Symbols: experiment, lines: model calculation. Data from [21, 23]

6.4 Estimation of EOL Degradation

Figure 6.11 shows the model calculated ΔV_T at the end of life (EOL) value of 10 years under normal operating bias (use condition) and the underlying ΔV_{IT}, ΔV_{HT}, and ΔV_{OT} subcomponents under (a) DC and (b) Mode-B AC stress in D1 and D2 devices. Note that both devices show very similar projected ΔV_T for either DC or Mode-B AC condition, and in all cases, ΔV_T is dominated by the ΔV_{IT} subcomponent. This is because ΔV_{HT} contribution is found to be negligible in these

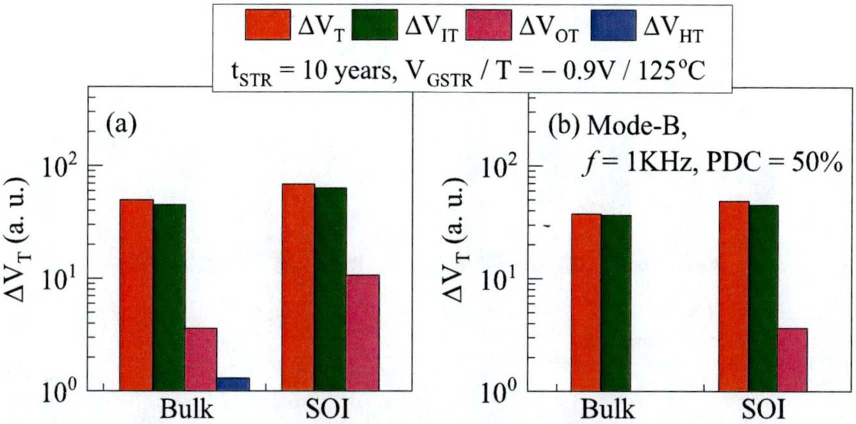

Fig. 6.11 Model based extrapolation of ΔV_T and underlying subcomponents to EOL at use condition in D1 and D2 devices during (a) DC and (b) mode-B AC stress

devices (due to production quality gate insulator stacks having low density of pre-existing defects), and ΔV_{OT}, although appreciable at high V_{GSTR} more so for the SOI device, is negligible at low V_G due to high VAF associated with this subcomponent.

As discussed in Chap. 1, Sect. 1.4, the conventional lifetime method fits the time kinetics of ΔV_T measured over short time and under high V_{GSTR} using a power-law time dependence and extrapolate to EOL. The extrapolated values at EOL, usually obtained at several V_{GSTR}, are then extrapolated to operating bias (V_{DD}) either using power law or exponential V_G dependence. The accuracy of the fit-based method is analyzed hereinafter.

Figure 6.12 compares the EOL ΔV_T at use condition calculated by the calibrated BAT framework to that obtained by the empirical methods (empirical methods are described in Chap. 1, Sect. 1.4), using either (a, c, e) power law or (b, d, f) exponential V_{GSTR} dependence. The empirical calculations are done for experimental data obtained using the full sweep MSM method for (a, b) different maximum t_{STR}, (c, d) different minimum V_{GSTR} (but fixed maximum V_{GSTR}), and (e, f) different ranges of minimum and maximum V_{GSTR} range.

The above exercise is done in devices having different processes, which results in higher (relative) contributions from the ΔV_{HT} (Device-B) and ΔV_{OT} (Device-C) subcomponents compared to the reference case (Device-A) during short time accelerated stress, although it is to be noted that ΔV_{IT} always dominates ΔV_T during short time stress (in Devices A and B) and also at EOL (in all devices). In Device-C, both ΔV_{IT} and ΔV_{OT} have similar contributions during short time accelerated stress.

Note that the difference between the modeled and empirically calculated EOL values depends on device type, the range of t_{STR} used for time regression, as well as the range of $V_{GSTR,}$ and the expression used for bias regression. The values are closer for Device-A. Some overestimation is obtained from the empirical methods for Device-B. For Device-C, the empirical methods sometimes overestimates and

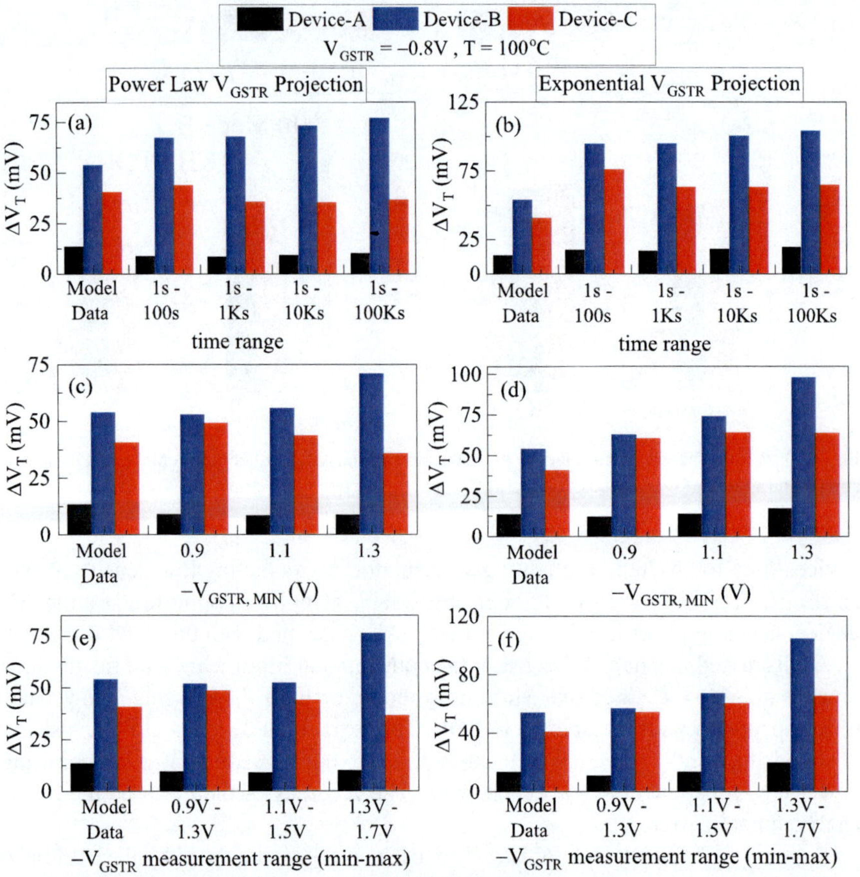

Fig. 6.12 Comparison of model based extrapolation and empirical methods for different devices (see text), for regression of measured data obtained for different maximum stress time (top panels), minimum stress bias (middle panels), and stress bias range (bottom panels)

sometimes underestimates. In general, the accuracy of the regression-based method depends on the range of V_{GSTR} (better if closer to V_{DD}) and t_{STR} (better if closer to EOL), although there are some exceptions. The relative dominance of different subcomponents at stress and use conditions based on VAF and their time kinetics, as well as the stress reduction effect (i.e., the reduction in effective stress at higher ΔV_T, see Chap. 4, Sect. 4.3) determine the empirically extrapolated ΔV_T at EOL. More such comparisons are shown in Chaps. 7, 10 and 11. Therefore, it is preferrable to use the model based approach for reliable estimation of ΔV_T at EOL under use condition.

6.5 Determination of Model Parameters

As mentioned before, in general, the measured time kinetics of ΔV_{T} during and after stress is due to uncorrelated contributions from ΔV_{IT}, $\Delta V_{\mathrm{HT,}}$ and ΔV_{OT} subcomponents. The determination of the underlying subcomponents and the related BAT model parameters are discussed in this section.

First, the overall time kinetics of ΔV_{T} during stress at different V_{GSTR} and T is empirically modeled by the following equation:

$$\Delta V_T = A * \left(V_{\mathrm{GSTR}} - \Delta V_{\mathrm{T,PRE}}\right)^{\Gamma} * e^{-(E_A/kT)} * t^n \tag{6.15}$$

where ΔV_{T} and $\Delta V_{\mathrm{T,PRE}}$ terms correspond to the values of ΔV_{T} respectively at the current and the previous time stamps, and the $\Delta V_{\mathrm{T,PRE}}$ term is used to model the impact of stress reduction effect. The parameters Γ, E_{A} and n are obtained by modeling data (preferably) from a $V_{\mathrm{GSTR}} \times T$ matrix. Note that only longer time data are used when ΔV_{T} shows power-law time dependence.

Next, the "stress reduction corrected" ΔV_{T} ($\Delta V_{\mathrm{T,COR}}$) time kinetics is constructed by dropping the $\Delta V_{\mathrm{T,PRE}}$ term in Eq. 6.15 and using the model parameters. This exercise is done using dataset at longer t_{STR} (>10 s) when the ΔV_{HT} term is saturated. Afterward, the $\Delta V_{\mathrm{T,COR}}$ time kinetics is modeled as follows:

$$\Delta V_{\mathrm{T,COR}} = A_{\mathrm{IT}} * t^{0.17} + A_{\mathrm{HT}} * t^{0} + A_{\mathrm{OT}} * t^{0.3} \tag{6.16}$$

and the parameters A_{IT}, $A_{\mathrm{HT,}}$ and A_{OT} corresponding to the "stress reduction corrected" ΔV_{IT}, $\Delta V_{\mathrm{HT,}}$ and ΔV_{OT} subcomponents are obtained at each V_{GSTR} and T.

Subsequently, the measured ΔV_{T} time kinetics is reconstructed using the following expression that contains the voltage and temperature dependencies of every subcomponent:

$$\begin{aligned} \Delta V_{\mathrm{T}} = {} & A_{\mathrm{IT}} * \left(V_{\mathrm{GSTR}} - \Delta V_{\mathrm{T,PRE}}\right)^{\Gamma_{\mathrm{IT}}} * e^{-(E_{\mathrm{AIT}}/kT)} * t^{0.17} \\ & + A_{\mathrm{HT}} * \left(V_{\mathrm{GSTR}} - \Delta V_{\mathrm{T,PRE}}\right)^{\Gamma_{\mathrm{HT}}} * e^{-(E_{\mathrm{AHT}}/kT)} * t^{0} \\ & + A_{\mathrm{OT}} * \left(V_{\mathrm{GSTR}} - \Delta V_{\mathrm{T,PRE}}\right)^{\Gamma_{\mathrm{OT}}} * e^{-(E_{\mathrm{AOT}}/kT)} * t^{0.3} \end{aligned} \tag{6.17}$$

The adjustable RD model parameters for BAT are obtained first using Eq. 6.17 and keeping $A_{\mathrm{HT}} = A_{\mathrm{OT}} = 0$. The ABDWT model parameter N_{0HT} is obtained next using Eq. 6.17 by keeping $A_{\mathrm{OT}} = 0$. The adjustable RDD model parameters are then obtained by using the full expression of Eq. 6.17. The measured ΔV_{T} kinetics from short to longer stress time is used to obtain the other adjustable ABDWT model parameters. Alternatively, the parameters can also be found using the sequence as RD and RD plus RDD using long-time data, and RD plus RDD and ABDWT using short to long-time data. Finally, the adjustable TTOM parameters are obtained using measured ΔV_T time kinetics during recovery after stress. As mentioned before, only the K_{F30} parameter is readjusted to model ΔV_T time kinetics during and after AC stress.

6.6 Summary

Uncorrelated contributions from ΔV_{IT}, ΔV_{HT}, and ΔV_{OT} subcomponents govern the stress and recovery time kinetics of measured ΔV_T at different V_{GSTR}, V_{GREC}, and T. The RD model calculates the generation and passivation of ΔN_{IT} during and after stress and is validated in Chap. 4. The dependence of ΔV_T on V_{GLOW} during AC stress, while ΔN_{IT} shows no such dependence, implies the importance of trap occupancy, which is calculated by TTOM in Chap. 5. The contributions due to ΔV_{HT} and ΔV_{OT}, when present, impact the time, bias, and T dependence of the stress and recovery kinetics, as shown in Chap. 3, and are calculated respectively using the ABDWT model in Chap. 5 and RDD model in this chapter.

The validations of the TTOM enabled RD, ABDWT plus TTOM enabled RD, and RDD plus TTOM enabled RD is done in Chap. 5 and in this chapter, by using specific stress conditions and/or devices, to have the ΔV_T contribution from either only ΔV_{IT} or from both ΔV_{IT} and ΔV_{HT} or from both ΔV_{IT} and ΔV_{OT} respectively, depending on the models under consideration. In Chap. 5, the TTOM enabled RD model is validated using Mode-B AC stress data when ΔV_T is dominated by ΔV_{IT} and other components are negligible. The TTOM enabled RD and ABDWT models are validated using data from Si channel, high N% device when ΔV_{OT} is negligible. In this chapter, the TTOM enabled RD and RDD models are validated using data from SiGe channel, low N% device when ΔV_{HT} is negligible. The complete BAT framework is used in this chapter to model the DC and AC (Mode-B) stress and recovery kinetics in bulk and SOI FinFETs. A step-by-step parameter calibration process is explained.

The framework is utilized to compare the model versus empirically (conventional fit-based method) calculated extrapolated ΔV_T at EOL under use condition in different devices. The empirical estimation is shown to be inaccurate, and the amount of inaccuracy depends on stress condition and fitting equations. The calibrated BAT framework can therefore be used for reliable estimation of NBTI at the end of device (product) life.

Acknowledgements All experimental data presented in this chapter are re-plotted from previously published reports. The authors acknowledge IBM for providing measurement facilities and FinFETs. One of the authors (Narendra Parihar) acknowledges Richard Southwick, Miaomiao Wang, and James Stathis for help with measurement facilities. Karansingh Thakor is acknowledged for help with manuscript preparation.

References

1. N. Kimizuka, K. Yamaguchi, K. Imai, T. Iizuka, C.T. Liu, R.C. Keller, T. Horiuchi, in *Symposium on VLSI Technology Digest of Technical Papers*, 92 (2000)
2. V. Huard, M. Denais, F. Perrier, N. Revil, C. Parthasarathy, A. Bravaix, E. Vincent, Microelectron. Reliab. **45**, 83 (2005)
3. A.T. Krishnan, C. Chancellor, S. Chakravarthi, P.E. Nicollian, V. Reddy, A. Varghese, R.B. Khamankar, S. Krishnan, in *IEEE International Electron Devices Meeting Technical Digest*, 688 (2005)
4. D. Varghese, D. Saha, S. Mahapatra, K. Ahmed, F. Nouri, M. Alam, in *IEEE International Electron Devices Meeting Technical Digest*, 684 (2005)
5. S. Mahapatra, K. Ahmed, D. Varghese, A.E. Islam, G. Gupta, L. Madhav, D. Saha, M.A. Alam, in *IEEE International Reliability Physics Symposium Proceedings*, 1 (2007)
6. Y. Mitani, H. Satake, A. Toriumi, IEEE Trans. Device Mater. Reliab. **8**, 6 (2008)
7. S. Pae, M. Agostinelli, M. Brazier, R. Chau, G. Dewey, T. Ghani, M. Hattendorf, J. Hicks, J. Kavalieros, K. Kuhn, M. Kuhn, J. Maiz, M. Metz, K. Mistry, C. Prasad, S. Ramey, A. Roskowski, J. Sandford, C. Thomas, J. Thomas, C. Wiegand, J. Wiedemer, in *IEEE International Reliability Physics Symposium Proceedings*, 352 (2008)
8. K. Joshi, S. Hung, S. Mukhopadhyay, V. Chaudhary, N. Nanaware, B. Rajamohanan, T. Sato, M. Bevan, A. Wei, A. Noori, B. McDougal, C. Ni, G. Saheli, C. Lazik, P. Liu, D. Chu, L. Date, S. Datta, A. Brand, J. Swenberg, S. Mahapatra, in *IEEE International Reliability Physics Symposium Proceedings*, 4C.2.1 (2013)
9. J. Franco, B. Kaczer, P.J. Roussel, J. Mitard, M. Cho, L. Witters, T. Grasser, G. Groeseneken, IEEE Trans. Electron Devices **60**, 396 (2013)
10. P. Srinivasan, J. Fronheiser, K. Akarvardar, A. Kerber, L.F. Edge, R.G. Southwick, E. Cartier, H. Kothari, in *IEEE International Reliability Physics Symposium Proceedings*, 6A.3.1 (2014)
11. N. Parihar, N. Goel, S. Mukhopadhyay, S. Mahapatra, IEEE Trans. Electron Devices **65**, 392 (2018)
12. N. Parihar, S. Mahapatra, in *IEEE International Reliability Physics Symposium Proceedings*, TX.5.1 (2018)
13. V. Huard, C. Ndiaye, M. Arabi, N. Parihar, X. Federspiel, S. Mhira, S. Mahapatra, A. Bravaix, in *IEEE International Reliability Physics Symposium Proceedings*, TX.4.1 (2018)
14. N. Parihar, R. Tiwari, C. Ndiaye, M. Arabi, S. Mhira, H. Wong, S. Motzny, V. Moroz, V. Huard, S. Mahapatra, in *International Conference on Simulation of Semiconductor Processes and Devices*, 167 (2018)
15. S. Ramey, A. Ashutosh, C. Auth, J. Clifford, M. Hattendorf, J. Hicks, R. James, A. Rahman, V. Sharma, A.St. Amour, C. Wiegand, in *IEEE International Reliability Physics Symposium Proceedings*, 4C.5.1 (2013)
16. K.T. Lee, W. Kang, E.-A. Chung, G. Kim, H. Shim, H. Lee, H. Kim, M. Choe, N.-I. Lee, A. Patel, J. Park, J. Park, in *IEEE International Reliability Physics Symposium Proceedings*, 2D.1.1 (2013)
17. J. Franco, B. Kaczer, A. Chasin, H. Mertens, L.-A. Ragnarsson, R. Ritzenthaler, S. Mukhopadhyay, H. Arimura, P.J. Roussel, E. Bury, N. Horiguchi, D. Linten, G. Groeseneken, A. Thean, in *IEEE International Reliability Physics Symposium Proceedings*, 4B.2.1 (2016)
18. G. Jiao, M. Toledano-Luque, K.-J. Nam, N. Toshiro, S.-H. Lee, J.-S. Kim, T. Kauerauf, E. Chung, D. Bae, G. Bae, D.-W. Kim, K. Hwang, in *IEEE International Electron Devices Meeting Technical Digest*, 31.2.1 (2016)
19. N. Parihar, R.G. Southwick, U. Sharma, M. Wang, J.H. Stathis, S. Mahapatra, in *IEEE International Reliability Physics Symposium Proceedings*, 2D.4.1 (2017)
20. N. Parihar, R. Southwick, M. Wang, J.H. Stathis, S. Mahapatra, in *IEEE International Electron Devices Meeting Technical Digest*, 7.3.1 (2017)
21. N. Parihar, U. Sharma, R.G. Southwick, M. Wang, J.H. Stathis, S. Mahapatra, IEEE Trans. Electron Devices **65**, 23 (2018)

22. N. Parihar, R.G. Southwick, M. Wang, J.H. Stathis, S. Mahapatra, IEEE Trans. Electron Devices **65**, 1699 (2018)
23. N. Parihar, R.G. Southwick, M. Wang, J.H. Stathis, S. Mahapatra, IEEE Trans. Electron Devices **65**, 1707 (2018)
24. N. Parihar, R. Tiwari, S. Mahapatra, in *International Conference on Simulation of Semiconductor Processes and Devices*, 176 (2018)
25. R. Tiwari, N. Parihar, K. Thakor, H.Y. Wong, S. Motzny, M. Choi, V. Moroz, S. Mahapatra, IEEE Trans. Electron Devices **66**, 2086 (2019)
26. R. Tiwari, N. Parihar, K. Thakor, H.Y. Wong, S. Motzny, M. Choi, V. Moroz, S. Mahapatra, IEEE Trans. Electron Devices **66**, 2093 (2019)
27. N. Parihar, U. Sharma, R.G. Southwick, M. Wang, J.H. Stathis, S. Mahapatra, in *IEEE International Reliability Physics Symposium Proceedings* (2019)
28. M. Wang, J. Zhang, H. Zhou, R.G. Southwick, R. Hsin, K. Chao, X. Miao, V.S. Basker, T. Yamashita, D. Guo, G. Karve, H. Bu, in *IEEE International Reliability Physics Symposium Proceedings* (2019)
29. N. Choudhury, U. Sharma, H. Zhou, R.G. Southwick, M. Wang, S. Mahapatra, in *IEEE International Reliability Physics Symposium Proceedings* (2020)
30. H. Zhou, M. Wang, J. Zhang, K. Watanabe, C. Durfee, S. Mochizuki, R. Bao, R. Southwick, M. Bhuiyan, B. Veeraraghavan, *IEEE International Reliability Physics Symposium Proceedings* (2020)
31. N. Choudhury, T. Samadder, R. Tiwari, H. Zhou, R.G. Southwick, M. Wang, S. Mahapatra, in *IEEE International Reliability Physics Symposium Proceedings* (2021)
32. S. Mahapatra, N. Goel, S. Desai, S. Gupta, B. Jose, S. Mukhopadhyay, K. Joshi, A. Jain, A.E. Islam, M.A. Alam, IEEE Trans. Electron Devices **60**, 901 (2013)
33. A.E. Islam, N. Goel, S. Mahapatra, M.A. Alam, in *Fundamentals of Bias Temperature Instability in MOS Transistors* (Springer India, 2015), pp. 181–207
34. N. Choudhury, N. Parihar, N. Goel, A. Thirunavukkarasu, S. Mahapatra, IEEE J. Electron Devices Society **8**, 1281 (2020)
35. T. Samadder, N. Choudhury, S. Kumar, D. Kochar, N. Parihar, S. Mahapatra, IEEE Trans. Electron Devices **68**, 485 (2021)
36. J.W. McPherson, H.C. Mogul, J. Appl. Phys. **84**, 1513 (1998)
37. M.A. Alam, J. Bude, A. Ghetti, in *IEEE International Reliability Physics Symposium Proceedings*, 21 (2000)
38. E.Y. Wu, J. Sune, W. Lai, IEEE Trans. Electron Devices **49**, 2141 (2002)
39. E. Wu, J. Sune, C. LaRow, R. Dufresne, in *2012 International Electron Devices Meeting Technical Digest*, 28.5.1 (2012)
40. K. Okada, H. Kubo, A. Ishinaga, K. Yoneda, in *Symposium on VLSI Technology Digest of Technical Papers*, 158 (1998)
41. T. Nigam, R. Degraeve, G. Groeseneken, M.M. Heyns, H.E. Maes, in *IEEE International Reliability Physics Symposium Proceedings*, 381 (1999)
42. A. Ghetti, J. Bude, G. Weber, in *Symposium on VLSI Technology Digest of Technical Papers*, 218 (2000)
43. M.A. Alam, IEEE Trans. Electron Devices **49**, 226 (2002)
44. S. Mahapatra, P. Bharath Kumar, M.A. Alam, IEEE Trans. Electron Devices **51**, 1371 (2004)
45. P.E. Nicollian, A.T. Krishnan, C. Bowen, S. Chakravarthi, C.A. Chancellor, R.B. Khamankar, in *IEEE International Electron Devices Meeting Technical Digest*, 392 (2005)
46. S. Mahapatra, D. Saha, D. Varghese, P. Bharath Kumar, IEEE Trans. Electron Devices **53**, 1583 (2006)
47. W.L. Chang, J.H. Stathis, E. Cartier, in *IEEE International Reliability Physics Symposium Proceedings*, 787 (2010)
48. J. Yang, M. Masuduzzaman, K. Joshi, S. Mukhopadhyay, J. Kang, S. Mahapatra, M.A. Alam, in *IEEE International Reliability Physics Symposium Proceedings*, 5D.4.1 (2012)
49. S. Kumar, R. Anandkrishnan, N. Parihar, S. Mahapatra, IEEE Trans. Electron Devices **67**, 4741 (2020)

50. T. Grasser, M. Waltl, Y. Wimmer, W. Goes, R. Kosik, G. Rzepa, H. Reisinger, G. Pobegen, A. El-Sayed, A. Shluger, B. Kaczer, in *IEEE International Electron Devices Meeting Technical Digest*, 20.1.1 (2015)
51. G. Rzepa, J. Franco, B. O'Sullivan, A. Subirats, M. Simicic, G. Hellings, P. Weckx, M. Jech, T. Knobloch, M. Waltl, P.J. Roussel, D. Linten, B. Kaczer, T. Grasser, Microelectron. Reliab. **85**, 49 (2018)
52. K.F. Schuegraf, C. Hu, IEEE Trans. Electron Devices **41**, 761 (1994)
53. L. Zhou, Q. Zhang, H. Yang, Z. Ji, Z. Zhang, Q. Liu, H. Xu, B. Tang, E. Simoen, X. Ma, X. Wang, Y. Li, H. Yin, J. Luo, C. Zhao, W. Wang, IEEE Electron Device Lett. **41**, 965 (2020)
54. R. Ranjan, Y. Liu, T. Nigam, A. Kerber, B. Parameshwaran, in *IEEE International Reliability Physics Symposium Proceedings*, DG.10.1 (2017)
55. M. Rafik, A.P. Nguyen, X. Garros, M. Arabi, X. Federspiel, C. Diouf, in *IEEE International Reliability Physics Symposium Proceedings*, 4A.3.1 (2018)

Chapter 7
BAT Framework Modeling of Gate First HKMG Si Channel MOSFETs

Souvik Mahapatra, Narendra Parihar, Nilesh Goel, Nilotpal Choudhury, and Tarun Samadder

7.1 Introduction

As described in the earlier chapters, Negative Bias Temperature Instability (NBTI) became an important issue with the migration from Silicon Dioxide (SiO_2) to Silicon Oxynitride (SiON) gate insulator MOSFETs [1–4]. It continues to remain so in dual layer (SiO_2 or SiON Interlayer (IL) and Hafnium Dioxide (HfO_2) High-K layer) High-K Metal Gate (HKMG) gate insulator based planar bulk [5–9] and Fully Depleted Silicon On Insulator (FDSOI) [10, 11] MOSFETs, bulk and SOI FinFETs [12–25], as well as Gate All Around Stacked Nanosheet (GAA-SNS) FETs [26–29], having either Silicon (Si) or Silicon Germanium (SiGe) channel. The key NBTI features are listed below (reproduced from Chap. 3, Sect. 3.1):

NBTI results in the accumulation of positive gate insulator charges when the gate bias (V_G) of the device is held at a negative value ($V_G = V_{GSTR}$). These charges result in device parametric drift, *e.g.*, threshold voltage shift (ΔV_T), over time. The accrued positive charges and the associated ΔV_T reduce over time if the magnitude of V_G is lowered to $V_G = V_{GREC}$ or removed ($V_G = 0$ V). As a consequence, AC stress results in lower NBTI compared to DC stress. However, NBTI recovery also necessitates the use of ultra-fast methods having measurement delay (t_M) in ~ microseconds for recovery artifact free NBTI data acquisition. As discussed in Chap. 1, Sect. 1.2, NBTI is measured by either full sweep or One Point Drop Down (OPDD) Measure-Stress-Measure (MSM) ultra-fast methods in this book.

As discussed in Chap. 1, Sect. 1.3, ΔV_T increases at higher magnitude of V_{GSTR} and at elevated temperature (T) during DC stress, and such dependencies are respectively determined using the Voltage Acceleration Factor (VAF) and Arrhenius T activation energy (E_A). The pulse high (V_{GHIGH}) and low (V_{GLOW}) values, Pulse

S. Mahapatra (✉) · N. Parihar · N. Goel · N. Choudhury · T. Samadder
Department of Electrical Engineering, Indian Institute of Technology Bombay, Powai, Mumbai 400076, India
e-mail: souvik@ee.iitb.ac.in

S. Mahapatra (ed.), *Recent Advances in PMOS Negative Bias Temperature Instability*,
https://doi.org/10.1007/978-981-16-6120-4_7

Duty Cycle (PDC) and frequency (f), as well as the stress mode (Mode-A or Mode-B) impact the ΔV_T magnitude during AC stress. As discussed in Chap. 2, the ΔV_T magnitude increases with higher Nitrogen content (N%) in the gate insulator but reduces with higher Germanium content (Ge%) in the channel. The time kinetics of ΔV_T during and after stress, *i.e.*, the power-law time slope n at longer stress time (t_{STR}) and the Fraction Remaining (FR) at a particular recovery time (t_{REC}) after stress (FR is the ratio of ΔV_T at t = t_{REC} during recovery to that at the end of stress at $t = t_{STR}$), as well as VAF, E_A, and T dependence of VAF (the reduction of VAF at higher T) are also impacted by N% and Ge%. The values of n, FR, VAF, and E_A as well as the T dependence of VAF reduce at higher N% but increase at higher Ge%. The impact of N% changes is studied in this chapter and in Chaps. 9 and 11, while the impact of Ge% changes is studied in Chaps. 8, 9 and 11 in different device architectures. Furthermore, changes in the layout and device dimensions also impact NBTI, and are discussed in Chaps. 9, 12 and 13. The impact of PDC, V_{GLOW}, f, and AC stress mode (A or B) are analyzed in Chaps. 14.

Figure 7.1 illustrates the BTI Analysis Tool (BAT) framework [8], described in Chap. 4 through Chap. 6, which is used to model the time kinetics of ΔV_T during and after DC and AC stress throughout this book (*e.g.*, the features summarized in Chap. 3, Sect. 3.1). BAT uses uncorrelated contributions from generated interface (ΔV_{IT}) and bulk (ΔV_{OT}) gate insulator traps, and hole trapping in process related pre-existing

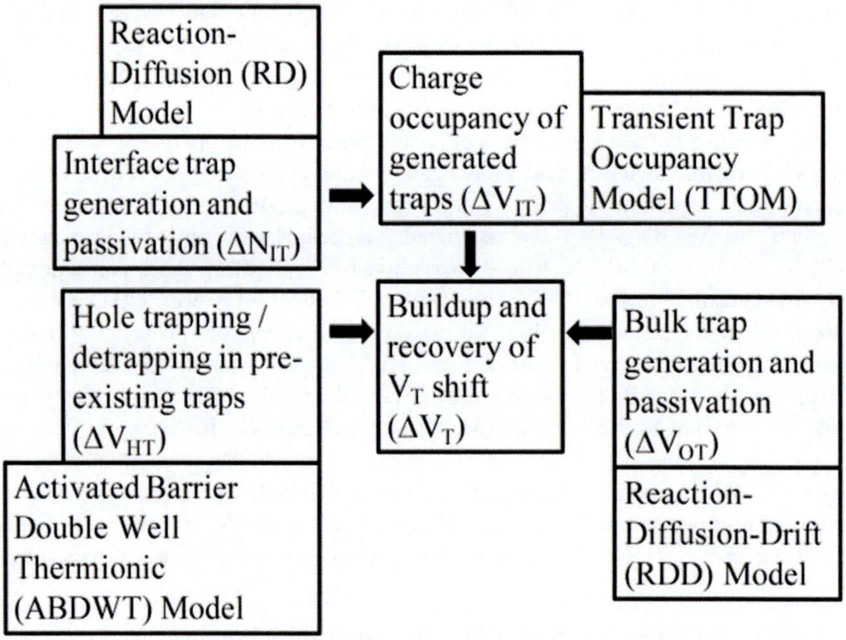

Fig 7.1 Schematic of the BTI Analysis Tool (BAT) framework used in this book to model measured ΔV_T kinetics during and after DC and AC NBTI stress, reproduced from Chap. 4

bulk gate insulator traps (ΔV_{HT}). The Reaction Diffusion (RD) model is used to calculate the time kinetics of interface traps (density ΔN_{IT}) [8, 30, 31]. RD model is described and independently validated using measured data from Direct Current IV (DCIV) method [32] in Chap. 4. Transient Trap Occupancy Model (TTOM) is used for the calculation of the occupancy of generated interface traps and their contribution (ΔV_{IT}) [8], which is described and validated in Chap. 5, Sect. 5.3. The ΔV_{HT} and ΔV_{OT} kinetics are modeled by the Activated Barrier Double Well Thermionic (ABDWT) model [33] and Reaction Diffusion Drift (RDD) model [34] respectively throughout this book, and these are described and validated in Chap. 5, Sect. 5.4 and Chap. 6, Sect. 6.2.

7.2 Device Details and Model Parameters

Table 7.1 lists the process description of the Gate First (GF) HKMG devices used in this chapter [6, 8]. Large area (width (W) = 10μm and length (L) = 1μm) devices with different N% in the gate stack (D1 and D2), different interlayer (IL) thickness (D1 and D4), and different High-K (HK) thickness (D1 and D3) are used. The IL of these devices is thermally grown by a Rapid Thermal Oxidation (RTO) based method, without (D1, D3, and D4) and with (D2) Nitrogen treatment before the IL growth. Atomic Layer Deposition (ALD) based Hafnium Dioxide (HfO_2) is used for the HK, and no air break is used between the IL and HK processes. Note, Remote Plasma Nitridation (RPN) followed by Post Nitridation Anneal (PNA) is used for Nitrogen incorporation in D1, D3, and D4 devices. Moreover, small area (W/L = 90 nm/70 nm) D1 devices are also used, where additional Nitrogen gets incorporated into the gate oxide from Spacers during the high temperature Source-Drain activation anneal in a GF process, these devices are denoted as D5.

The IL and HK thicknesses are verified by cross section Transmission Electron Microscopy (TEM), and the Equivalent Oxide Thickness (EOT) is determined by the quantum corrected CV measurements. All process details are provided in [6]. All data in large area devices are measured by the full sweep MSM method unless specifically mentioned otherwise. For small area devices, stress kinetics is measured

Table 7.1 Description of the GF HKMG MOSFETs analyzed in this chapter. The D1 and D2 devices are same as the D1 and D2 devices of Chap. 4, Table 4.2. The D5 device is same as D1 device, except smaller W/L values and Nitrogen incorporation from Spacer (see text)

Device	IL (nm)	High-K (nm)	Nitrogen
D1	0.3	2.3	Low
D2	0.2	2.3	High
D3	0.3	1.8	Low
D4	0.5	2.3	Low
D5	0.3	2.3	High

during OPDD MSM method, while the recovery kinetics is measured using a constant bias (similar to the On-The-Fly (OTF) method), see Chap. 1, Sect. 1.2. It is to be noted that the small area recovery kinetics measured in this way shows some unique features, which are described in Sect. 7.7.

As mentioned before, the measured ΔV_T time kinetics is modeled using uncorrelated contributions from ΔV_{IT}, $\Delta V_{HT,}$ and ΔV_{OT} subcomponents by the framework illustrated in Fig. 7.1, and the device specific model parameters are listed in Table 7.2. As described in Chap. 6, Sect. 6.3, the process dependent RD model parameters are related to the pre-factor (K_{F10}), T independent field acceleration (Γ_0), bond polarization (α), and T activation energy (E_{AKF1}) of the inversion layer hole assisted bond dissociation process, see Chap. 4, Sect. 4.3. Note, K_{F10}, $\Gamma_{0,}$ and E_{AKF1} remain fixed

Table 7.2 Process dependent RD, TTOM, ABDWT, and RDD parameters (as per classification done in Chap. 4 through Chap. 6) used in this chapter. Other adjustable model parameters are E_{AOT} = 1.14 eV, Γ_{0OT} = 0.13 cm/MV, α_{OT} = 3.6 qÅ and K_{F50} = 23 cm³/s for RDD, for D1, D3 and D4 devices. Bulk trap generation is negligible for D2 and D5. Recovery is not modeled for D4. See Table 7.1 for device details

Device	Unit	D1	D3	D4
K_{F10}	cm/Vs	8	8	8
E_{AKF1}	eV	0.4	0.4	0.4
Γ_0	cm/MV	0.38	0.38	0.38
α	qÅ	1.2	1.2	1.2
f_{FAST}	–	0.78	0.78	–
τ_{EC}	s	2.0×10^{-2}	2.0×10^{-2}	–
N_{0HT}	1/cm²	6.0×10^{11}	8.0×10^{11}	1.1×10^{12}
E_{BM}	eV	1.18	1.18	1.37
γ_B	C.cm	6.0×10^{-9}	6.0×10^{-9}	6.0×10^{-9}
m	–	3.2	2.4	3.2
K_{F30}	1/s	8.0×10^4	4.0×10^4	8.0×10^5

Device	Unit	D2	D5
K_{F10}	cm/Vs	5.8	8
E_{AKF1}	eV	0.18	0.35
Γ_0	cm/MV	0.10	0.34
α	qÅ	1.2	1.2
f_{FAST}	–	0.60	0.55
τ_{EC}	S	8.0×10^{-5}	9.0×10^{-1}
N_{0HT}	1/cm²	4.64×10^{12}	1.5×10^{12}
E_{BM}	eV	1.25	1.25
γ_B	C.cm	3.6×10^{-9}	6.0×10^{-9}
m	–	3.5	4
K_{F30}	1/s	–	–

for D1, D3, and D4 devices, while α is same for all devices, and these parameters for D1 and D2 devices are independently validated using DCIV measurements in Chap. 4, Sect. 4.4. Due to differences in the energy zone scanned by V_T and DCIV measurements, the K_{F10} parameter is higher than that shown in Chap. 4, Table 4.2. The process dependent TTOM parameters are related to fast fraction of traps that undergo electron capture (f_{FAST}) and the electron capture time constant (τ_{EC}), see Chap. 5, Sect. 5.3. Similar values are obtained for D1 and D3 devices (recovery is not modeled for D4). As described in Chap. 5, Sect. 5.4, the process dependent ABDWT model parameters are related to the density of pre-existing defects (N_{0HT}), the energy barrier (E_{BM}), and terms associated with the oxide electric field (E_{OX}) dependence of the barrier (γ_B) and trap energy level ($\gamma_{E2}= m\gamma_B$), see Chap. 5, Fig. 5.10. N_{0HT} is varied for all devices and determine the saturated ΔV_{HT} component, while the other parameters determine the kinetics of hole trapping and detrapping and may or may not vary with processes. The process dependent RDD model parameters are related to the pre-factor (K_{F30}), T independent field acceleration (Γ_{0OT}), and T activation energy (E_{AOT}) of the Anode Hole Injection (AHI) assisted bond dissociation process (note, the bond polarization factor (α_{OT}) is not varied across devices but listed for completeness), and the forward reaction rate for the ions (K_{F50}), see Chap. 6, Sect. 6.2. Only K_{F30} is changed across devices. All the other model parameters are process agnostic and are listed in the respective sections of Chap. 4 through Chap. 6 (Tables 4.1, 5.1, 5.2 and 6.1).

7.3 DC Stress Kinetics—Impact of Nitrogen Incorporation

Figure 7.2 shows the time evolution of measured and modeled ΔV_T along with the underlying model subcomponents for DC stress at a fixed V_{GSTR} and T in (a) D1 and (b) D2 devices. ΔV_T evolves fast at the initiation of stress and shows power-law time dependence at a longer time with slope n ~ 0.13 for D1 and n ~ 0.1 for D2 devices. The ΔV_{IT} subcomponent evolves fast at the initiation of stress and shows power-law time dependence at a long time with n ~ 1/6. The DCIV measured ΔV_{IT} (= $q^*\Delta N_{IT}/C_{OX}$, where q is the electronic charge and C_{OX} is the gate capacitance), after required delay and bandgap corrections (see Chap. 3, Sect. 3.2), is used to verify the correctness of the RD model. Note that the measured and delay corrected DCIV time kinetics data at multiple V_{GSTR} and T are used to validate the RD model and its parameters for the D1 and D2 devices, see Chap. 4, Sect. 4.4. Except for the pre-factor K_{F10}, all the other adjustable RD model parameters remain identical between DCIV calibrated data without and with bandgap correction.

The ΔV_{HT} subcomponent evolves even faster in time for shorter stress time but saturates at longer time (n ~ 0 in a log–log plot). The saturation value is higher for device D2 and suggests larger pre-existing defect density in the gate insulator due to higher N%, which is verified by flicker noise measurements (see related discussion in Chap. 3, Sect. 3.3). The ΔV_{OT} subcomponent is only visible for D1 and not for D2 devices, and that too only when V_{GSTR} and/or T are high, and shows power-law time

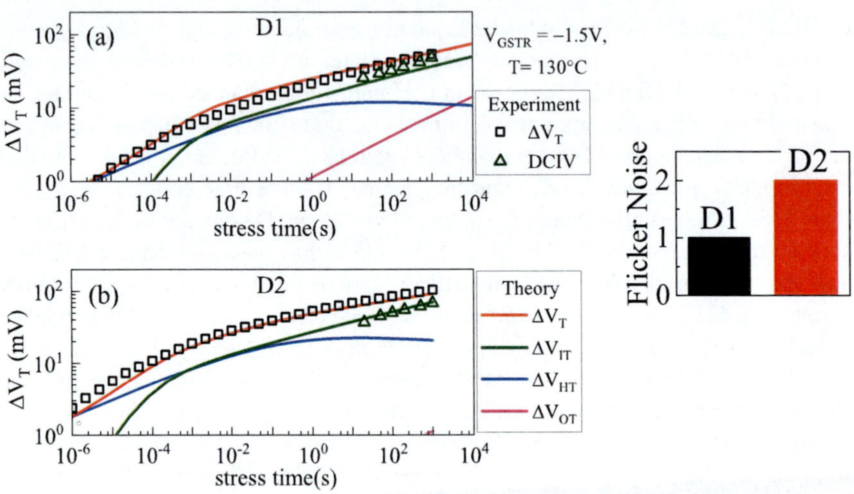

Fig 7.2 Measured and modeled ΔV_T time kinetics in D1 and D2 devices during stress. The underlying model subcomponents are also plotted. Delay and bandgap corrected DCIV data are used to verify the ΔV_{IT} subcomponent. Flicker noise is used to verify relative contribution from the ΔV_{HT} subcomponent (see [6]). Symbols: experiment, lines: model calculation. Data from [8]

dependence with $n \sim 1/3$ at longer time. Note that measured ΔV_T at very short time is dominated by ΔV_{HT}, although ΔV_{IT} clearly dominates at longer time. The long-time ΔV_{IT} domination holds at different V_{GSTR} and T conditions used in this chapter. The contribution from ΔV_{OT} is generally small in these devices, unless very high V_{GSTR} and T values are used in stress, and is negligible for D2 and D5. As ΔV_{HT} saturates ($n \sim 0$) while ΔV_{OT} shows power-law time dependence with $n \sim 1/3$ at longer time, the relatively higher ΔV_{HT} and negligible ΔV_{OT} reduces the long-time slope n of the overall ΔV_T for the D2 device, see Chap. 3, Sect. 3.3.

Figure 7.3 shows the time evolution of measured and modeled ΔV_T in D1 (left panels) and D2 (right panels) devices at various T, multiple V_{GSTR} values are used in each panel at a particular T.

Table 7.2 lists the process dependent parameters to model the process dependence of ΔV_T kinetics from very short to long stress time under such $V_{GSTR} \times T$ matrix. Of these, two parameters governing the RD model (E_{AKF1} and Γ_0, since α is same) are independently verified using DCIV measured and delay corrected ΔN_{IT} kinetics in Chap. 4, Fig. 4.6. As mentioned before, the pre-factor K_{FI0} would not be the same to model the delay corrected DCIV measurements (for ΔN_{IT} in Chap. 4) and the ΔV_{IT} contribution to overall ΔV_T (from I-V measurements) due to difference in the energy bandgap scanned by these methods, as is explained in Chap. 3, Sect. 3.2. The necessary bandgap correction is done by assuming donor like generated interface trap generation uniformly throughout the substrate bandgap. The delay and bandgap corrected DCIV measured ΔV_{IT} (and K_{FI0}) matches that needed to simulate the ΔV_{IT} contribution to ΔV_T using the RD model and used in Fig. 7.3.

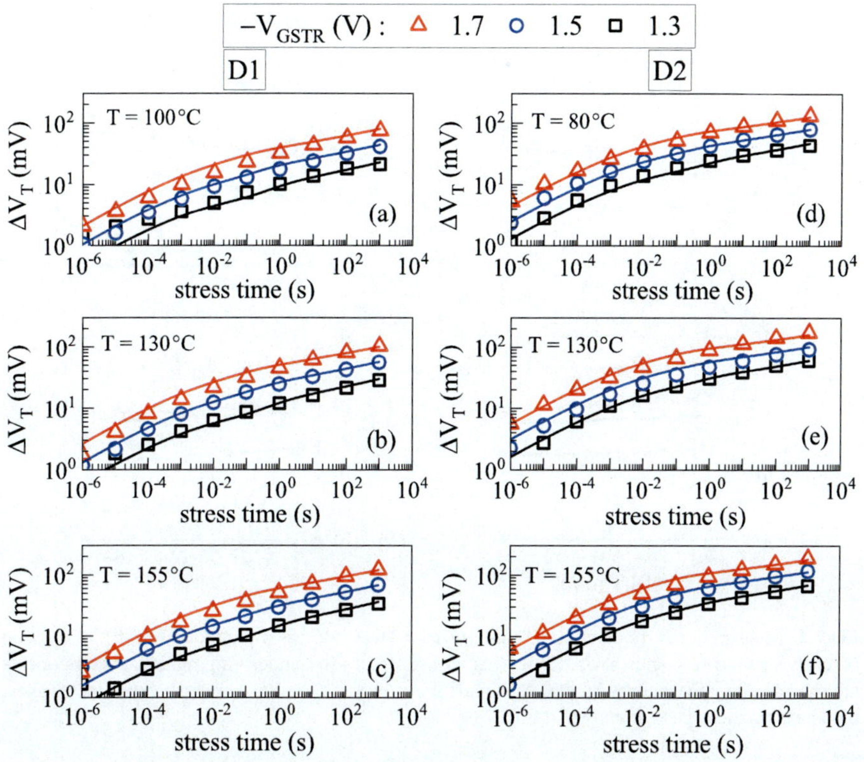

Fig 7.3 Measured and modeled ΔV_T time kinetics in (a–c) D1 and (d–f) D2 devices during stress at different V_{GSTR} and T ($V_{GSTR} \times T$ matrix). Symbols: experiment, lines: model calculation. Data from [8]. The corresponding DCIV data are shown in Chap. 4, Fig. 4.6

The product of pre-factor K_{F10} and $\exp(-E_{AKF1}/kT)$ for the ΔV_{IT} subcomponent is higher but E_{AIT} and Γ_0 values are lower at higher N%, the reason for this is discussed later in this section, also see Chap. 4, Sect. 4.3. Higher N% also results in higher ΔV_{HT} subcomponent due to higher pre-existing trap density, and the other ABDWT model parameters indicate faster trapping. Higher N% reduces the pre-factor K_{F30} for the ΔV_{OT} subcomponent, which is consistent with the reduced Time Dependent Dielectric Breakdown (TDDB) reported in gate insulators having higher N% [35]. Note that the generated bulk gate insulator traps (density ΔN_{OT}, contribution ΔV_{OT} to overall ΔV_T) are also responsible for Stress Induced Leakage Current (SILC) and TDDB in gate insulators, see Chap. 3, Sect. 3.4 for related discussion [36].

Figure 7.4 shows the measured and modeled ΔV_T at fixed time (t_{STR} = 1Ks) as a function of (a, b) V_{GSTR} at fixed T and (c, d) T at fixed V_{GSTR}, for (a, c) D1 and (b, d) D2 devices. The underlying ΔV_{IT}, $\Delta V_{HT,}$ and ΔV_{OT} subcomponents are shown in every panel.

The device D2 has higher N% in the IL compared to D1 which results in larger pre-existing trap density as verified using flicker noise measurements, Fig. 7.2. The

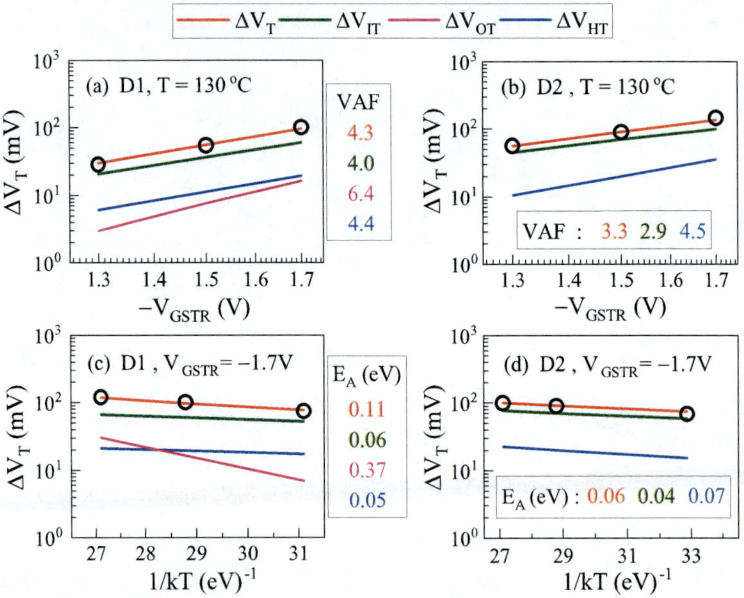

Fig 7.4 Measured and modeled fixed time ($t_{STR} = 1$Ks) ΔV_T as a function of (a, b) V_{GSTR} and (c, d) T for (a, c) D1 and (b, d) D2 devices during stress. The underlying model subcomponents are shown. The power-law VAF and Arrhenius E_A values are shown. Symbols: experiment, lines: model calculation. Data from [8]

contributions from ΔV_{HT} and ΔV_{IT} increase while contribution from ΔV_{OT} reduces with increase in N%. However, ΔV_{IT} dominates ΔV_T for the range of V_{GSTR} and T studied in these experiments, while ΔV_{OT} becomes negligible for the D2 device (also see associated discussion in Chap. 5, Sect. 5.4).

Note that the V_{GSTR} dependent slope or VAF (~E_{OX} dependence) for the ΔV_{IT} and ΔV_{HT} is similar in these devices, and it is highest for the ΔV_{OT} subcomponent. However, the Arrhenius T activation E_A is lower for ΔV_{HT} compared to ΔV_{IT} (see related discussion in Chap. 3, Sect. 3.3), while it is highest for the ΔV_{OT} subcomponent. Note, the VAF and E_A values do not change across processes for the ΔV_{HT} and ΔV_{OT} subcomponents, but they reduce at higher N% for the ΔV_{IT} subcomponent. Therefore, larger relative contribution from ΔV_{HT} (although ΔV_{IT} dominates) and negligible ΔV_{OT} reduce the VAF and E_A of overall ΔV_T for the D2 device. Moreover, the reduction in VAF and E_A for the ΔV_{IT} subcomponent at higher N% (which is explained below) also contributes. In addition, the relatively higher ΔV_T in the D2 device also reduces the overall VAF due to the stress reduction effect (reduction in the effective electrical stress at higher ΔV_T).

Figure 7.5 shows the measured and modeled ΔV_T at fixed time ($t_{STR} = 1$Ks) as a function of (a, b) V_{GSTR} for different T and (c, d) T for different V_{GSTR}, for (a, c) D1 and (b, d) D2 devices. The VAF remains similar across different T, and as a consequence, the E_A also remains similar across various V_{GSTR} for both devices. It

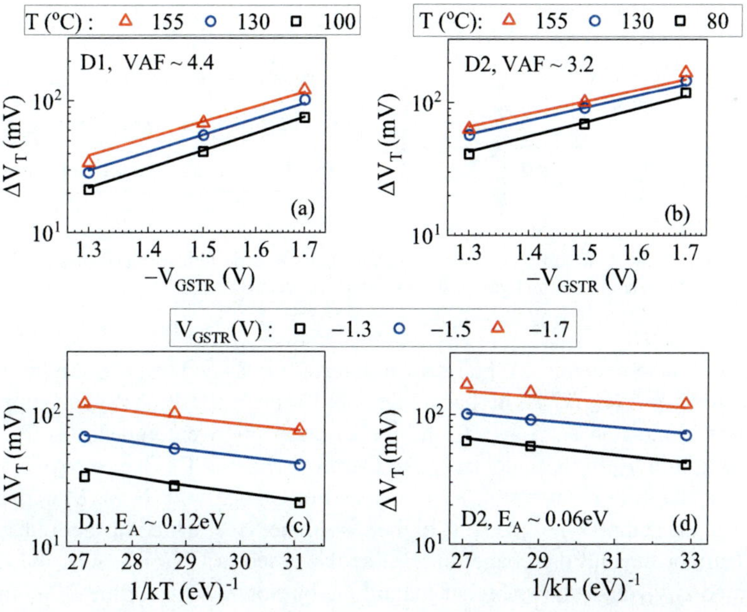

Fig 7.5 Measured and modeled fixed time ΔV_T as a function of (a, b) V_{GSTR} at different T and (c, d) T at different V_{GSTR} for (a, c)D1 and (b, d) D2 devices during stress. The power-law VAF and Arrhenius E_A values are shown. Symbols: experiment, lines: model calculation. Data from [8]

is important to remark that the invariance of VAF at different T (and hence of E_A at different V_{GSTR}) is not universally observed, refer to Chap. 11 for contradictory results, *i.e.*, the reduction of VAF at higher T and reduction in E_A at higher V_{GSTR}. Note that the longer time slope n is similar across different V_{GSTR} and T in these devices, see Fig. 7.3, although lower n is observed for D2 compared to D1 due to the reasons mentioned before. This aspect is discussed before in Chap. 1, Sect. 1.3, and the invariance of n across V_{GSTR} and T imply the VAF and E_A values are not sensitive to the value of t_{STR} at which they are evaluated. It also implies the VAF and E_A values at fixed t_{STR} are related to their fixed ΔV_T level counterparts (*i.e.*, the VAF and E_A of lifetime) by the slope n. However, the above feature is rather an exception and is not necessarily always true, contradictory results (*i.e.*, variation in n with V_{GSTR} and/or T) are shown in Chap. 10.

The impact of N% on the ΔV_{IT} parameters K_{FI0} and Γ_0 can be understood from the inversion layer hole and oxide electric field (E_{OX}) induced X–H bond dissociation process, see Chap. 4, Sect. 4.3 and Fig. 4.5 (reproduced in Fig. 7.6). The inversion layer holes tunnel to the interfacial X–H bonds, get captured and make them weak, and the weak bonds subsequently get dissociated by thermal process. Note that higher N% reduces the hole tunneling barrier (φ_B) in the valence band as SiON has smaller bandgap than pure Silicon Dioxide (SiO_2) gate insulator. However, although the impact of higher N% on tunnel effective mass (m_T) is debated, it is found to reduce at higher N% in these devices. The reduction in φ_B and m_T increases the pre-factor K_{FI0}

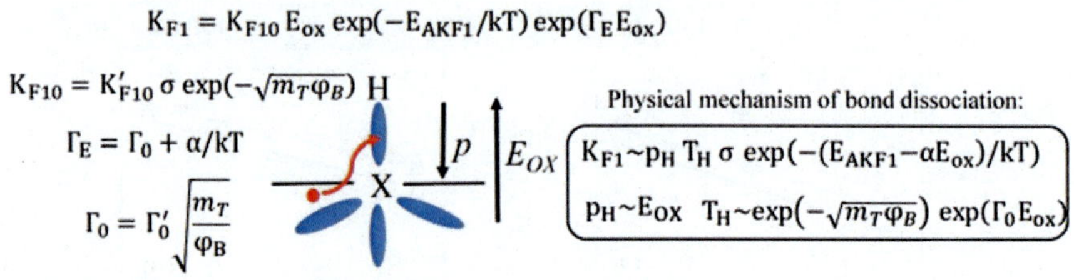

Fig 7.6 Schematic of the inversion layer hole and oxide electric field-induced dissociation of H passivated defects at the channel/gate insulator interface, details in Chap. 4, Sect. 4.3

(note, K_{F10} also depends on the intrinsic interface quality, however, the product of K_{F10} and $\exp(-E_{AKF1}/kT)$ is higher at higher N%) and hence the ΔV_{IT} subcomponent increases at larger N%. However, the reduction in m_T overcompensates the effect of reduction in φ_B in these devices, and therefore, Γ_0 and Γ_E (hence VAF) reduces at higher N% (the polarization term α does not change with N%). Moreover, the dielectric constant (ε_{IL}) of the IL is higher for higher N%, which further reduces the VAF (further beyond the change in Γ_E) as discussed in Chap. 4, Sect. 4.3. DCIV measured ΔN_{IT} also shows higher magnitude but lower VAF for the D2 compared to the D1 device, see Chap. 4, Fig. 4.6.

7.4 DC Stress Kinetics—Impact of Thickness Scaling

Figure 7.7 shows the time evolution of measured and modeled ΔV_T during stress to study the impact of IL and HK thickness scaling. Overall ΔV_T and the underlying ΔV_{IT}, ΔV_{HT}, and ΔV_{OT} subcomponents at fixed V_{GSTR} and T are shown in the left panels while modeling at multiple V_{GSTR} and T are shown in the right panels.

Figure 7.8 shows the measured and modeled ΔV_T at fixed time ($t_{STR} = 1$Ks) as a function of (a, c, e) V_{GSTR} at fixed T and (b, d, f) T at fixed V_{GSTR}, for (a, b) D1, (c, d) D3 and (e, f) D4 devices. The underlying ΔV_{IT}, ΔV_{HT}, and ΔV_{OT} subcomponents are shown in every panel. Overall ΔV_T is dominated by ΔV_{IT} for all devices and stress conditions, although the relative ΔV_{OT} contribution increases at highest V_{GSTR} and/or T, especially for the thicker IL device. The values of VAF and E_A for these devices are similar to each other (the modeled values are same, small changes in simulated results are due to the stress reduction effect), due to ΔV_{IT} domination of overall degradation, and no significant change of the channel/IL interface quality.

As listed in Table 7.1, D1 is the reference device, the HK thickness is reduced in D3 while the IL thickness is increased in D4. The N content is similar (low) in these devices (especially when compared to D2 and D5). The adjustable model parameters are listed in Table 7.2. Note that the channel/IL interface quality and N content do not change between D1, D3, and D4 devices. Therefore, all four parameters related to the ΔV_{IT} subcomponent remain fixed. The N_{0HT} term for ΔV_{HT} increases for thinner HK (D1–D3) and thicker IL t_{STR} (D1–D4). The K_{F30} pre-factor for ΔV_{OT} reduces

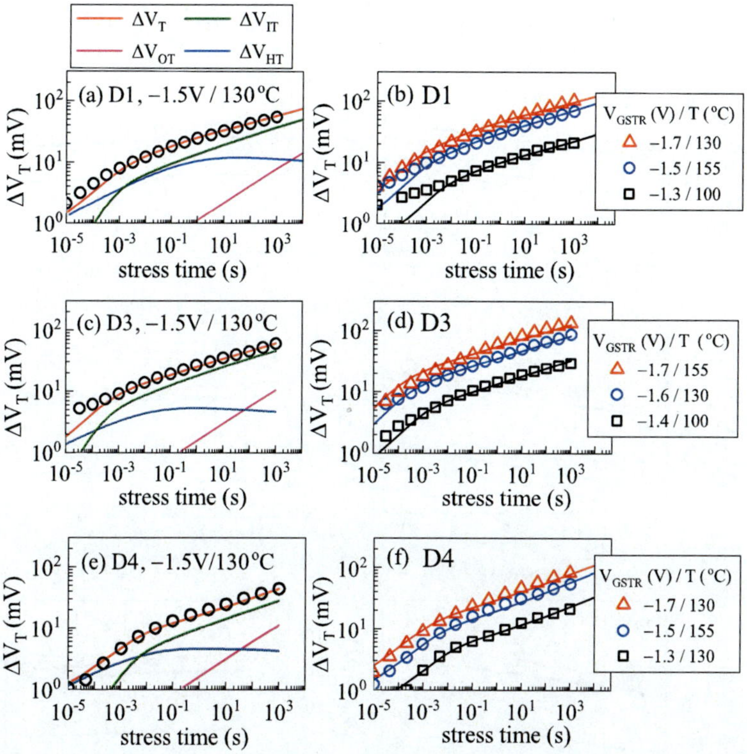

Fig 7.7 Measured and modeled ΔV_{T} time kinetics in (a, b) D1, (c, d) D3, and (e, f) D4 devices at different V_{GSTR} and T during stress. Symbols: experiment, lines: model calculation. The underlying subcomponents are shown in the left panels

slightly for thinner HK but increases for thicker IL. Note that the ΔV_{IT} subcomponent dominates overall ΔV_{T}, however, the relative contributions from ΔV_{HT} and ΔV_{OT} change as the IL and HK thicknesses are varied.

As discussed in [37], incorporation of Hafnium (Hf) and Nitrogen (N) in the IL gives rise to defects near the valence band, which act as hole trapping centers. The slightly poorer IL quality for D3 and D4 devices (higher $N_{0\mathrm{HT}}$ ~ more pre-existing defects for ΔV_{HT}) can be explained as follows. For thinner HK, N incorporation in the IL during the post-HK RPN step would be higher, although for an optimized RPN-PNA step the majority of the N in the IL would be near the IL/HK interface and there is not much difference in N% at the channel/IL interface. Moreover, for thicker IL, the aerial density of traps would be higher due to higher IL volume.

Moreover, for a fixed V_{GSTR}, the voltage drop across the IL would increase, and across the HK would reduce for thinner HK or thicker IL devices. This would increase generation of bulk traps in the IL but reduce it in the HK, however, the IL traps contribute more to ΔV_{OT} due to their proximity to the channel. Since the RDD

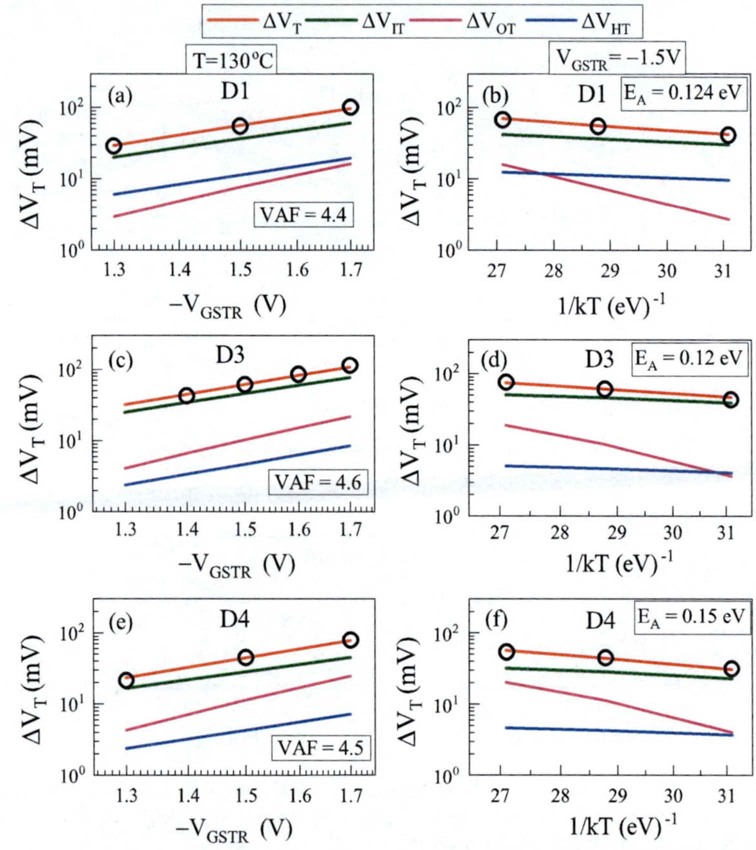

Fig 7.8 Measured and modeled ΔV_T at fixed time in (a, b) D1, (c, d) D3 and (e, f) D4 devices as a function of V_{GSTR} (left panels) and T (right panels) during stress, Symbols: experiment, lines: model calculation. The underlying subcomponents are shown. The power-law VAF and Arrhenius E_A values are listed for the overall ΔV_T

model uses overall E_{OX} (and not individually in the IL and HK layers), the K_{F30} term is varied to reflect the above changes in ΔV_{OT}.

7.5 Recovery Kinetics After DC Stress

Figure 7.9 shows the time evolution of measured and modeled ΔV_T during recovery after DC stress in D1 device. The modeling of overall ΔV_T along with underlying ΔV_{IT}, ΔV_{HT}, and ΔV_{OT} subcomponents are shown in Fig. 7.9 (a) at fixed V_{GSTR} and T, while modeling at various V_{GSTR} and T are shown in Fig. 7.9 (b). The ΔV_{HT} and ΔV_{OT} subcomponents recover fast and slow (~negligible) respectively.

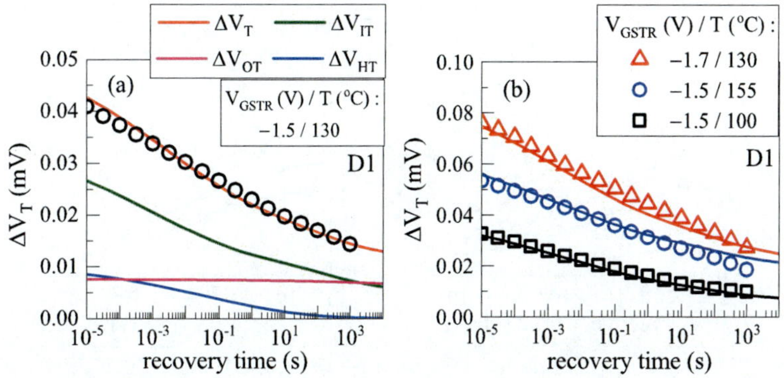

Fig 7.9 Measured and modeled ΔV_T recovery time kinetics in D1 device at different V_{GSTR} and T. Symbols: experiment, lines: model calculation. the underlying subcomponents are shown in the left panel. Data from [8]

The ΔV_{IT} subcomponent recovers via a combination of two sub-processes, with a fast recoverable component (ΔV_{IT_FAST}) due to capture of electrons in the generated interface traps that go below the Fermi level as the magnitude of V_G is reduced from V_{GSTR} to V_{GREC} (handled by TTOM), and a slow recoverable component (ΔV_{IT_SLOW}) due to the re-passivation of generated interface traps triggered by Hydrogen (H and H_2) back diffusion and reverse reaction (handled by RD model), see Chap. 5, Sect. 5.3 for details.

Note that during recovery, the H_2 molecules during backward diffusion have to hop to find an un-passivated defect to passivate, which slows down the rate of recovery at longer time (when available defects for recovery become less). Moreover, a certain fraction of H_2 molecules can get trapped or lost out of the diffusion framework and results in some sort of lock-in effect [38, 39], which also slows down the recovery. The hopping and lock-in of H_2 molecules and related slowing down of recovery have been verified using the Kinetic Monte Carlo (KMC) based stochastic RD model [40]. In the macroscopic RD model, the recovery slow down (due to hopping and lock-in of H_2) is modeled by reducing the H_2 diffusivity only during recovery, which is explained in Chap. 4, Sect. 4.3 [8, 30].

Figure 7.10 shows the time evolution of measured and modeled ΔV_T recovery in D1 devices at (a, c, e) $V_{GREC} = 0$ V and (b, d, f) $V_{GREC} \neq 0$ V that would respectively reflect digital and analog cases, for (a, b) changes in T at fixed V_{GSTR} and t_{STR}, (c, d) changes in V_{GSTR} at fixed T and t_{STR}, and (e, f) changes in t_{STR} at fixed V_{GSTR} and T. Note that the recovery is modeled after stress at different $V_{GSTR} \times T$ values (the same matrix shown in Fig. 7.3), only a few cases are explicitly plotted in panels (a) through (d). The fractional recovery at a given time reduces, hence the FR becomes higher, for higher magnitude of V_{GSTR} and V_{GREC}, higher T, and higher t_{STR}. In particular, the start of recovery is delayed for the $V_{GREC} \neq 0$ V case for all stress conditions. The stress (Fig. 7.3, left panels) and recovery (Fig. 7.10) time kinetics of measured ΔV_T under different V_{GSTR}, T, V_{GREC}, and t_{STR} conditions for the D1 device are modeled

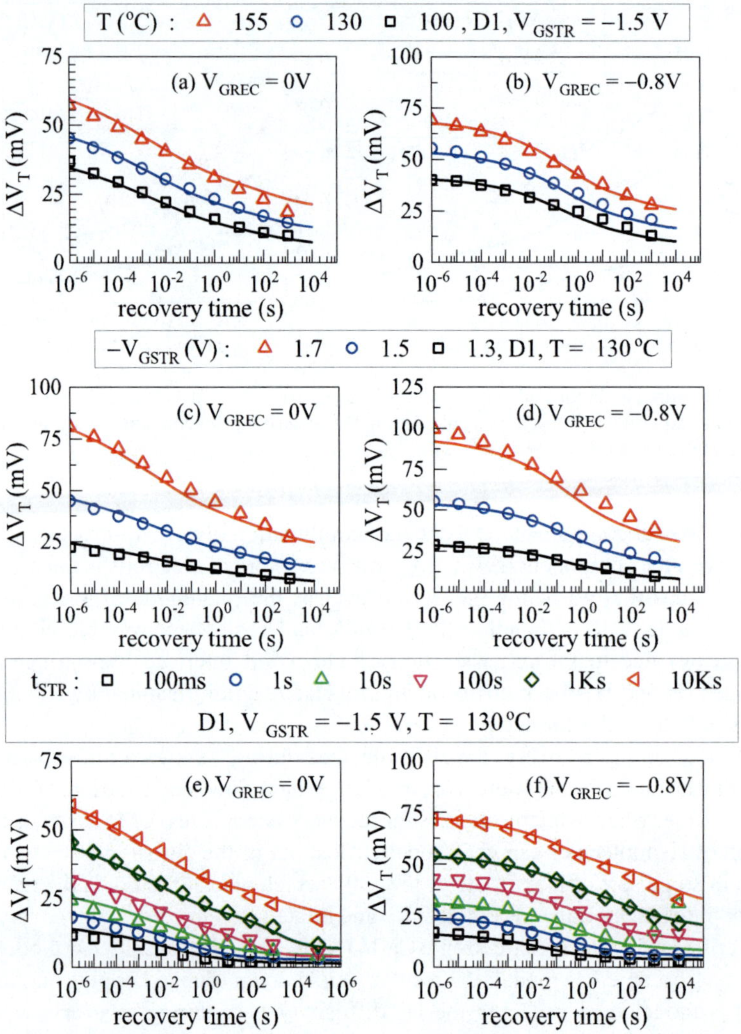

Fig 7.10 Measured and modeled ΔV_T recovery time kinetics in D1 device at different (a, b) T, (c, d) V_{GSTR} and (e, f) stress time, for zero (left panels) and non-zero (right panels) recovery bias. Symbols: experiment, lines: model calculation. Data from [8]

using the adjustable parameters listed in Table 7.2. Only three TTOM parameters are varied with V_{GREC}, which is shown in Chap. 5, Sect. 5.3, Fig. 5.6. Detailed stress and recovery modeling for D2 device is shown in Chap. 5, Sects. 5.4 and 5.5.

Figure 7.11 shows the time evolution of measured and modeled ΔV_T recovery in D1 devices after (a) short and (b) long t_{STR}. The recovery kinetics (onset of recovery and the recovery fraction at a given time) slows down at higher magnitude of V_{GREC}. Importantly, the gap between the end of stress and start of recovery ΔV_T

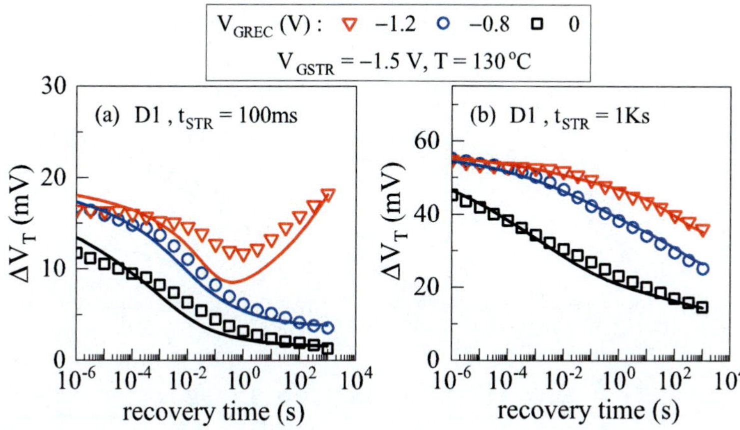

Fig 7.11 Measured and modeled ΔV_T recovery time kinetics in D1 device at different values of V_{GREC} after stress and for (a) short and (b) long t_{STR} values. Symbols: experiment, lines: model calculation. Data from [8]

values reduces and is absent at higher V_{GREC} magnitude. Moreover, re-stressing of the device (increase in ΔV_T) is observed when short time stress is followed by long recovery, and the difference between the stress and recovery biases is small. Except for the V_{GREC} dependence of three TTOM parameters, no other parameters are adjusted to model the measured data. Similar results are also analyzed in Chap. 10 for SOI FinFET devices.

7.6 Impact of Measurement Delay

As shown in Chap. 1, Sect. 1.2, the measurement delay (t_M) is an important parameter for proper determination of the ΔV_T time kinetics using the MSM method. It is shown that higher t_M results in lower ΔV_T magnitude but higher long-time slope n during DC stress, although t_M does not impact the Mode-B AC stress. As shown in the previous section, ΔV_T starts to recover immediately after the DC stress bias is removed. Based on the analysis of ΔV_T recovery time kinetics in Chap. 5 and also in the previous section, the delay artifact can be primarily attributed to the recovery of ΔV_{IT_FAST} and ΔV_{HT} subcomponents for not very large t_M values. However, for very large t_M, the recovery of ΔV_{IT_SLOW} would also contribute.

Figure 7.12 shows the time evolution of measured ΔV_T during DC stress in D3 device using the OPDD MSM method having delay of (a) $t_M = 10\mu s$ and (b) $t_M = 1\mu s$. Overall modeled ΔV_T and the underlying ΔV_{IT}, $\Delta V_{HT,}$ and ΔV_{OT} subcomponents are also shown. As expected, lower magnitude of ΔV_T and higher long time (t_{STR} > 1s) power-law slope n is observed for delayed measurements. The default t_M = $10\mu s$ data are modeled as pure DC with no recovery. To model t_M = 1 ms data,

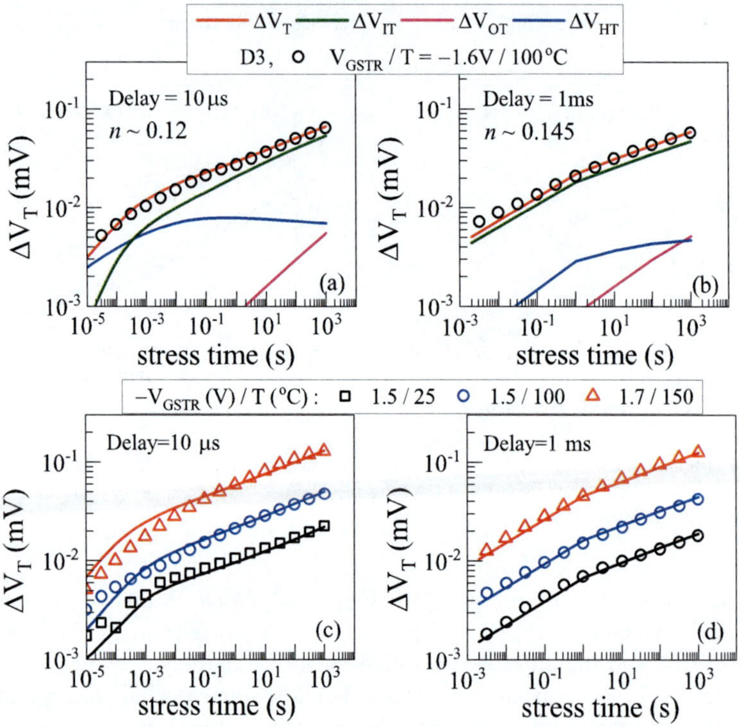

Fig 7.12 Modeling of ΔV_{T} kinetics in the D3 device during stress at different V_{GSTR} and T, measured using different delays. Symbols: experiment, lines: model calculation. The underlying subcomponents are shown in panels (a) and (b)

recovery phases are inserted at identical stress-off (for measurement) intervals as in experiment (see Chap. 1, Sect. 1.2 for MSM method), and ΔV_{T} at the end of the recovery phase is matched with measured data. The model parameters are listed in Table 7.2. The analysis of measured stress data in Chap. 8 is also done with inserted delay (as slow MSM method is used for measurements).

Note, the ΔV_{IT} (primarily due to $\Delta V_{\mathrm{IT\_FAST}}$) and ΔV_{HT} subcomponents are both impacted at $t_{\mathrm{M}} = 1\mathrm{ms}$, while there is negligible impact on ΔV_{OT} since it is semi-permanent. However, the overall ΔV_{T} is still dominated by the ΔV_{IT} subcomponent. Also note that the $\Delta V_{\mathrm{IT\_SLOW}}$ subcomponent does not change, unless the delay is very large (which is not the case in these experiments). The accuracy of the delay simulations is verified by modeling ΔV_{T} kinetics at different V_{GSTR} and T, as shown in Fig. 7.12 for (c) $t_{\mathrm{M}} = 10 \mu\mathrm{s}$ and (d) $t_{\mathrm{M}} = 1 \mu\mathrm{s}$ delay (by the OPDD MSM method). No parameters are re-adjusted to model data at different t_{M} and different V_{GSTR} and T.

7.7 Measurement of Small Area Devices

As mentioned in Chap. 2, Sect. 2.5, ΔV_T measurements in individual small area devices show variation due to variation in the as-fabricated device and also stochastic effects. It is expected that the mean ΔV_T stress-recovery kinetics of multiple small area device measurements would be the same as that of a large area device, provided the gate oxide quality remains the same. This is shown in Chap. 10 for SOI FinFETs. However, this is not true for the D5 devices studied in this chapter, due to additional Nitrogen incorporation discussed in Sect. 7.2. Hence, the D5 devices are rather similar to the D2 devices with higher N%, although the method of Nitrogen incorporation is different between the two devices.

Figure 7.13 shows the time evolution of measured (a) individual ΔV_T kinetics and their mean at fixed V_{GSTR} and T, and (b) mean ΔV_T kinetics at multiple V_{GSTR} and T during stress. The modeling of the mean is done using the deterministic framework, and the underlying subcomponents (ΔV_{IT} and ΔV_{HT}, since ΔV_{OT} is negligible) are shown for one stress condition in Fig. 7.13 (a). The time kinetics of ΔV_{IT} and ΔV_{HT} is the same as shown before for large area devices. The short time degradation is dominated by ΔV_{HT}, however, ΔV_{IT} dominates at longer time since ΔV_{HT} saturates. The saturated ΔV_{HT} contribution is higher in these devices due to higher N% (like D2), when compared to the longer L counterpart (D1) having identically processed gate insulator stack. As mentioned before, this is due to Nitrogen incorporation during the Source-Drain activation anneal, after the gate stack formation, in this case.

Figure 7.14 shows the time evolution of measured individual ΔV_T kinetics and their mean during recovery after stress for variation in: (a, b) V_{GSTR}, (c, d) T and (e, f) t_{STR}, all other stress and recovery conditions (other than the one varied) are kept same between different panels. As mentioned in Sect. 7.2, the individual kinetics is

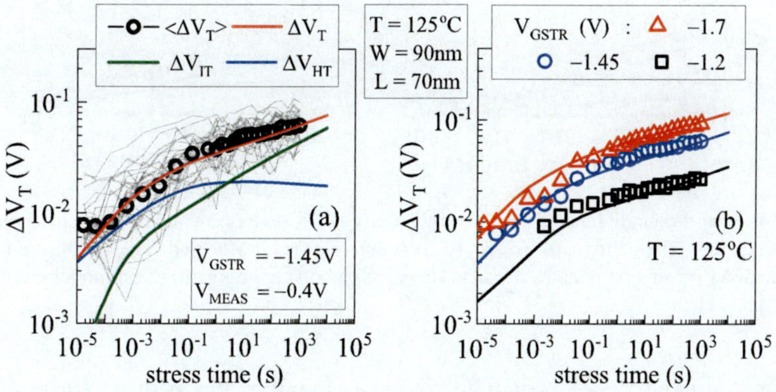

Fig 7.13 Measured ΔV_T kinetics in D5 device during stress: (a) multiple measurements and their mean at fixed V_{GSTR} and T, (b) mean of measured data at multiple V_{GSTR} and T. Mean of the measured data is modeled in both panels, subcomponents are shown in (a). Symbols: experiment, lines: model calculation. Data from [41]

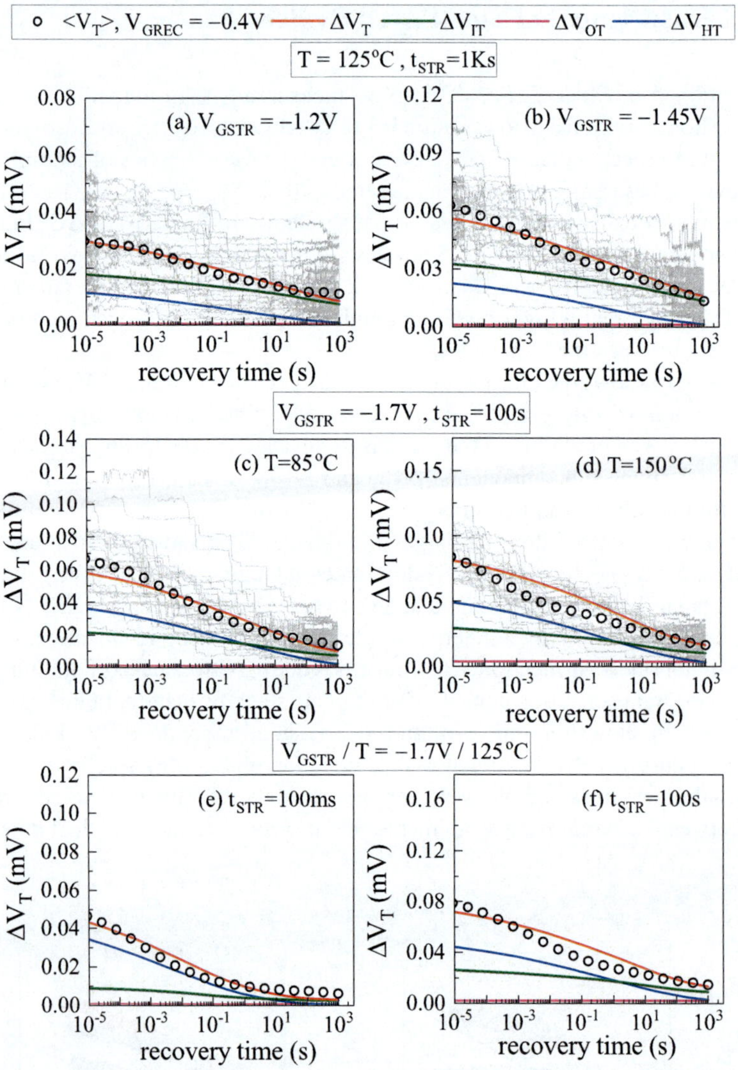

Fig 7.14 Measured individual ΔV_T kinetics in multiple D5 devices and their mean during recovery after stress, for variation in (a, b) V_{GSTR}, (c, d) T, and (e, f) t_{STR}. Modeled ΔV_T and the underlying ΔV_{IT} and ΔV_{HT} subcomponents are also shown. Symbols: experiment, lines: model calculation. Data from [41]

measured at fixed V_{GREC}, which continuously monitors the recovery traces. Therefore, each discrete step indicates a recovery event, either hole detrapping, or electron capture in generated interface traps or passivation of generated interface traps. However, the mean ΔV_T recovery kinetics shows smooth characteristics like large area device data shown earlier. The modeled ΔV_T and underlying subcomponents are shown in every panel. The recovery of ΔV_{HT} is slightly slower than the D2 device (see Chap. 5, Sect. 5.4), but nevertheless, gets over in ~ 10s (same as that for D1 device). The ΔV_{IT} recovery spans over several decades in time and is due to the combinations of ΔV_{IT_FAST} and ΔV_{IT_SLOW} subcomponents as discussed in Sect. 7.5, also see Chap. 5, Sect. 5.3. The relative ΔV_{HT} contribution is larger at the onset of recovery (*i.e.*, at the end of stress) for lower T and/or shorter t_{STR}. This is because of ΔV_{HT} domination at shorter t_{STR}, lower E_A of ΔV_{HT} than that of the ΔV_{IT} subcomponent, and saturation of ΔV_{HT} at longer time (and hence ΔV_{IT} dominates overall ΔV_T in this case). These hole trapping features are discussed in detail in Chap. 3, Sect. 3.3.

7.8 Estimation of EOL Degradation

Figure 7.15 shows the time evolution of measured and modeled ΔV_T for (a, b) D1 and (c, d) D2 devices during (a, c) DC and (b, d) Mode-B AC stress at different T and V_{GSTR} (V_{GHIGH} for AC) values (see Chap. 1, Sect. 1.2 for definition of Mode-B AC stress). The data at higher V_{GSTR} and T are measured for a relatively shorter

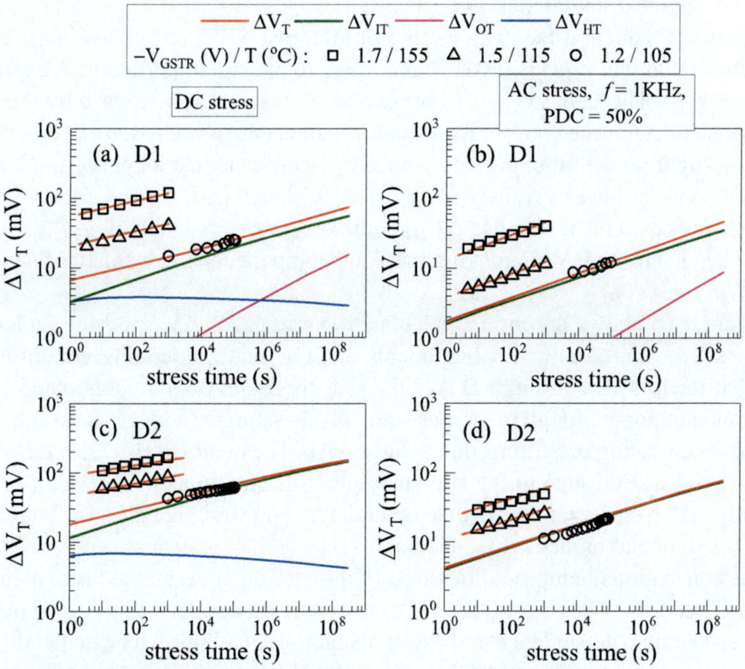

Fig 7.15 Time evolution of measured and modeled ΔV_T kinetics in **a**, **b** D1 and **c**, **d** D2 devices during **a**, **c** DC and **b**, **d** Mode-B AC stress at different V_{GSTR} and T. Symbols: experiment, lines: model calculation. The model subcomponents are shown. Data from [8]

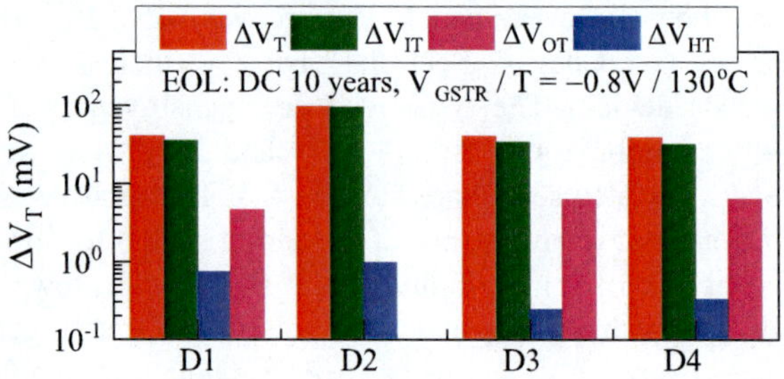

Fig 7.16 Projected ΔV_T and underlying subcomponents at EOL (using calibrated BAT framework) for different devices

duration, but at the lowest V_{GSTR} and T till $t_{STR} = 100$Ks. The model parameters are shown in Table 7.2. Note, only the K_{F30} parameter of the RDD model is reduced for AC stress, due to the reasons mentioned in Chap. 6, Sect. 6.3. No other parameters are adjusted between DC and AC stress for different devices.

The modeled ΔV_T is extrapolated to the end of life (EOL) value of 10 years for the lowest stress condition, and the underlying ΔV_{IT}, $\Delta V_{HT,}$ and ΔV_{OT} subcomponents are shown. Note, it is computationally expensive to calculate ΔV_T till long t_{STR} values during AC stress. Therefore, the cycle-by-cycle AC simulation is done for relatively shorter t_{STR} (the computational overhead depends on the product of f and t_{STR}), and the resulting ΔV_T and underlying subcomponents are then fitted by the DC model and extrapolated to EOL.

The stress reduction factor (lowering of effective NBTI stress due to high ΔV_T magnitude at longer t_{STR}) is invoked and found to play an important role, especially at longer t_{STR}. The EOL ΔV_T is lower due to "soft saturation" caused by the stress reduction effect if turned on, compared to the default (stress reduction off) case; it also causes a slight reduction in the saturated ΔV_{HT} component at longer t_{STR}. Note that the EOL ΔV_T at lower V_G use condition is dominated by the ΔV_{IT} subcomponent for both devices. This is because ΔV_{HT} saturates for $t_{STR} > 1$ s, while ΔV_{OT} is small at low V_G due to high VAF, making these subcomponents much smaller than ΔV_{IT} as shown.

Figure 7.16 shows the model calculated extrapolated ΔV_T and the underlying ΔV_{IT}, $\Delta V_{HT,}$ and ΔV_{OT} subcomponents at EOL under operating conditions for different devices (D1 through D4). Note that the EOL ΔV_T is dominated by the ΔV_{IT} subcomponent for all processes since ΔV_{HT} saturates a longer time and ΔV_{OT} reduces at operating conditions due to large VAF. The devices D1, D3, and D4 show similar ΔV_T at EOL due to the similar channel/IL quality and ΔV_{IT} see Sect. 7.3. Although ΔV_{HT} is larger for D2 than other devices due to larger N%, the larger EOL ΔV_T is still due to higher ΔV_{IT}, see Sect. 7.3.

The conventional, empirical fit based EOL determination method is explained in Chap. 1, Sect. 1.4. The measured ΔV_T time kinetics at accelerated V_{GSTR} is fitted by a power-law time dependence and extrapolated to EOL. This is repeated at different V_{GSTR}, and the extrapolated EOL ΔV_T is fitted using either an exponential or a power-law V_{GSTR} dependence and extrapolated to use ($V_G = V_{DD}$) condition.

Figure 7.17 compares the BAT calculated and empirically estimated EOL ΔV_T for

Fig 7.17 Comparison of BAT and empirically estimated EOL ΔV_T for different devices

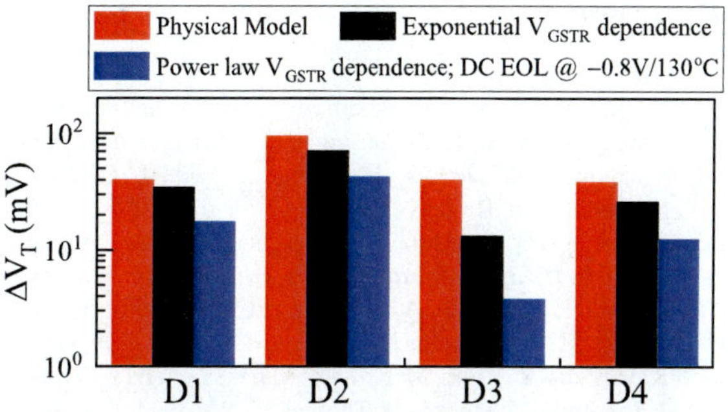

different devices (D1 through D4). In these devices, the power-law method results in lower EOL ΔV_T than the exponential method. Interestingly, both empirical methods result in lower ΔV_T than that obtained by the calibrated model. It is important to remark that the empirical method can lead to underestimated (as shown) or overestimated (as shown in later chapters) EOL ΔV_T when compared to the modeled data, see Chap. 6, Sect. 6.4. Therefore, the use of model-based extrapolation is advisable for reliable determination of NBTI limited device lifetime.

7.9 Summary

The ultra-fast measured ΔV_T kinetics during and after DC stress and during Mode-B AC stress in GF HKMG Si p-MOSFETs is modeled by a comprehensive physics-based framework, having uncorrelated contributions from the underlying ΔV_{IT}, $\Delta V_{HT,}$ and ΔV_{OT} subcomponents. The subcomponents have different time kinetics of stress and recovery, VAF and E_A, and their relative dominance controls these quantities of the overall ΔV_T for different experimental conditions, such as stress at various V_{GSTR} and T, as well as recovery at various V_{GREC} and T, after stress at different V_{GSTR} and t_{STR}. The gate insulator process dependence, such as changes in N%, as well as the IL and High-K thicknesses are modeled. The framework can model the impact of measurement delay. It can also model the mean ΔV_T stress-recovery time kinetics from multiple small area devices. The ΔV_{IT} subcomponent dominates the overall ΔV_T for all processes and stress conditions (especially so at longer t_{STR}), including when projected to the EOL under operating conditions. Empirical fit based methods result in different EOL ΔV_T compared to that projected via the calibrated BAT framework. Hence, the model-based extrapolation should be preferred for reliable estimation of NBTI lifetime.

Acknowledgements Most experimental data presented in this chapter are re-plotted from previously published reports. The authors acknowledge Applied Materials for providing the planar MOSFETs. Ankush Chaudhary is acknowledged for data shown in Fig. 7.14. One of the authors (Souvik Mahapatra) acknowledges Steven Hung and Johanes Swenberg for useful discussions. Karansingh Thakor is acknowledged for help with manuscript preparation.

References

1. N. Kimizuka, K. Yamaguchi, K. Imai, T. Iizuka, C.T. Liu, R.C. Keller, T. Horiuchi, in *Symposium on VLSI Technology Digest of Technical Papers*, p. 92 (2000).
2. V. Huard, M. Denais, F. Perrier, N. Revil, C. Parthasarathy, A. Bravaix, E. Vincent, Microelectron. Reliab. **45**, 83 (2005)
3. S. Mahapatra, K. Ahmed, D. Varghese, A. E. Islam, G. Gupta, L. Madhav, D. Saha, M. A. Alam, in *IEEE International Reliability Physics Symposium Proceedings*, p. 1 (2007).
4. Y. Mitani, H. Satake, A. Toriumi, IEEE Trans. Device Mater. Reliab. **8**, 6 (2008)
5. S. Pae, M. Agostinelli, M. Brazier, R. Chau, G. Dewey, T. Ghani, M. Hattendorf, J. Hicks, J. Kavalieros, K. Kuhn, M. Kuhn, J. Maiz, M. Metz, K. Mistry, C. Prasad, S. Ramey, A. Roskowski, J. Sandford, C. Thomas, J. Thomas, C. Wiegand, J. Wiedemer, in *IEEE International Reliability Physics Symposium Proceedings*, p. 352 (2008).
6. K. Joshi, S. Hung, S. Mukhopadhyay, V. Chaudhary, N. Nanaware, B. Rajamohanan, T. Sato, M. Bevan, A. Wei, A. Noori, B. McDougal, C. Ni, G. Saheli, C. Lazik, P. Liu, D. Chu, L. Date, S. Datta, A. Brand, J. Swenberg, S. Mahapatra, in *IEEE International Reliability Physics Symposium Proceedings*, 4C.2.1 (2013).
7. J. Franco, B. Kaczer, P.J. Roussel, J. Mitard, M. Cho, L. Witters, T. Grasser, G. Groeseneken, IEEE Trans. Electron Devices **60**, 396 (2013)
8. N. Parihar, N. Goel, S. Mukhopadhyay, S. Mahapatra, IEEE Trans. Electron Devices **65**, 392 (2018)
9. N. Parihar, S. Mahapatra, in *IEEE International Reliability Physics Symposium Proceedings*, TX.5.1 (2018).
10. V. Huard, C. Ndiaye, M. Arabi, N. Parihar, X. Federspiel, S. Mhira, S. Mahapatra, A. Bravaix, in *IEEE International Reliability Physics Symposium Proceedings*, TX.4.1 (2018).
11. N. Parihar, R. Tiwari, C. Ndiaye, M. Arabi, S. Mhira, H. Wong, S. Motzny, V. Moroz, V. Huard, S. Mahapatra, in *International Conference on Simulation of Semiconductor Processes and Devices*, p. 167 (2018).
12. S. Ramey, A. Ashutosh, C. Auth, J. Clifford, M. Hattendorf, J. Hicks, R. James, A. Rahman, V. Sharma, A. St. Amour, C. Wiegand, in *IEEE International Reliability Physics Symposium Proceedings*, 4C.5.1 (2013).
13. K. T. Lee, W. Kang, E-A Chung, G. Kim, H. Shim, H. Lee, H. Kim, M. Choe, N-I Lee, A. Patel, J. Park, J. Park, in *IEEE International Reliability Physics Symposium Proceedings*, 2D.1.1 (2013).
14. J. Franco, B. Kaczer, A. Chasin, H. Mertens, L.-A. Ragnarsson, R. Ritzenthaler, S. Mukhopadhyay, H. Arimura, P.J. Roussel, E. Bury, N. Horiguchi, D. Linten, G. Groeseneken, A. Thean, in *IEEE International Reliability Physics Symposium Proceedings*, 4B.2.1 (2016).
15. G. Jiao, M. Toledano-Luque, K.-J. Nam, N. Toshiro, S.-H. Lee, J.-S. Kim, T. Kauerauf, E. Chung, D. Bae, G. Bae, D.-W. Kim, K. Hwang, in *IEEE International Electron Devices Meeting Technical Digest*, 31.2.1 (2016).
16. N. Parihar, R.G. Southwick, U. Sharma, M. Wang, J.H. Stathis, S. Mahapatra, in *IEEE International Reliability Physics Symposium Proceedings*, 2D.4.1 (2017).
17. N. Parihar, R. Southwick, M. Wang, J.H. Stathis, S. Mahapatra, in *IEEE International Electron Devices Meeting Technical Digest*, 7.3.1 (2017).
18. A. Rahman, J. Dacuna, P. Nayak, G. Leatherman, S. Ramey, in *IEEE International Reliability Physics Symposium Proceedings*, 6F.4.1 (2018).
19. N. Parihar, U. Sharma, R.G. Southwick, M. Wang, J.H. Stathis, S. Mahapatra, IEEE Trans. Electron Devices **65**, 23 (2018)
20. N. Parihar, R.G. Southwick, M. Wang, J.H. Stathis, S. Mahapatra, IEEE Trans. Electron Devices **65**, 1699 (2018)
21. N. Parihar, R.G. Southwick, M. Wang, J.H. Stathis, S. Mahapatra, IEEE Trans. Electron Devices **65**, 1707 (2018)
22. N. Parihar, R. Tiwari, S. Mahapatra, in *International Conference on Simulation of Semiconductor Processes and Devices*, p. 176 (2018).

23. R. Tiwari, N. Parihar, K. Thakor, H.Y. Wong, S. Motzny, M. Choi, V. Moroz, S. Mahapatra, IEEE Trans. Electron Devices **66**, 2086 (2019)
24. R. Tiwari, N. Parihar, K. Thakor, H.Y. Wong, S. Motzny, M. Choi, V. Moroz, S. Mahapatra, IEEE Trans. Electron Devices **66**, 2093 (2019)
25. N. Parihar, U. Sharma, R.G. Southwick, M. Wang, J.H. Stathis, S. Mahapatra, in *IEEE International Reliability Physics Symposium Proceedings* (2019).
26. M. Wang, J. Zhang, H. Zhou, R.G. Southwick, R. Hsin, K. Chao, X. Miao, V.S. Basker, T. Yamashita, D. Guo, G. Karve, H. Bu, in *IEEE International Reliability Physics Symposium Proceedings* (2019).
27. N. Choudhury, U. Sharma, H. Zhou, R.G. Southwick, M. Wang, S. Mahapatra, in *IEEE International Reliability Physics Symposium Proceedings* (2020).
28. H. Zhou, M. Wang, J. Zhang, K. Watanabe, C. Durfee, S. Mochizuki, R. Bao, R. Southwick, M. Bhuiyan, B. Veeraraghavan, *IEEE International Reliability Physics Symposium Proceedings* (2020).
29. N. Choudhury, T. Samadder, R. Tiwari, H. Zhou, R.G. Southwick, M. Wang, S. Mahapatra, in *IEEE International Reliability Physics Symposium Proceedings* (2021).
30. S. Mahapatra, N. Goel, S. Desai, S. Gupta, B. Jose, S. Mukhopadhyay, K. Joshi, A. Jain, A.E. Islam, M.A. Alam, IEEE Trans. Electron Devices **60**, 901 (2013)
31. A.E. Islam, N. Goel, S. Mahapatra, M.A. Alam, in *Fundamentals of Bias Temperature Instability in MOS Transistors* (Springer India, 2015), pp. 181–207.
32. S. Mukhopadhyay, K. Joshi, V. Chaudhary, N. Goel, S. De, R.K. Pandey, K.V.R.M. Murali, S. Mahapatra, in *IEEE International Reliability Physics Symposium Proceedings*, GD 3.1 (2014).
33. N. Choudhury, N. Parihar, N. Goel, A. Thirunavukkarasu, S. Mahapatra, IEEE J. Electron Devices Soc. **8**, 1281 (2020)
34. T. Samadder, N. Choudhury, S. Kumar, D. Kochar, N. Parihar, S. Mahapatra, IEEE Trans. Electron Devices **68**, 485 (2021)
35. L. K. Han, S. Crowder, M. Hargrove, E. Wu, S.H. Lo, F. Guarin, E. Crabbe, and L. Su, in IEEE International Electron Devices Meeting Technical Digest, 643 (1997).
36. M.A. Alam, IEEE Trans. Electron Devices **49**, 226 (2002)
37. S. Mahapatra, S. De, K. Joshi, S. Mukhopadhyay, R.K. Pandey, K.V.R.M. Murali, IEEE Electron Device Lett. **34**, 963 (2013)
38. K.O. Jeppson, C.M. Svensson, J. Appl. Phys. **48**, 2004 (1977)
39. S. Rangan, N. Mielke, E.C.C. Yeh, in *IEEE International Electron Devices Meeting Technical Digest*, 14.3.1 (2003).
40. S. Kumar, R. Anandkrishnan, N. Parihar, S. Mahapatra, IEEE Trans. Electron Devices **67**, 4741 (2020)
41. N. Parihar, N. Goel, A. Chaudhary, S. Mahapatra, IEEE Trans. Electron Devices **63**, 946 (2016)

Chapter 8
BAT Framework Modeling of Gate First HKMG Si-Capped SiGe Channel MOSFETs

Narendra Parihar, Tarun Samadder, and Souvik Mahapatra

8.1 Introduction

As described in the earlier chapters, Negative Bias Temperature Instability (NBTI) became an important concern during the migration from Silicon Dioxide (SiO_2) to Silicon Oxynitride (SiON) gate insulator MOSFETs[1–4]. It continues to impact the dual layer (SiO_2 or SiON Interlayer (IL) and Hafnium Dioxide (HfO_2) high-K layer) High-K Metal Gate (HKMG) gate insulator-based planar bulk [5–11] and Fully Depleted Silicon On Insulator (FDSOI) [12, 13] MOSFETs, bulk and SOI FinFETs [13–26], and Gate All Around Stacked Nanosheet (GAA-SNS) FETs[27–30], having either Silicon (Si) or Silicon Germanium (SiGe) channel.

As discussed in Chap. 1, Sect. 1.3, NBTI results in a gradual buildup of positive gate insulator charges and results in device parametric shift, *e.g.*, threshold voltage shift (ΔV_{T}), over time, when the gate bias (V_{G}) is held at a negative value ($V_{\mathrm{G}} = V_{\mathrm{GSTR}}$) compared to other terminals of the device. The ΔV_{T} magnitude increases at higher magnitude of V_{GSTR} and temperature (T) during stress, and is governed by the Voltage Acceleration Factor (VAF) and Arrhenius T activation energy (E_{A}), respectively. However, the charges accrued during stress and ΔV_{T} reduce when the magnitude of V_G is reduced or removed ($V_{\mathrm{G}} = V_{\mathrm{GREC}}$ or 0 V). Therefore, AC stress results in lower ΔV_{T} than DC stress. The key NBTI features are summarized hereinafter (reproduced from Chap. 3, Sect. 3.1):

As discussed earlier in Chap. 2, the magnitude of NBTI increases with nitrogen content (N%) in the gate insulator stack, while it reduces with germanium content (Ge%) in the channel. The N% impact is shown for both SiON [1–4] and HKMG devices fabricated using the Gate First (GF) [7, 10, 13] and Replacement Metal Gate

N. Parihar · T. Samadder · S. Mahapatra (✉)
Department of Electrical Engineering, Indian Institute of Technology Bombay, Powai, Mumbai 400076, India
e-mail: souvik@ee.iitb.ac.in

S. Mahapatra (ed.), *Recent Advances in PMOS Negative Bias Temperature Instability*,
https://doi.org/10.1007/978-981-16-6120-4_8

(RMG) [18, 19, 21, 22] flows. The Ge% impact is observed in HKMG gate insulator-based planar [6, 8, 9, 11], FDSOI [12, 13] and FinFET [16–19, 21, 22] devices. Besides the ΔV_T magnitude, N% in the gate insulator and Ge% in the channel impact the time kinetics during and after stress, *i.e.*, the power-law time slope *n* at long stress time (t_{STR}) and Fraction Remaining (FR) at a particular recovery time (t_{REC}) after stress (FR is defined as the ratio of ΔV_T at t = t_{REC} to that at the end of stress at t = t_{STR}). These processes also impact VAF, E_A, and *T* dependence of VAF (the reduction of VAF at higher *T*). Note that increase in N% reduces *n* at long t_{STR}, FR for a particular ratio of t_{REC} to t_{STR} (*i.e.*, recovery becomes faster), VAF, E_A, and the *T* dependence of VAF (lower reduction in VAF at higher T). However, increase in Ge% increases *n* at long t_{STR}, FR for a particular ratio of t_{REC} to t_{STR} (*i.e.*, recovery becomes slower), VAF, E_A, and the *T* dependence of VAF (larger VAF reduction at higher *T*).

The above aspects are analyzed in Chap. 7 for Si channel bulk MOSFETs, in Chap. 9 for Si and SiGe channel FDSOI MOSFETs, and in Chap. 11 for SiGe channel FinFETs. In addition, changes in the compressive stress in the channel due to changes in the layout or device dimension also impact NBTI, which is discussed in Chap. 9 for FDSOI MOSFETs, as well as in Chaps. 12 and 13 for FinFETs and GAA-SNS FETs.

The impact of Si-capped SiGe channel process changes in planar MOSFETs is studied in this chapter. The process dependencies listed above are analyzed and modeled using the BTI Analysis Tool (BAT) framework in this book. The framework is described in Chap. 4 through Chap. 6 and is briefly summarized below (reproduced from Chap. 6, Sect. 6.3).

Figure 8.1 illustrates the BAT framework [10] used to model the time kinetics of ΔV_T during and after stress. It uses uncorrelated contributions from generated interface (ΔV_{IT}) and bulk (ΔV_{OT}) gate insulator traps, and hole trapping in process-related preexisting bulk gate insulator traps (ΔV_{HT}). The Reaction Diffusion (RD) model is used to calculate the time kinetics of interface traps (density ΔN_{IT}) [10, 31,

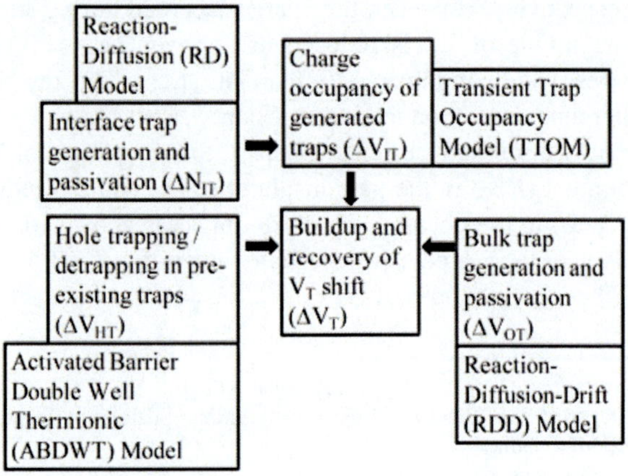

Fig. 8.1 Schematic of the BTI Analysis Tool (BAT) framework used in this book to model measured ΔV_T kinetics during and after DC and AC NBTI stress, reproduced from Chap. 4

32]. In Chap. 4, the RD model is described and independently validated by measured data from Direct Current IV (DCIV) method [33]. Transient Trap Occupancy Model (TTOM) is used to calculate the occupancy of generated interface traps and their contribution (ΔV_{IT}) to ΔV_T [10], which is described and validated in Chap. 5, Sect. 5.3. The ΔV_{HT} and ΔV_{OT} kinetics are, respectively, modeled by the Activated Barrier Double Well Thermionic (ABDWT) model [34] and Reaction Diffusion Drift (RDD) model [35], and these models are described and validated in Chap. 5, Sect. 5.4 and Chap. 6, Sect. 6.2.

In Chap. 7, the impact of N% on ΔV_T is shown to be due to increase in ΔV_{IT} and ΔV_{HT} (with relatively higher increase in ΔV_{HT}) and reduction in ΔV_{OT} subcomponents (although ΔV_{IT} dominates overall ΔV_T across V_{GSTR} and T). However, the mechanism responsible for the impact of Ge% on ΔV_T is debated. Lower NBTI is suggested to be due to the higher pre-stress interface trap density that results in lower interface trap generation during stress [6], unfavorable valence band alignment resulting in either lower hole trapping [8, 16], or lower contribution from interface traps [9], and lowering of oxide electric field (E_{OX}) due to large negative charge density near the valence band edge [17]. However, none of the above reports have modeled the measured ΔV_T time kinetics across different V_{GSTR} and T, which is a prerequisite of any successful model.

Recently, the impact of Ge% on ΔV_T stress and recovery kinetics is shown to be due to reduction in ΔV_{IT}, ΔV_{HT}, and ΔV_{OT} subcomponents (with relatively higher reduction in ΔV_{HT} and lower reduction in ΔV_{OT}), although ΔV_{IT} dominates the overall ΔV_T across V_{GSTR} and T (unless for very high Ge% devices, where both ΔV_{IT} and ΔV_{OT} similarly contribute) [12, 13, 18, 19, 21, 22, 24]. This is also shown to be valid for the Si-capped SiGe devices analyzed in this chapter.

8.2 Device Details and Model Parameters

Table 8.1 lists the devices analyzed in this chapter (the measured data are obtained from [8], provided by the original authors). The reference device D1 is Si channel planar MOSFET having GF HKMG gate insulator stack. All other devices (D2 through D8) are Si-capped SiGe channel planar MOSFETs. The illustration is for a Si-capped SiGe device, showing SiGe Quantum Well (QW), Si cap and HKMG gate insulator stack having IL, high-K, and metal gate. The gate insulator for these devices have slightly different IL thickness but identical high-K thickness, refer to [8] for details. D2 and D4 devices have different Ge% in the QW; D6, D7, and D8 devices have different SiGe QW thickness; and D3, D4, D5, and D6 devices have different Si cap thickness.

In the original reference [8], the time kinetics of ΔV_T is measured using the extended Measure-Stress-Measure (e-MSM) method, explained in Chap. 1, Sect. 1.2. Note that a measurement delay (t_M) of 1 ms is used during stress, and continuous sampling (from 1 ms and beyond) is used for recovery after stress. The stress time

Table 8.1 Process description of the Si reference and Si-capped SiGe planar MOSFETs, from [8]. The channel and gate stack for the Si-capped SiGe device is illustrated. Process changes regarding Ge% in the QW, QW thickness, and Si cap thickness are studied

Device	Ge% in QW (%)	SiGe QW thickness (nm)	Si cap thickness (nm)
D1	0	0	0
D2	45	5	1.3
D3	55	5	2
D4	55	5	1.3
D5	55	5	1
D6	55	5	0.65
D7	55	3	1.3
D8	55	7	1.3

kinetics is measured at different V_{GSTR} and T, while the recovery kinetics is measured after stress at different V_{GSTR}, T, and t_{STR}.

As mentioned before, the measured ΔV_T time kinetics is modeled using uncorrelated contributions from ΔV_{IT}, ΔV_{HT}, and ΔV_{OT} subcomponents using the framework illustrated in Fig. 8.1. The model parameters are listed in Table 8.2, and their

Table 8.2 Process-dependent RD, TTOM, ABDWT, and RDD parameters (as per classification done in Chap. 4 through Chap. 6) used in this chapter. See Table 8.1 for device details. The other adjustable parameters (see text) are as follows: $\tau_{EC} = 5 \times 10^{-3}$ s for TTOM, $E_{BM} = 1.1$ eV, $\gamma_B = 6.5 \times 10^{-9}$ C.cm, and $m = 4.5$ for ABDWT, and $E_{AOT} = 1.14$ eV, $\Gamma_{0OT} = 0.13$ cm/MV, $\alpha_{OT} = 3.6$ qÅ, and $K_{F50} = 23$ cm^3/s for RDD, for all devices

Device	Unit	D1	D2	D3	D4
K_{F10}	cm/Vs	50	8	9	10
E_{AKF1}	eV	0.54	0.58	0.60	0.64
Γ_0	cm/MV	0.90	0.49	0.88	0.45
α	qÅ	1.8	2.8	2.8	2.85
f_{FAST}	–	0.40	0.43	0.51	0.53
N_{0HT}	1/cm^2	8.0×10^{11}	6.0×10^{11}	5.0×10^{11}	1.8×10^{11}
K_{F30}	1/s	9.0×10^5	8.0×10^5	1.2×10^6	4.0×10^5
Device	Unit	D5	D6	D7	D8
K_{F10}	cm/Vs	18	20	15	4
E_{AKF1}	eV	0.68	0.70	0.64	0.64
Γ_0	cm/MV	0.36	0.10	0.45	0.45
α	qÅ	2.90	3.60	2.85	2.85
f_{FAST}	–	0.57	0.70	0.53	0.53
N_{0HT}	1/cm^2	2.0×10^{10}	1.0×10^{10}	1.8×10^{11}	1.8×10^{11}
K_{F30}	1/s	2.0×10^4	2.5×10^3	4.0×10^5	4.0×10^5

description is as follows (reproduced from Chap. 6, Sect. 6.3).

As described in Chap. 6, Sect. 6.3, the process-dependent RD model parameters are related to the pre-factor (K_{F10}), T-independent field acceleration (Γ_0), bond polarization (α), and T activation energy (E_{AKF1}) of the inversion layer hole-assisted bond dissociation process, see Chap. 4, Fig. 4.5. The process-dependent TTOM parameters are related to fast fraction of traps that undergo electron capture (f_{FAST}) and the electron capture time constant (τ_{EC}), see Chap. 5, Sect. 5.3. However, τ_{EC} is not varied for these devices. As described in Chap. 5, Sect. 5.4, the process-dependent ABDWT model parameters are related to the density of preexisting defects (N_{0HT}), the energy barrier (E_{BM}), and terms associated with E_{OX} dependence of the barrier (γ_B) and trap energy level ($\gamma_{E2} = m\gamma_B$), see Chap. 5, Fig. 5.10. Only N_{0HT} is varied across devices. The process-dependent RDD model parameters are related to the pre-factor (K_{F30}), T-independent field acceleration (Γ_{0OT}), and T activation energy (E_{AOT}) of the Anode Hole Injection (AHI)-assisted bond dissociation (note, the bond polarization factor (α_{OT}) is not varied across devices but listed for completeness), and the forward reaction rate for ions (K_{F50}), see Chap. 6, Sect. 6.2. Only K_{F30} is varied across devices, while $\Gamma_{0,OT}$ is different between Si and all SiGe devices. All the other model parameters are process agnostic and are listed in the respective sections of Chap. 4 through Chap. 6 (Tables 4.1, 5.1, 5.2, and 6.1).

8.3 Modeling of Stress Kinetics

Figure 8.2(a) shows the time evolution of simulated ΔV_T and the underlying ΔV_{IT}, ΔV_{HT}, and ΔV_{OT} subcomponents during stress for the device D1, using the model parameters listed in Table 8.2. Note, ΔV_{IT} evolves rapidly at the initiation of stress

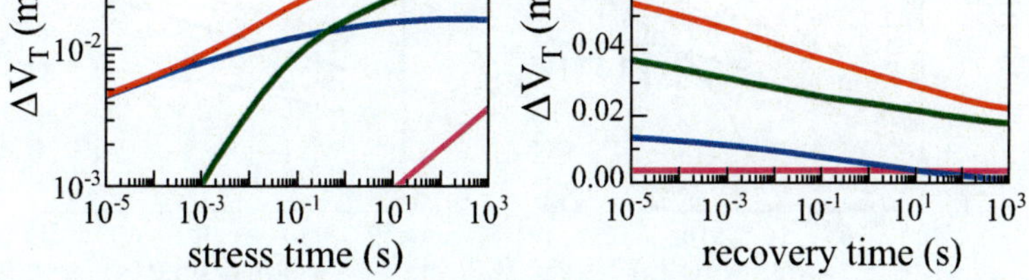

Fig. 8.2 Time evolution of overall ΔV_T and model subcomponent during (a) stress and (b) recovery from short to long time. The parameters of device D1 are used for simulation

and asymptotically settles into a power-law dependence with slope n ~ 1/6 at long time (t_{STR} > 1 s). The ΔV_{HT} magnitude depends on the gate insulator quality, see Chap. 5, Sect. 5.4, which evolves even rapidly than ΔV_{IT} at the beginning of stress but saturates at long time (n ~ 0 in a log–log plot). The initial buildup of ΔV_{OT} is smaller than the other components, but it shows power-law dependence with slope n ~ 1/3 at long time. The VAF and E_A of ΔV_{OT} is largest and of ΔV_{HT} is smallest among all the subcomponents for this device (which is usually the case across all devices studied in this book).

As discussed in Chaps. 1 and 7, a measurement delay of 1 ms would cause significant recovery of ΔV_T. As discussed in Chap. 7, the relative contributions at the end of stress of the underlying ΔV_{IT}, ΔV_{HT} and ΔV_{OT} subcomponents govern the recovery time kinetics of ΔV_T after stress. Figure 8.2(b) shows the time evolution of simulated ΔV_T and the underlying ΔV_{IT}, ΔV_{HT}, and ΔV_{OT} subcomponents during recovery after stress for the D1 device. ΔV_{HT} recovers fast while ΔV_{OT} is semi-permanent and recovers slowly. The ΔV_{IT} recovery is over several decades in time and is due to two different processes: ΔV_{IT_FAST}, for the fraction of traps that go below the Fermi level (f_{FAST}) as the magnitude of V_G reduces from V_{GSTR} to V_{GREC} or 0 V during recovery and becomes neutral by capturing electrons (handled by the TTOM framework), and ΔV_{IT_SLOW}, for the remaining traps that get re-passivated by hydrogen back diffusion (handled by the RD model), see Chaps. 4 and 7 for details. Note, the recovery of ΔV_{IT_FAST} and ΔV_{HT} would impact stress measurements with 1 ms delay, see Chap. 7, Sect. 7.5 for related discussion.

Figure 8.3(a) shows the time evolution of simulated ΔV_T along with the underlying ΔV_{IT}, ΔV_{HT}, and ΔV_{OT} subcomponents for device D1 during DC stress. Note that a delay of 1 ms is inserted in the simulations at times when the stress is paused for measurement (see Chap. 1, Sect. 1.2 for MSM method details), to mimic actual experimental conditions. The ΔV_{IT} (owing to ΔV_{IT_FAST}) and ΔV_{HT} subcomponents recover during the measurement interval and reduce overall ΔV_T. The recovery of

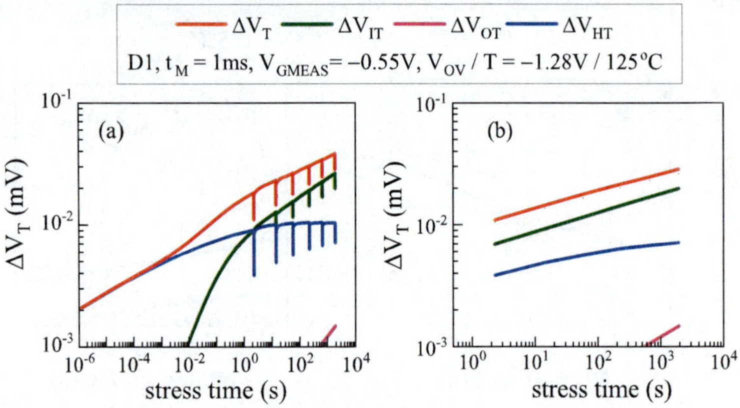

Fig 8.3 (a) Model calculated ΔV_T stress time kinetics with the underlying subcomponents with 1 ms inserted delay. (b) The lower envelope of ΔV_T and subcomponents from panel (a)

ΔV_{OT} is negligible. Figure 8.3(b) shows the time evolution of the lower envelope (end points after recovery during measurement intervals in Fig. 8.3(a) of simulated ΔV_T and the underlying subcomponents. Note that unlike the modeling of ultra-fast measured data in Chap. 7, the time evolution of the lower envelope of ΔV_{HT} does not saturate within $t_{STR} \sim 1$ s. This is due to the hole detrapping process, which delays the time to saturation. Furthermore, the slope of ΔV_{IT} is slightly higher than $n \sim 1/6$ due to recovery (see Chap. 1, Sect. 1.2). All stress kinetics in this chapter is modeled using 1 ms inserted delay, and the lower envelope of ΔV_T is compared to measured data.

Figure 8.4 shows the time evolution of measured and modeled ΔV_T at different V_{OV} (overdrive voltage, $V_{OV} = V_{GSTR} - V_{T0}$, where V_{T0} is the pre-stress threshold voltage) but fixed T for the devices D1 through D8 listed in Table 8.1. The time kinetics is measured for longer stress time ($t_{STR} > 1$ s) and shows power-law time dependence for all devices. ΔV_T is lower for SiGe compared to Si channel device under identical stress condition (V_{OV}, T, and t_{STR}), and among the different SiGe channel devices, it is lower for higher Ge% in the QW (Fig. 8.4(b),(d)), lower Si cap thickness (Fig. 8.4(c),(d),(e),(f)), and higher QW thickness (Fig. 8.4(d),(g),(h)). Although ΔV_T increases at higher V_{OV} for a particular device, there is no noticeable change in the time slope n as V_{OV} is increased. Note, although the relative ΔV_{OT} contribution increases at higher V_{OV} (which is described below), ΔV_{IT} dominates overall ΔV_T in the range of V_{GSTR} used in these devices. It is expected that higher (relative) ΔV_{OT} contribution would increase the long time slope n of overall ΔV_T, see Chap. 3, Sect. 3.4. However, the stress reduction effect (reduction in the effective electrical stress at higher ΔV_T) would offset the effect of higher relative ΔV_{OT} contribution at higher V_{OV}. This explains the invariance of n across V_{OV}.

Note that due to measurement delay of 1 ms, the measured slope n is slightly higher than that would have been obtained if ultra-fast measurements were used. As demonstrated using Fig. 8.3, the lower envelop of ΔV_T (after recovery in 1 ms) is shown as model lines for all the devices in this figure. Also note that although the recovery of both ΔV_{IT_FAST} and ΔV_{HT} impact the simulated lower envelope for D1 reference, only the recovery of ΔV_{IT_FAST} impacts that of D2 through D8 since ΔV_{HT} is negligible for SiGe channel devices. The ΔV_{IT_SLOW} and ΔV_{OT} subcomponents recover slowly and have no impact for 1 ms measurement delay.

Figure 8.5 shows the measured and modeled ΔV_T as a function of V_{GSTR} but at fixed T and $t_{STR} = 1$Ks, together with the underlying ΔV_{IT}, ΔV_{HT}, and ΔV_{OT} subcomponents for all devices. Note that ΔV_{IT} dominates the overall ΔV_T for different devices and for the entire range of V_{GSTR} (or V_{OV}) studied in these experiments. The contribution from ΔV_{HT} is negligible for all devices except the Si reference device. It is important to remark that this is in contrast to the original analysis of [8], where the V_{GSTR} dependence of fixed time ΔV_T is modeled exclusively by the ΔV_{HT} subcomponent. It is to be noted that the time kinetics of ΔV_T during stress is not modeled in the original reference, without which the relative contribution of different subcomponents cannot be properly ascertained.

Although both ΔV_{IT} and ΔV_{OT} reduce, the relative ΔV_{OT} contribution increases for SiGe compared to Si channel devices, especially for higher Ge% in the QW

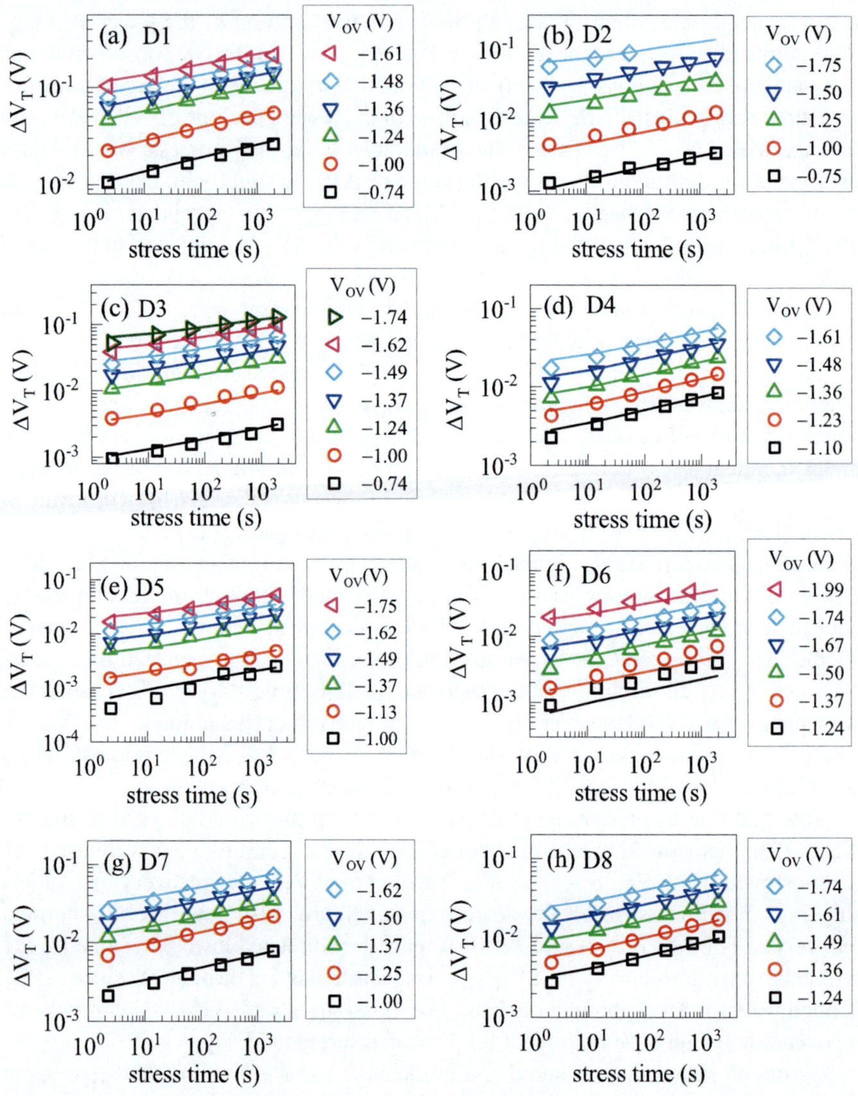

Fig 8.4 Measured and modeled time evolution of ΔV_T at different V_{OV} for different devices listed in Table 8.1. Stress T is 125 °C. The model lines represent the lower envelope of ΔV_T, as illustrated in Fig. 8.3. Symbols: experiment, lines: model calculation. Data from [8]

(Fig. 8.5(b),(d)), lower Si cap thickness (Fig. 8.5(c),(d),(e),(f)), and increase in the QW thickness (Fig. 8.5(d),(g),(h)). The VAF of the ΔV_{OT} and ΔV_{HT} subcomponents are highest and lowest, respectively, and show process independence (this is applicable for ΔV_{OT}, since ΔV_{HT} is negligible except in D1). However, the VAF for ΔV_{IT} is process dependent and is related to that of Γ_0 and α, see Table 8.2.

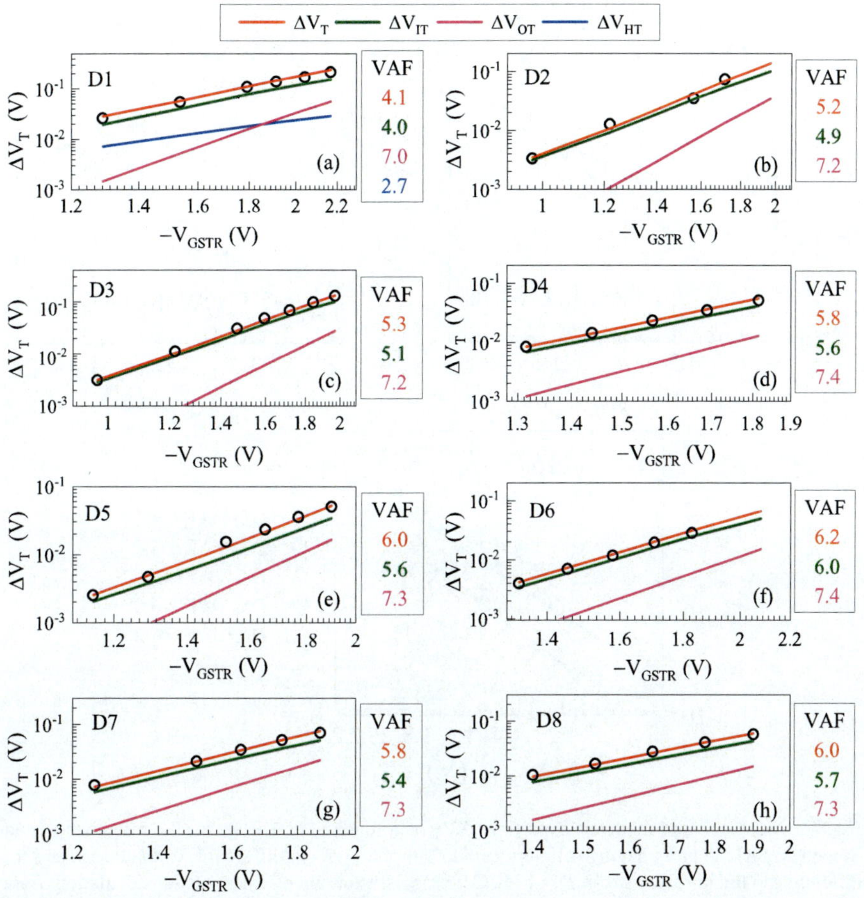

Fig 8.5 Measured and modeled fixed time ΔV_{T} as a function of V_{GSTR} for different devices; the underlying model subcomponents are also shown. Stress T is 125 °C. Except D1, the ΔV_{HT} subcomponent is almost negligible for other devices and is not plotted. Symbols: experiment, lines: model calculation. Data from [8]

Figure 8.6 shows the measured and modeled ΔV_{T} as a function of V_{GSTR} but at fixed T and $t_{\mathrm{STR}} = 1$ Ks, for changes in the (a) Ge% in the QW, (b) QW thickness, and (c) Si cap thickness. The ΔV_{T} magnitude reduces and VAF increases as Ge% increases in the SiGe QW, the Si cap thickness reduces, and the QW thickness increases (however, the QW thickness change impact is not large).

Note that ΔV_{T} is due to ΔV_{IT} and ΔV_{OT} subcomponents (except in D1 when the ΔV_{HT} subcomponent also contributes), although it is ΔV_{IT} that dominates the overall ΔV_{T} in all devices. The VAF is lower for the Si device due to relatively higher ΔV_{HT} (note, ΔV_{HT} has lower VAF) and relatively lower ΔV_{OT} (note, ΔV_{OT} has higher VAF) contribution. The higher VAF for SiGe devices is due to negligible ΔV_{HT} and relatively higher ΔV_{OT} contribution. Higher VAF of overall ΔV_{T} related

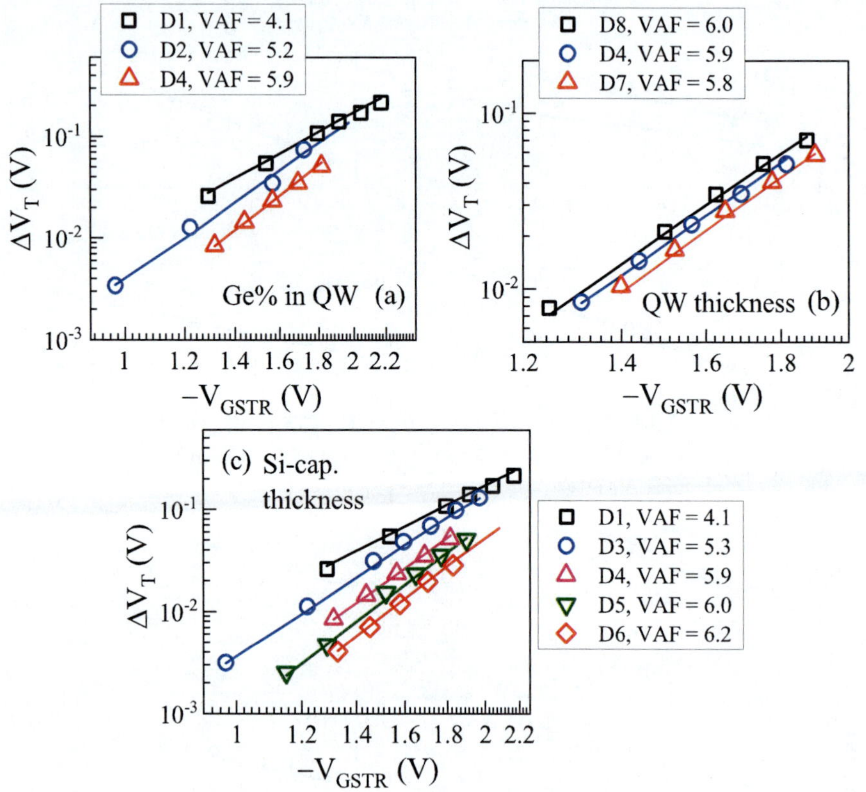

Fig 8.6 Measured and modeled fixed time ΔV_T as a function of V_{GSTR} for the reference Si and Si-capped SiGe devices having different (a) Ge% in the QW, (b) different QW thickness, and (c) different cap thicknesses. Stress T is 125 °C. Symbols: experiment, lines: model calculation. Data from [8]

to SiGe process changes is primarily due to higher relative ΔV_{OT} contribution. The reduction in ΔV_T magnitude (note, higher VAF is always linked to lower ΔV_T) reduces the stress reduction effect, which also increases the VAF of ΔV_{IT}, ΔV_{OT}, and overall ΔV_T.

Figure 8.7 shows the measured and modeled (a, c, e) time evolution of ΔV_T at fixed V_{OV} but different T, and (b, d, f) fixed time ($t_{STR} = 1$Ks) ΔV_T as a function of T but fixed V_{OV} along with the underlying ΔV_{IT}, ΔV_{HT}, and ΔV_{OT} subcomponents, for the reference Si (D1) and selected SiGe devices with different Si cap thickness (D3, D6). ΔV_{IT} dominates the overall ΔV_T over the complete T range and for all devices. ΔV_{HT} is negligible in SiGe devices and is a minor component for Si reference. Note that the E_A for ΔV_{OT} and ΔV_{HT} is highest and lowest, respectively, and are process independent. However, the E_A for ΔV_{IT} is process dependent and increases with reduction in the Si cap thickness. The E_A of overall ΔV_T increases with reduction in

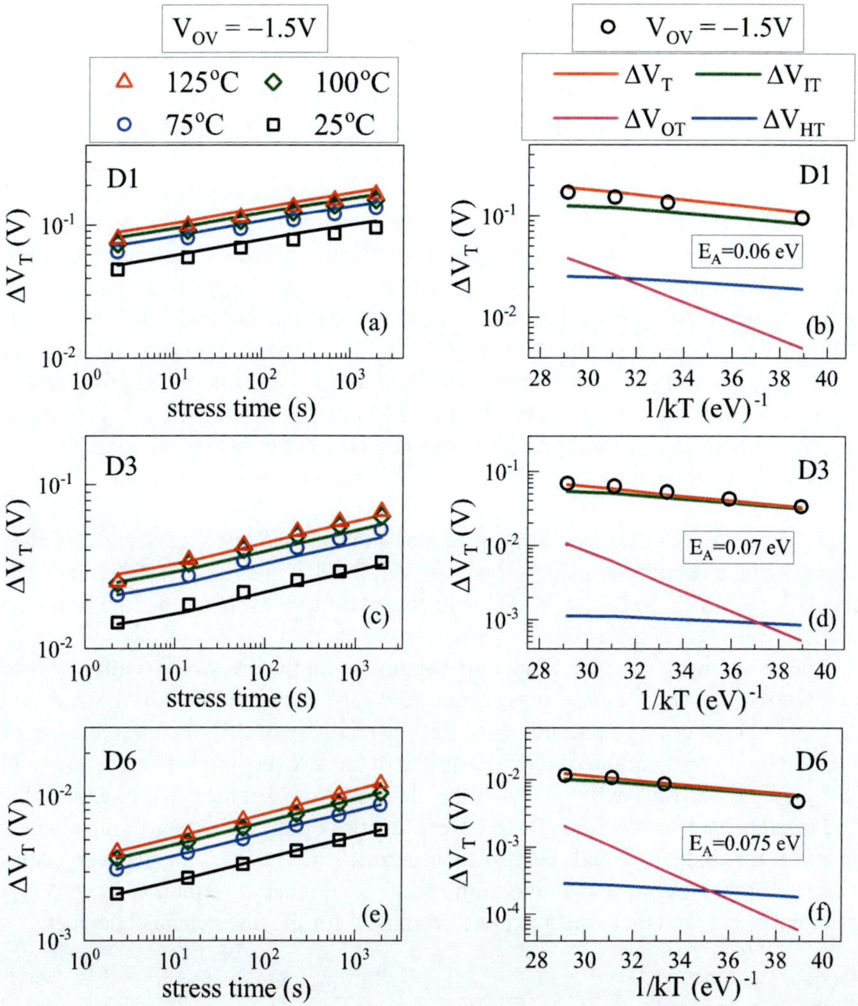

Fig 8.7 (a–c) Measured and modeled stress time kinetics of ΔV_{T} at different T, for Si and SiGe devices having different Si cap thickness. (d–f) Temperature dependence of fixed time ΔV_{T} for the same devices; also shown are different model subcomponents. Except D1, the ΔV_{HT} subcomponent is almost negligible for other devices and is not plotted. Symbols: experiment, lines: model calculation. Data from [8]

the Si cap thickness, due to increase of the same for ΔV_{IT} and the relatively higher ΔV_{OT} contribution.

8.4 Modeling of Recovery Kinetics

The recovery kinetics after DC stress is measured and modeled at different V_{GSTR}, T, and t_{STR} for the reference Si (D1) and SiGe (D3 and D6) channel devices having different Si cap thickness.

Figure 8.8 shows the time evolution of measured and modeled ΔV_T and the underlying ΔV_{IT}, ΔV_{HT}, and ΔV_{OT} subcomponents during recovery after stress in the D1 device, for fixed V_{GSTR} and T. The time kinetics of the subcomponents during recovery is explained in Fig. 8.2. ΔV_{HT} and ΔV_{OT}, respectively, show fast and slow recovery, while ΔV_{IT} recovery is distributed over an extended timescale. The overall ΔV_T recovery is governed by ΔV_{HT} and ΔV_{IT_FAST} at the short to medium time and by ΔV_{IT_SLOW} and ΔV_{OT} at longer time. Note that ΔV_{HT} is non-negligible for D1 but is negligible for the other (D2 through D8) devices. Moreover, ΔV_{OT} is small in D1 and is relatively larger for the other devices. Nevertheless, the overall ΔV_T recovery is dominated by ΔV_{IT} for all devices in the range of V_{GSTR} and T used in these experiments.

Figure 8.9 shows the time evolution of measured and modeled ΔV_T recovery after stress for (a, c, e) different V_{GSTR} (or V_{OV}) but fixed T and (b, d, f) different T but fixed V_{GSTR}, in (a, b) D1, (c, d) D3, and (e, f) D6 devices. Note that the first data point for all cases is measured at ~2 ms.

The modeling is done by simulating the stress and then recovery (with first data point at 1 μs), which takes into account the "gap" between the end of stress and start of actual recovery measurements, *i.e.*, the recovery of ΔV_T before the onset of measurement. As mentioned before, early part of the recovery is due to ΔV_{IT_FAST} and ΔV_{HT} for D1 and only ΔV_{IT_FAST} for other devices. Note that the gap is largest for D1 and smaller for other devices. The recovery kinetics of the ΔV_{IT_SLOW} component is identical for all devices, and ΔV_{OT} is semi-permanent. The long time recovery slows down at higher V_{GSTR} and T due to the relatively higher contribution from ΔV_{OT} subcomponent, and this aspect is more prominent for the SiGe channel devices.

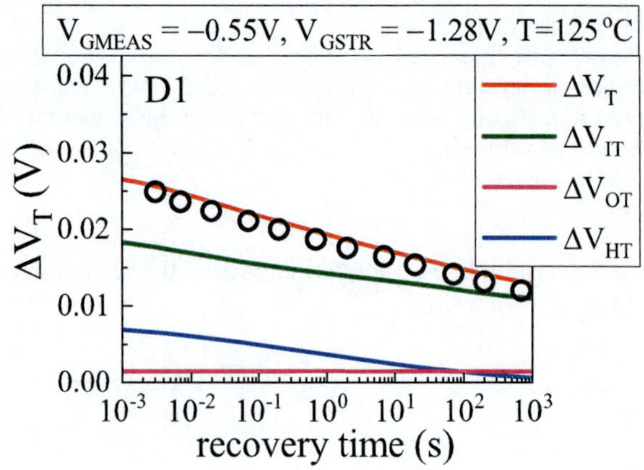

Fig 8.8 Measured and modeled ΔV_T recovery time kinetics with model subcomponents for the D1 device, stress and recovery T is 125 °C. Symbols: experiment, lines: model calculation. Data from [8]

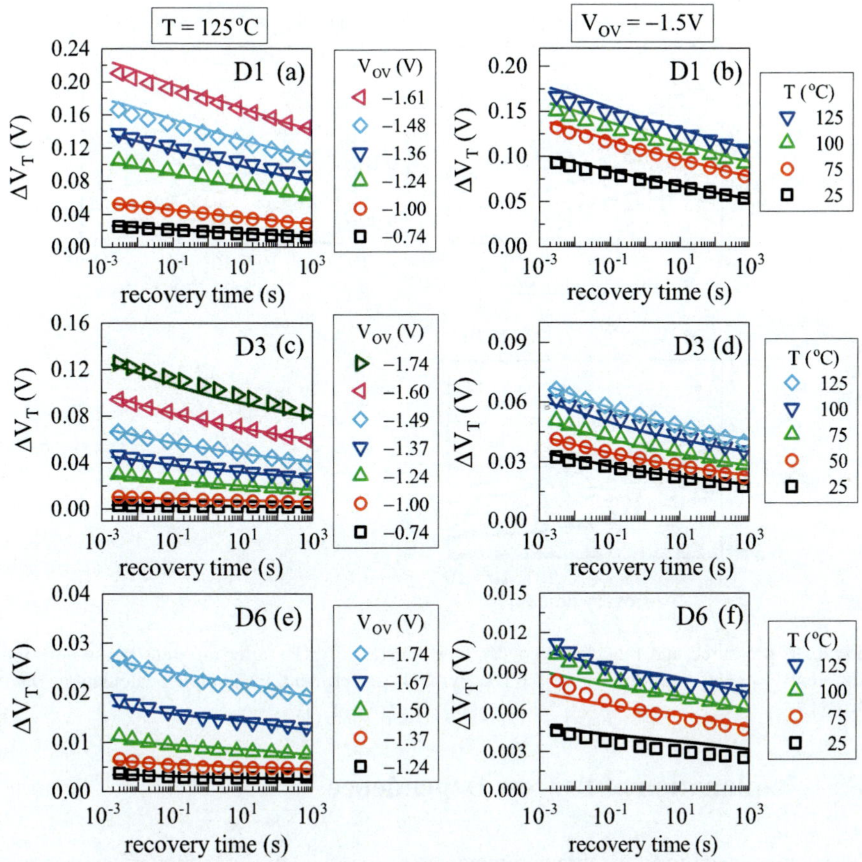

Fig 8.9 Measured and modeled recovery time kinetics of ΔV_T for different V_{OV} (left) and T (right), for Si and SiGe devices having different Si cap thickness. Symbols: experiment, lines: model calculation. Data from [8]

The time evolution of measured and modeled ΔV_T recovery at different t_{STR} is shown for the D1 (Fig. 8.10), D3 (Fig. 8.11), and D6 (Fig. 8.12) devices. In each of these plots, each of the subpanels (a) through (c) correspond to different V_{GSTR}/T conditions. As listed in Table 8.2, only two additional parameters are needed to explain the recovery kinetics for the range of V_{GSTR}, T, and t_{STR} conditions shown in Fig. 8.9 through Fig. 8.12. These process-dependent parameters are f_{FAST} and τ_{EC} related to the electron capture process (TTOM), also see Chap. 5, Sect. 5.3. All the other parameters are only adjusted during stress and are not re-adjusted during recovery.

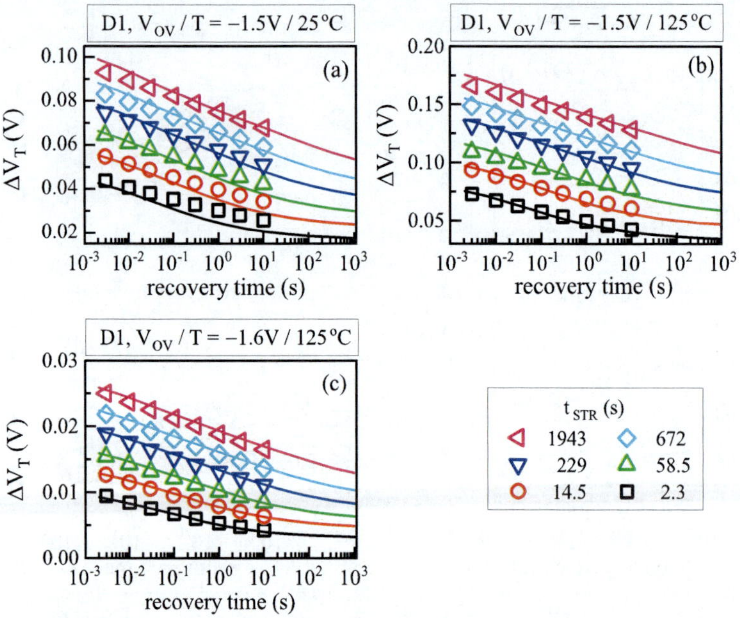

Fig 8.10 Measured and modeled recovery time kinetics of ΔV_T after different stress time for different V_{GSTR}/T conditions in device D1. Symbols: experiment, lines: model calculation. Data from [8]

8.5 Explanation of Process Dependence

The device-specific adjustable parameters used to model the DC stress-recovery time kinetics for different processes and diverse range of experimental conditions are listed in Table 8.2. Only the N_{0HT} parameter of the ABDWT model, which is related to the density of preexisting traps, is varied across processes. However, as mentioned before, ΔV_{HT} is only appreciable in the D1 reference Si device and is negligible in all SiGe devices. The relative contribution of ΔV_{OT} is small for D1 but increases for the other SiGe devices. Only the pre-factor K_{F30} is varied across processes, which is linked to the AHI mechanism, see Chap. 6, Sect. 6.2.

The measured ΔV_T is dominated by the ΔV_{IT} subcomponent for all processes and across different V_{GSTR} and T. The parameters related to the RD model (K_{F10}, E_{AFK1}, Γ_0 and α) and TTOM (f_{FAST}) are varied across processes. Note, the TTOM parameter τ_{EC} is same across devices, while f_{FAST} becomes higher for thinner Si cap and higher Ge% in the QW, while it does not vary with QW thickness.

The SiGe process dependence of RD model parameters impact the inversion layer hole and E_{OX} induced X–H bond dissociation at the Si cap/IL interface, see Chap. 4, Sect. 4.3 and Fig. 4.5 (reproduced below in Fig. 8.13). The inversion layer holes tunnel to the interfacial X–H bonds get captured and make them weak, and the weak bonds subsequently get dissociated by thermal activation. The pre-factor (K_{F10}) depends on

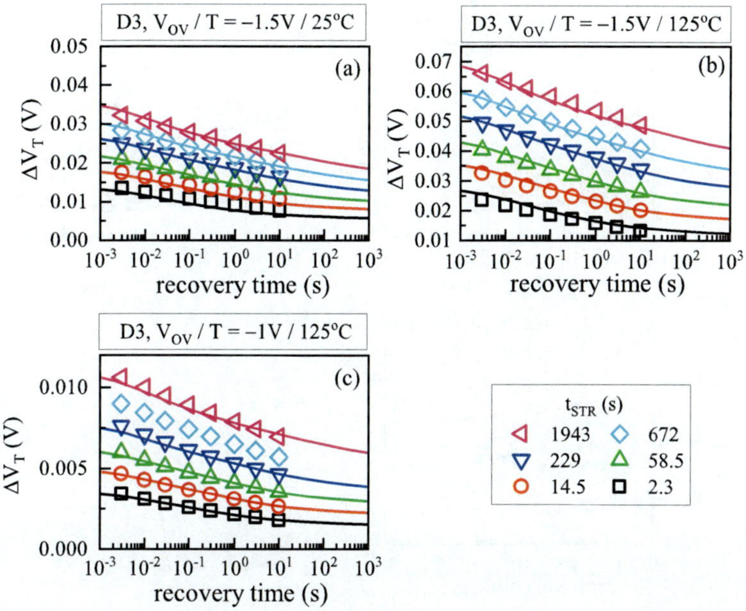

Fig 8.11 Measured and modeled recovery time kinetics of ΔV_T after different stress time for different V_{GSTR}/T conditions in device D3. Symbols: experiment, lines: model calculation. Data from [8]

the inversion layer hole density (p_H), tunneling coefficient (T_H), and capture cross section (σ); T_H depends on the tunneling effective mass (m_T) and tunneling barrier (φ_B). The T-independent field acceleration factor (Γ_0) depends on φ_B and m_T. The overall field acceleration factor (Γ_E) depends on both Γ_0 and α.

8.5.1 Impact of Si Cap Thickness

Note that the inversion holes are primarily confined to the SiGe QW in Si-capped SiGe devices. The inversion hole confinement is due to the valence band offset between the Si and SiGe layers (note, the bandgap of SiGe is smaller than Si), which is illustrated using the energy band diagram in Fig. 8.14 (the schematic of channel and HKMG gate insulator stack of Table 8.1 is reproduced for reference). Therefore, it is expected that the tunneling of holes from the SiGe QW to the Si cap/IL interface would be higher with reduction in the Si cap thickness, and ΔV_T should increase. However, the measured ΔV_T reduces for lower Si cap thickness, which is explained below.

Note that Ge out-diffusion from the SiGe QW toward the gate insulator during source-drain anneal in a GF integration would be higher in devices having thinner Si cap. As a consequence, the Si cap/IL interface would be more Ge rich for thinner

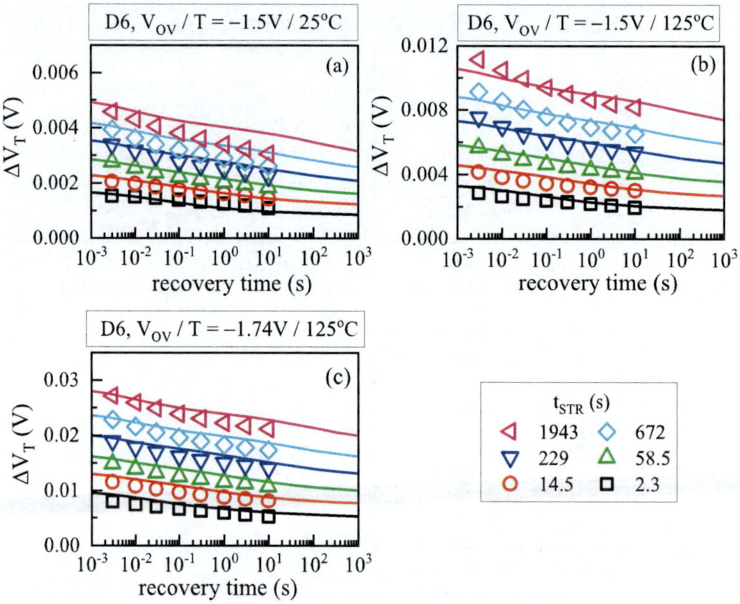

Fig 8.12 Measured and modeled recovery time kinetics of ΔV_T after different stress time for different V_{GSTR}/T conditions in device D6. Symbols: experiment, lines: model calculation. Data from [8]

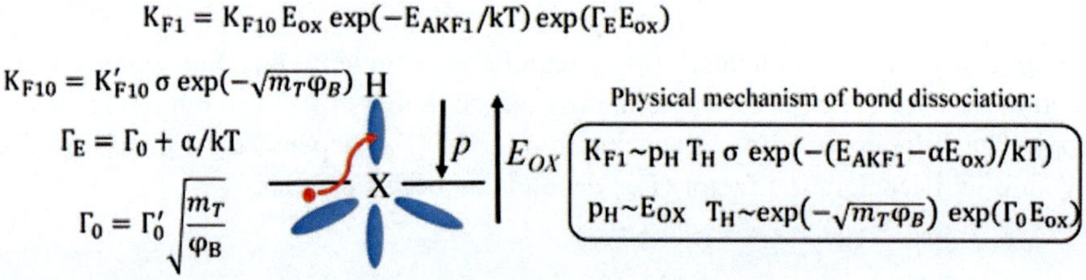

Fig 8.13 Schematic of the inversion layer hole and oxide electric field induced dissociation of H passivated defects at the Si cap/gate insulator interface, details in Chap. 4, Sect. 4.3

Fig 8.14 Schematic of the channel and gate stack and corresponding energy band diagram during stress for Si-capped SiGe device

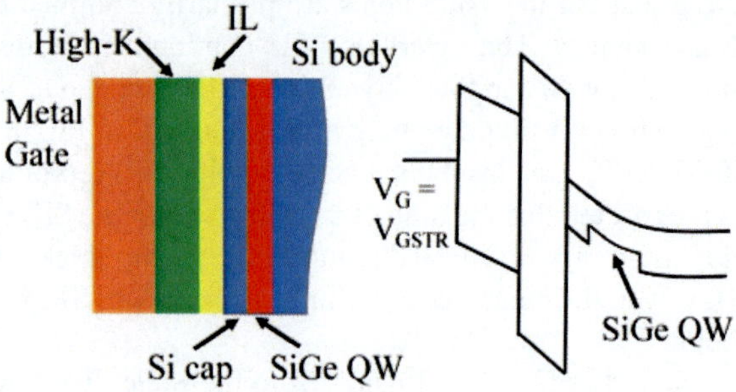

Si cap devices. Therefore, as per the discussion in Chap. 4, Sect. 4.3, the barrier φ_B would increase and the tunneling rate (T_H) would reduce according to Fig. 8.13. It is presumed that this aspect overcompensates the expected increase in T_H due to the thinner Si cap, and thereby lowers the magnitude of ΔV_{IT} hence ΔV_T.

Although E_{AKF1} increases, the T-activated ΔV_{IT} pre-factor K_{F10}*exp($-E_{AKF1}/kT$) reduces at lower Si cap thickness, due to the reduction in K_{F10} at higher φ_B. First principles calculations are needed to explain the impact of Ge% on E_{AKF1}, which is beyond the scope of this analysis. Note that higher φ_B also reduces the field dependence Γ_0 according to Fig. 8.13. However, higher Ge% in thinner Si cap devices would also affect other interfacial properties; *i.e.*, increase in the polarization factor α. Therefore, the overall field acceleration Γ_E ($= \Gamma_0 + \alpha/kT$) of ΔV_{IT} does not significantly change with Si cap thickness.

Although both ΔV_{IT} and ΔV_{OT} reduce at lower Si cap thickness, the reduction in ΔV_{IT} is much larger than ΔV_{OT}. Therefore, the relatively higher ΔV_{OT} contribution helps in increasing the VAF of overall ΔV_T. Furthermore, the stress reduction-related self-saturation effect becomes smaller since ΔV_T reduces for thinner Si cap devices, and results in higher VAF for both ΔV_{IT} and ΔV_{OT} (even if the related Γ_E values are not changed) and hence of overall ΔV_T, which experimentally demonstrated in Fig. 8.6(c).

8.5.2 Impact of SiGe QW Thickness

As mentioned before, the inversion holes are primarily confined to the SiGe QW in Si-capped SiGe devices. The subband energy levels inside the SiGe QW would shift up in energy (toward the valence band of Si) at lower QW thickness due to stronger quantization. This would increase the hole tunneling probability through the Si cap and therefore would increase the ΔV_{IT} subcomponent and overall ΔV_T. The T-activated ΔV_{IT} pre-factor K_{F10}*exp($-E_{AKF1}/kT$) increases at lower SiGe QW thickness due to lowering of the effective barrier height caused by the increase in the subband energy levels. However, the QW thickness change impact is smaller compared to other process changes, see Fig. 8.6(b). Moreover, the parameters Γ_0 and α remain constant across different QW thickness, which explains the invariance of Γ_E and therefore VAF, see Fig. 8.6(b). The E_{AKF1} also does not vary with variation in QW thickness.

8.5.3 Impact of Ge% in SiGe QW

Higher Ge% in the QW increases the valence band offset between Si and SiGe and lowers the hole tunneling probability and K_{F10}. Moreover, higher Ge% in the QW would also result in higher Ge out-diffusion into the Si cap and hence reduce K_{F10} (due to the reasons mentioned in Sect. 8.5.1). As a consequence, although the T

activation E_{AKF1} increases, the T-activated ΔV_{IT} pre-factor $K_{F10}*\exp(-E_{AKF1}/kT)$ reduces at higher Ge% in the QW. Moreover, higher Ge in the Si cap also reduces Γ_0 but increases α and therefore Γ_E remains almost unchanged. Note that although both ΔV_{IT} and ΔV_{OT} reduce at higher Ge% in the QW, the reduction in ΔV_{IT} is much larger than ΔV_{OT}. Therefore, the relatively higher ΔV_{OT} contribution helps in increasing the VAF of overall ΔV_T. Moreover, the stress reduction-related self-saturation effect is smaller since ΔV_T reduces in higher Ge% devices, which also helps in increasing the VAF of both ΔV_{IT} and ΔV_{OT} and overall ΔV_T, which experimentally demonstrated in Fig. 8.6(a). Note that the impact of higher Ge% in the QW on $\Delta V_{IT,}$ ΔV_{OT}, and overall ΔV_T is similar to that for the reduction in Si cap thickness described earlier in this section.

8.6 Estimation of EOL Degradation

Figure 8.15 shows the measured and modeled ΔV_T at fixed, short t_{STR} under accelerated stress as a function of (a) Ge% in the SiGe QW, (b) QW thickness, and (c) Si cap thickness. The model calculated ΔV_T projection to end of life (EOL) of 10 years under lower V_G operating condition is also shown for each of the cases. The EOL ΔV_T is highest for the Si reference (D1) and lowest for the thinnest Si cap SiGe (D6) devices. In SiGe devices, the EOL ΔV_T reduces with increase in Ge% in the QW,

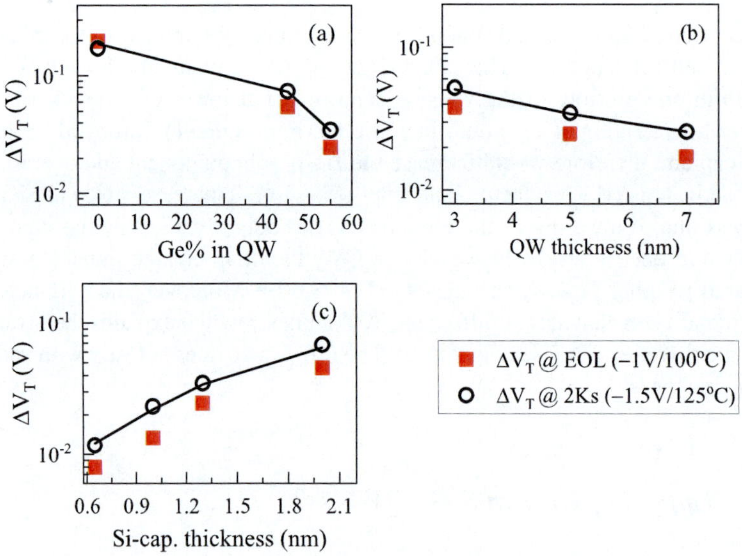

Fig 8.15 Measured and modeled fixed time ΔV_T at short time under accelerated stress and model projected ΔV_T at EOL under operating condition for Si-capped SiGe devices having different (a) Ge% in QW, (b) QW thickness, and (c) Si cap thickness. Data from [8]

increase in QW thickness, and reduction in Si cap thickness. Note that higher VAF of SiGe (compared to Si) channel devices, observed for some process changes as discussed above, aids in reducing the projected ΔV_T at low-V_G operating condition.

8.7 Summary

The experimental ΔV_T kinetics (from [8]) during and after DC NBTI stress in GF HKMG Si and Si-capped SiGe p-MOSFETs is modeled using a comprehensive physics-based framework having uncorrelated contributions from ΔV_{IT}, ΔV_{HT}, and ΔV_{OT} subcomponents. All the subcomponents and hence ΔV_T reduce, while VAF increases, for SiGe devices (compared to Si reference), especially so for lower Si cap thickness and higher Ge% in the SiGe QW, while the impact of QW thickness change remains small. Overall ΔV_T is dominated by ΔV_{IT} for different processes and stress conditions (V_{GSTR} and T), including that at EOL under operating condition. The ΔV_{HT} contribution remains negligible, while the fractional ΔV_{OT} contribution increases for SiGe devices. The impact of process changes on NBTI can be attributed to Ge segregation at the Si cap/IL interface and the related changes in ΔV_{IT} and ΔV_{OT} subcomponents, which exacerbated for lower Si cap thickness and higher Ge% in the QW. Increased hole tunneling for thinner QW devices also slightly increases ΔV_{IT} and overall ΔV_T.

Acknowledgements All experimental data presented in this chapter are re-plotted from a previously published report. The authors would like to acknowledge IMEC for providing the measurement data used in this chapter, and Jacopo Franco for useful discussion. Karansingh Thakor is acknowledged for help with manuscript preparation.

References

1. N. Kimizuka, K. Yamaguchi, K. Imai, T. Iizuka, C.T. Liu, R.C. Keller, T. Horiuchi, in *Symposium on VLSI Technology Digest of Technical Papers*, p. 92 (2000).
2. V. Huard, M. Denais, F. Perrier, N. Revil, C. Parthasarathy, A. Bravaix, E. Vincent, Microelectron. Reliab. **45**, 83 (2005)
3. S. Mahapatra, K. Ahmed, D. Varghese, A.E. Islam, G. Gupta, L. Madhav, D. Saha, M.A. Alam, in *IEEE International Reliability Physics Symposium Proceedings*,p. 1 (2007).
4. Y. Mitani, H. Satake, A. Toriumi, IEEE Trans. Device Mater. Reliab. **8**, 6 (2008)
5. S. Pae, M. Agostinelli, M. Brazier, R. Chau, G. Dewey, T. Ghani, M. Hattendorf, J. Hicks, J. Kavalieros, K. Kuhn, M. Kuhn, J. Maiz, M. Metz, K. Mistry, C. Prasad, S. Ramey, A. Roskowski, J. Sandford, C. Thomas, J. Thomas, C. Wiegand, J. Wiedemer, in *IEEE International Reliability Physics Symposium Proceedings*, p. 352 (2008).
6. S. Krishnan, U. Kwon, N. Moumen, M.W. Stoker, E.C.T. Harley, S. Bedell, D. Nair, B. Greene, W. Henson, M. Chowdhury, D.P. Prakash, E. Wu, D. Ioannou, E. Cartier, M.-H. Na, S. Inumiya, K. Mcstay, L. Edge, R. Iijima, J. Cai, M. Frank, M. Hargrove, D. Guo, A. Kerber, H. Jagannathan, T. Ando, J. Shepard, S. Siddiqui, M. Dai, H. Bu, J. Schaeffer, D. Jaeger, K. Barla, T.

Wallner, S. Uchimura, Y. Lee, G. Karve, S. Zafar, D. Schepis, Y. Wang, R. Donaton, S. Saroop, P. Montanini, Y. Liang, J. Stathis, R. Carter, R. Pal, V. Paruchuri, H. Yamasaki, J.-H. Lee, M. Ostermayr, J.-P. Han, Y. Hu, M. Gribelyuk, D.-G. Park, X. Chen, S. Samavedam, S. Narasimha, P. Agnello, M. Khare, R. Divakaruni, V. Narayanan, M. Chudzik, in *IEEE International Electron Devices Meeting Technical Digest*, 28.1.1 (2011).
7. K. Joshi, S. Hung, S. Mukhopadhyay, V. Chaudhary, N. Nanaware, B. Rajamohanan, T. Sato, M. Bevan, A. Wei, A. Noori, B. McDougal, C. Ni, G. Saheli, C. Lazik, P. Liu, D. Chu, L. Date, S. Datta, A. Brand, J. Swenberg, S. Mahapatra, in *IEEE International Reliability Physics Symposium Proceedings*, 4C.2.1 (2013).
8. J. Franco, B. Kaczer, P.J. Roussel, J. Mitard, M. Cho, L. Witters, T. Grasser, G. Groeseneken, IEEE Trans. Electron Devices **60**, 396 (2013)
9. P. Srinivasan, J. Fronheiser, K. Akarvardar, A. Kerber, L.F. Edge, R.G. Southwick, E. Cartier, H. Kothari, in *IEEE International Reliability Physics Symposium Proceedings*, 6A.3.1 (2014).
10. N. Parihar, N. Goel, S. Mukhopadhyay, S. Mahapatra, IEEE Trans. Electron Devices **65**, 392 (2018)
11. N. Parihar, S. Mahapatra, in *IEEE International Reliability Physics Symposium Proceedings*, TX.5.1 (2018).
12. V. Huard, C. Ndiaye, M. Arabi, N. Parihar, X. Federspiel, S. Mhira, S. Mahapatra, A. Bravaix, in *IEEE International Reliability Physics Symposium Proceedings*, TX.4.1 (2018).
13. N. Parihar, R. Tiwari, C. Ndiaye, M. Arabi, S. Mhira, H. Wong, S. Motzny, V. Moroz, V. Huard, S. Mahapatra, in *International Conference on Simulation of Semiconductor Processes and Devices*, p. 167 (2018).
14. S. Ramey, A. Ashutosh, C. Auth, J. Clifford, M. Hattendorf, J. Hicks, R. James, A. Rahman, V. Sharma, A. St. Amour, C. Wiegand, in *IEEE International Reliability Physics Symposium Proceedings*, 4C.5.1 (2013).
15. K. T. Lee, W. Kang, E-A Chung, G. Kim, H. Shim, H. Lee, H. Kim, M. Choe, N-I Lee, A. Patel, J. Park, J. Park, in *IEEE International Reliability Physics Symposium Proceedings*, 2D.1.1 (2013).
16. J. Franco, B. Kaczer, A. Chasin, H. Mertens, L.-A. Ragnarsson, R. Ritzenthaler, S. Mukhopadhyay, H. Arimura, P.J. Roussel, E. Bury, N. Horiguchi, D. Linten, G. Groeseneken, A. Thean, in *IEEE International Reliability Physics Symposium Proceedings*, 4B.2.1 (2016).
17. G. Jiao, M. Toledano-Luque, K.-J. Nam, N. Toshiro, S.-H. Lee, J.-S. Kim, T. Kauerauf, E. Chung, D. Bae, G. Bae, D.-W. Kim, K. Hwang, in *IEEE International Electron Devices Meeting Technical Digest*, 31.2.1 (2016).
18. N. Parihar, R. G. Southwick, U. Sharma, M. Wang, J. H. Stathis, S. Mahapatra, in *IEEE International Reliability Physics Symposium Proceedings*, 2D.4.1 (2017).
19. N. Parihar, R. Southwick, M. Wang, J.H. Stathis, S. Mahapatra, in *IEEE International Electron Devices Meeting Technical Digest*, 7.3.1 (2017).
20. N. Parihar, U. Sharma, R.G. Southwick, M. Wang, J.H. Stathis, S. Mahapatra, IEEE Trans. Electron Devices **65**, 23 (2018)
21. N. Parihar, R.G. Southwick, M. Wang, J.H. Stathis, S. Mahapatra, IEEE Trans. Electron Devices **65**, 1699 (2018)
22. N. Parihar, R.G. Southwick, M. Wang, J.H. Stathis, S. Mahapatra, IEEE Trans. Electron Devices **65**, 1707 (2018)
23. N. Parihar, R. Tiwari, S. Mahapatra, in *International Conference on Simulation of Semiconductor Processes and Devices*, p. 176 (2018).
24. R. Tiwari, N. Parihar, K. Thakor, H.Y. Wong, S. Motzny, M. Choi, V. Moroz, S. Mahapatra, IEEE Trans. Electron Devices **66**, 2086 (2019)
25. R. Tiwari, N. Parihar, K. Thakor, H.Y. Wong, S. Motzny, M. Choi, V. Moroz, S. Mahapatra, IEEE Trans. Electron Devices **66**, 2093 (2019)
26. N. Parihar, U. Sharma, R. G. Southwick, M. Wang, J. H. Stathis, S. Mahapatra, in *IEEE International Reliability Physics Symposium Proceedings* (2019).
27. M. Wang, J. Zhang, H. Zhou, R.G. Southwick, R. Hsin, K. Chao, X. Miao, V.S. Basker, T. Yamashita, D. Guo, G. Karve, H. Bu, in *IEEE International Reliability Physics Symposium Proceedings* (2019).

28. N. Choudhury, U. Sharma, H. Zhou, R.G. Southwick, M. Wang, S. Mahapatra, in *IEEE International Reliability Physics Symposium Proceedings* (2020).
29. H. Zhou, M. Wang, J. Zhang, K. Watanabe, C. Durfee, S. Mochizuki, R. Bao, R. Southwick, M. Bhuiyan, B. Veeraraghavan, *IEEE International Reliability Physics Symposium Proceedings* (2020).
30. N. Choudhury, T. Samadder, R. Tiwari, H. Zhou, R.G. Southwick, M. Wang, S. Mahapatra, in *IEEE International Reliability Physics Symposium Proceedings* (2021).
31. S. Mahapatra, N. Goel, S. Desai, S. Gupta, B. Jose, S. Mukhopadhyay, K. Joshi, A. Jain, A.E. Islam, M.A. Alam, IEEE Trans. Electron Devices **60**, 901 (2013)
32. A. E. Islam, N. Goel, S. Mahapatra, M. A. Alam, in *Fundamentals of Bias Temperature Instability in MOS Transistors* (Springer India, 2015), pp. 181–207.
33. S. Mukhopadhyay, K. Joshi, V. Chaudhary, N. Goel, S. De, R.K. Pandey, K.V.R.M. Murali, S. Mahapatra, in *IEEE International Reliability Physics Symposium Proceedings*, GD 3.1 (2014).
34. N. Choudhury, N. Parihar, N. Goel, A. Thirunavukkarasu, S. Mahapatra, IEEE J. Electron Devices Soc. **8**, 1281 (2020)
35. T. Samadder, N. Choudhury, S. Kumar, D. Kochar, N. Parihar, S. Mahapatra, IEEE Trans. Electron Devices **68**, 485 (2021)

Chapter 9
BAT Framework Modeling of Gate First HKMG Si and SiGe Channel FDSOI MOSFETs

Narendra Parihar, Tarun Samadder, Nilotpal Choudhury, Vincent Huard, and Souvik Mahapatra

9.1 Introduction

As shown in the previous chapters, negative-bias temperature instability (NBTI) became a crucial reliability issue during the migration from Silicon Dioxide (SiO_2) to Silicon Oxynitride (SiON) MOSFETs [1–4], and it continues to remain so for the dual-layer (SiO_2 or SiON Interlayer (IL) and Hafnium Dioxide (HfO_2) High-K layer) High-K metal gate (HKMG) gate insulator -based bulk [5–11] and Fully Depleted Silicon On Insulator (FDSOI) [12, 13] planar MOSFETs, bulk and SOI FinFETs [13–26], and Gate All Around Stacked Nanosheet (GAA-SNS) FETs [27–30], with either Silicon (Si) and Silicon Germanium (SiGe) channel.

The planar Fully Depleted Silicon On Insulator (FDSOI) technology is of interest for Internet of Things (IoT), Analog, and Radio Frequency (RF) applications, because of lower cost, ultra-low power dissipation, and excellent dynamic performance control [31–35]. Reduced source-drain capacitance and dynamic threshold voltage adjustment using the body bias are two important advantages of this technology in lowering the dynamic power dissipation. However, mechanical stress as a performance booster is necessary to take full advantage of this architecture down to 10nm node. Mechanical stress due to SiGe source-drain and SiGe channel is shown to improve the performance of FDSOI p-MOSFETs.

The important NBTI features are summarized below (reproduced from Chap. 3, Sect. 3.1). As shown earlier in Chap. 1, Sect. 1.3, NBTI results in a gradual buildup of positive gate insulator charges and shifts transistor parameters, *e.g.*, threshold voltage shift (ΔV_T), over time. It gets accelerated at more negative gate bias (V_G) during

N. Parihar · T. Samadder · N. Choudhury · S. Mahapatra (✉)
Department of Electrical Engineering, Indian Institute of Technology Bombay, Powai, Mumbai 400076, India
e-mail: souvik@ee.iitb.ac.in

V. Huard
Dolphin Design, 1 bis À chemin du pré carré, 38240 Meylan, France

S. Mahapatra (ed.), *Recent Advances in PMOS Negative Bias Temperature Instability*,
https://doi.org/10.1007/978-981-16-6120-4_9

stress ($V_G = V_{GSTR}$) and at higher temperature (T), governed by the voltage acceleration factor (VAF) and Arrhenius T activation energy (E_A), respectively. However, the positive charges accrued during stress reduce after the magnitude of V_G is reduced or removed ($V_G = V_{GREC}$ or 0V), which reduces ΔV_T. Therefore, NBTI for AC stress results in lower ΔV_T than DC stress. The AC to DC ratio depends on the pulse duty cycle (PDC) and pulse low bias (V_{GLOW}), and may or may not depend on the pulse frequency (f), since the f (in) dependence depends on the AC stress mode. On the other hand, NBTI recovery necessitates the use of ultra-fast methods for artifact-free measurements. As discussed in Chap. 1, Sect. 1.2, the ultra-fast Measure Stress Measure (MSM) method is used throughout this book.

Note that various reports have suggested reduced NBTI in SiGe channel-based planar bulk, FDSOI, and FinFET technologies, with increased germanium content (Ge%) in the channel [6, 8–11, 13, 18, 19, 21, 22]. The nitrogen content (N%) in the gate stack, on the other hand, increases NBTI, which is shown for Si channel bulk MOSFETs, SiGe channel FDSOI MOSFETs, and SiGe channel bulk FinFETs[10, 12, 13, 18, 19, 21, 22].

As shown earlier in Chap. 2, besides changing the magnitude of ΔV_T, Ge% in the channel and N% in the gate insulator impact the time dependence during and after stress, *i.e.*, the power-law time slope n at longer stress time (t_{STR}) and Fraction Remaining (FR) at a particular recovery time (t_{REC}) after stress (FR is defined as the ratio of ΔV_T at $t = t_{REC}$ to that at the end of stress at $t = t_{STR}$). These processes also impact the VAF, E_A, and T dependence of VAF. Increase in Ge% increases n at longer t_{STR}, FR for a particular ratio of t_{REC} to t_{STR} (*i.e.*, recovery becomes slower), VAF, E_A, and the T dependence of VAF (higher reduction in VAF at higher T). However, increase in N% reduces n at longer t_{STR}, FR for a particular ratio of t_{REC} to t_{STR} (*i.e.*, recovery becomes faster), VAF, E_A, and the T dependence of VAF (lower reduction in VAF at higher T). These features are universally observed in different device architectures.

In addition, changes in the compressive mechanical stress in the channel, due to changes in the layout or channel dimensions (length and width), have been shown to impact NBTI, in FDSOI MOSFETs [12, 13], FinFETs [23, 25], and GAA-SNS FETs [30]. Note, NBTI reduces with increased spacing between the Shallow Trench Isolation (STI) and device active, with reduction in fin length and width in FinFETs and sheet length in GAA-SNS FETs; however, it increases with the reduction in sheet width in GAA-SNS FETs.

As mentioned in Chap. 8, the physical mechanism responsible for the benefit of SiGe channel is debated. The benefit is suggested due to higher pre-stress interface trap density that reduces interface trap generation during stress [6], unfavorable valence band alignment resulting in either lower hole trapping [8, 16] or lower contribution from generated interface traps [9], and the lowering of oxide electric field (E_{OX}) due to large negative charge density near the valence band edge [17]. However, none of the reports have modeled the measured time kinetics of ΔV_T during and after stress, which is a prerequisite for the validation of any physical model.

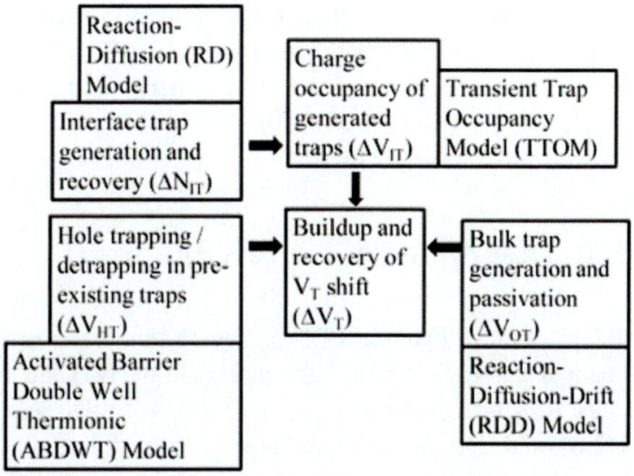

Fig. 9.1 Schematic of the BTI analysis tool (BAT) framework used in this book to model measured ΔV_T kinetics during and after DC and AC NBTI stress, reproduced from Chap. 4

Such modeling has been done recently, by using the BTI analysis tool (BAT) framework described in Chap. 4 through Chap. 6. The benefit of the SiGe channel is discussed in Chap. 8 for Si-capped SiGe planar bulk MOSFETs, in this chapter for SiGe FDSOI MOSFETs, and in Chap. 11 for SiGe bulk FinFETs. The impact of N% is discussed in Chap. 7 for Si channel bulk MOSFETs and in Chap. 11 for SiGe channel bulk FinFETs. The impact of mechanical stress due to changes in the layout is discussed in this chapter for SiGe FDSOI MOSFETs, and due to changes in the device dimensions in Chaps. 12 and 13 for SOI and SiGe bulk FinFETs and Si GAA-SNS FETs. The BAT model framework is briefly described hereinafter (reproduced from Chap. 6, Sect. 6.3).

Figure 9.1 illustrates the BAT framework [10] used to model the time kinetics of ΔV_T during and after stress. It uses uncorrelated contributions from generated interface (ΔV_{IT}) and bulk (ΔV_{OT}) gate insulator traps, and hole trapping in process-related preexisting bulk gate insulator traps (ΔV_{HT}). The reaction diffusion (RD) model is used to calculate the time kinetics of interface traps (density ΔN_{IT}) [10, 36, 37]. In Chap. 4, the RD model is described and independently validated by measured data from direct current IV (DCIV) method [38]. The transient trap occupancy model (TTOM) is used to calculate the occupancy of generated interface traps and their contribution (ΔV_{IT}) [10]. TTOM is described and validated in Chap. 5, Sect. 5.3. The ΔV_{HT} and ΔV_{OT} kinetics are modeled by the activated barrier double well thermionic (ABDWT) model [39] and reaction diffusion drift (RDD) model [40], respectively, and these are described and validated in Chap. 5, Sect. 5.4 and Chap. 6, Sect. 6.2.

In Chap. 7, the impact of N% on ΔV_T is shown to be due to increase in ΔV_{IT} and ΔV_{HT} (with relatively higher increase in ΔV_{HT}) and reduction in ΔV_{OT} subcomponents, although ΔV_{IT} dominates overall ΔV_T across V_{GSTR} and T. In Chap. 8, the impact of Ge% on ΔV_T is shown to be due to reduction in ΔV_{IT}, ΔV_{HT}, and ΔV_{OT} subcomponents (with relatively higher reduction in ΔV_{HT} and lower reduction in

ΔV_{OT}), although once again, ΔV_{IT} dominates overall ΔV_T across V_{GSTR} and T. The impacts of Ge% and N% changes on ΔV_T in FDSOI devices are also explained by the above mechanisms in this chapter. Furthermore, the impact of changes in mechanical strain (due to layout changes) on ΔV_T is also explained using the above framework.

9.2 Description of Process Splits

Table 9.1 describes the FDSOI devices analyzed in this chapter. The Si and SiGe channel devices are fabricated using 28nm (D0) and 14nm (D1 through D6) proprietary STMicroelectronics processes [31–35]. D0 and D1 feature Si channel, while D2 through D6 are SiGe channel devices. All devices have epitaxial SiGe source-drain. The resulting uniaxial compressive stress (UCS) in the channel is varied using different spacing (SA) between the device active (source-drain edge) and the STI, illustrated using Fig. 9.2 (a) [12, 13]. All devices feature HKMG gate insulator consisting of SiON based IL, Hafnium Silicate (HfSiO) for D0 and HfO_2 for D1–D6 based High-K layer, and Titanium Nitride (TiN) High-K cap, and fabricated using Gate First (GF) integration. Different TiN cap thicknesses (10 Å–45 Å) are used for the 14 nm node devices. Note that higher N% is introduced in the gate stack for thicker TiN capped devices during high T source-drain thermal annealing in the GF integration scheme.

Note that the use of SiGe channel and epitaxial SiGe source-drain causes UCS in the channel. Figure 9.2 (b) shows the stress in SiGe channel measured using the Nanobeam Diffraction (NBD) technique [12, 41]. It is observed that the magnitude of UCS reduces as the transistor channel comes closer to the edge of STI. As a further evidence, Fig. 9.2 (c) shows the stress profile along the channel for different SA, obtained from Technology CAD (TCAD) simulations [12]. It can be seen that UCS in channel reduces with the reduction in SA. Therefore, due to stress relaxation near the STI edge, lower SA devices have lower compressive strain (due to lower UCS) in the channel.

Table 9.1 Process description of the Si reference and SiGe channel FDSOI MOSFETs analyzed in this chapter, from [12, 13]

Device	Ge% in channel	N% in gate stack
D0	0	Low
D1	0	Low
D2	25	High
D3	30	High
D4	30	Medium
D5	30	Low
D6	34	Low

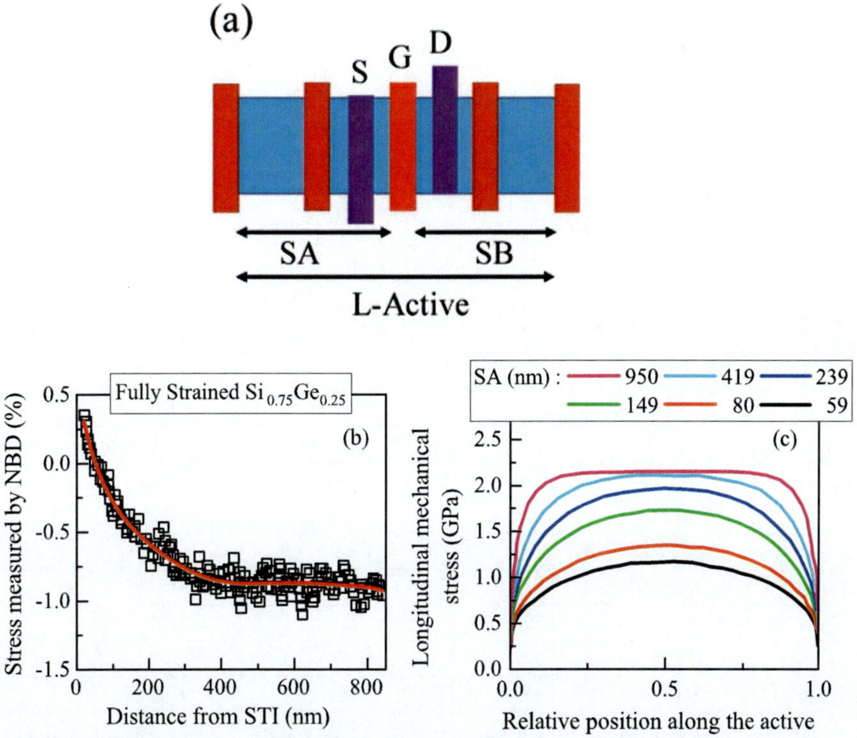

Fig. 9.2 (a) Schematic to illustrate the spacing (SA) between STI edge and active (S, G and D represent source, gate, and drain, L-active is total length of active). (b) Measured channel stress as a function of STI distance from NBD method. (c) Stress profiles along the active for various SA from TCAD simulation. Data from [12]

9.3 Modeling of Measured Data, Model Parameters

The time kinetics of ΔV_T during and after stress, measured using the one point drop down (OPDD) measure-stress-measure (MSM) method with measurement delay of 1 μs (see Chap. 1, Sect. 1.2), is analyzed in this section. The impact of extended stress T range on the stress and recovery time kinetics and VAF is shown for the device D0. The stress and recovery time kinetics are compared for different channel Ge% (0, 25, 30, and 34%), gate stack N% (low, medium, and high), and channel strain (SA: 59nm–419nm) for different processes (Ge% and N%) in the devices D1 through D6 (D1 is the reference).

As mentioned before, the measured ΔV_T time kinetics is modeled using uncorrelated contributions from ΔV_{IT}, ΔV_{HT}, and ΔV_{OT} subcomponents using the framework illustrated in Fig. 9.1. The model parameters are listed in Table 9.2, and their description is as follows (reproduced from Chap. 6, Sect. 6.3).

Table 9.2 Process-dependent RD, TTOM, ABDWT, and RDD parameters (as per classification done in Chap. 4 through Chap. 6) used in this chapter

Device	Unit	D0	D1	D2	D3
K_{F10}[a]	cm/Vs	–	1.0	0.53	1.67
E_{AKF1}	eV	0.41	0.41	0.45	0.50
Γ_0	cm/MV	0.22	0.35	0.20	0.15
α	qÅ	1.8	1.8	2.3	2.5
N_{0HT}[a]	1/cm^2	–	1.0	0.80	0.39
E_{BM}	eV	1.0	1.3	1.3	1.3
γ_B	C.cm	4.5×10^{-9}	5.0×10^{-9}	5.0×10^{-9}	5.0×10^{-9}
K_{F30}[a]	1/s	–	1	30	30

Device	Unit	D4	D5	D6
K_{F10}[a]	cm/Vs	13.3	66.7	6.7×10^3
E_{AKF1}	eV	0.55	0.64	0.80
Γ_0	cm/MV	0.12	0.08	0.03
α	qÅ	2.5	2.5	2.7
N_{0HT}[a]	1/cm^2	0.56	0.53	0.33
E_{BM}	eV	1.3	1.3	1.3
γ_B	C.cm	5.0×10^{-9}	5.0×10^{-9}	3.5×10^{-9}
K_{F30}[a]	1/s	35	36	3.5

See Table 9.1 for device details. [a]The parameters K_{F10}, N_{0HT}, and K_{F30} are not mentioned for D0, while for D1–D6 are normalized to the D1 device to maintain confidentiality, and they depend on SA. The other adjustable parameters are: $f_{FAST} = 0.44$ and $\tau_{EC} = 3 \text{ x } 10^{-2}$ s for TTOM, $m = 2.4$ for ABDWT for all devices, Γ_{0OT} (in cm/MV) = 0.13 for D0 and 0.31 for others, E_{AOT} (in eV) = 0.55 for D0 and 0.9 for others, $\alpha_{OT} = 3.6$ qÅ and $K_{F50} = 23$ cm^3/s for all devices, for RDD. See text for details

As described in Chap. 4, Sect. 4.3, the process-dependent RD model parameters are related to the pre-factor (K_{F10}), T-independent field acceleration (Γ_0), bond polarization (α), and T activation energy (E_{AKF1}) of the inversion layer hole-assisted bond dissociation process, see Chap. 4, Fig. 4.5. E_{AKF1} increases with higher Ge% and reduces with higher N%, the term K_{F10}*exp($-E_{AKF1}/kT$) reduces with higher Ge% and increases with higher N% for the SiGe devices, Γ_0 reduces with increase in Ge% but increases with increase in N%, while α increases at higher Ge% but does not change with N%. The process-dependent TTOM parameters are related to fast fraction of traps that undergo electron capture (f_{FAST}) and the electron capture time constant (τ_{EC}), see Chap. 5, Sect. 5.3, and do not change for process changes.

As described in Chap. 5, Sect. 5.4, the process-dependent ABDWT model parameters are related to the density of preexisting defects (N_{0HT}), the energy barrier (E_{BM}), and terms associated with E_{OX} dependence of the barrier (γ_B) and trap energy level ($\gamma_{E2} = m\gamma_B$), see Chap. 5, Fig. 5.10. The process-dependent RDD model parameters are related to the pre-factor (K_{F30}), T-independent field acceleration (Γ_{0OT}), and T

activation energy (E_{AOT}) of the anode hole injection (AHI)-assisted bond dissociation (note, the bond polarization factor (α_{OT}) is not varied across devices but listed for completeness), and the forward reaction rate for ions (K_{F50}), see Chap. 6, Sect. 6.2. Only N_{0HT} (also γ_B for D6) and K_{F30} are changed across different 14 nm node devices. Other model parameters are process agnostic and are listed in the respective sections of Chap. 4 through Chap. 6 (Tables 4.1, 5.1, 5.2, and 6.1).

The following aspects are considered in the modeling framework. Note that the bandgap, flatband voltage, and interface trap precursor density are different in different devices and depend on Ge% in the channel. The bandgap is also dependent on confinement due to the ultrathin body. Due to the difference in flatband voltage, the E_{OX} for a particular V_{GSTR} becomes different in different devices. However, the bulk trap precursor density is same for all devices. Note, although K_{F30} is higher for SiGe compared to Si channel devices, due to difference in E_{OX} at a given V_{GSTR}, the ΔV_{OT} contribution is lower for SiGe compared to the Si reference at low to moderate V_{GSTR} and T.

9.3.1 Stress and Recovery Kinetics Over Extended T *Range*

Extended T (–40°C to 165°C) stress and recovery measurements are performed on the device D0. Figure 9.3 shows the time evolution of measured and modeled ΔV_T and the underlying ΔV_{IT}, ΔV_{HT}, and ΔV_{OT} subcomponents during (left panels) and after (right panels) DC stress at low, medium, and high stress T but fixed V_{GSTR}. The ΔV_T stress time kinetics is usually plotted in a log–log scale, and it evolves rapidly at the initiation of stress and asymptotically settles into a power-law dependence with slope n at long time ($t_{STR} > 1s$). The ΔV_T recovery kinetics is usually plotted in a semi-log scale, and it shows logarithmic time dependence.

Note that the ΔV_{IT} and ΔV_{OT} subcomponents show power-law time dependence at longer stress time with slope n of ~ 1/6 and ~ 1/3, respectively, while ΔV_{HT} saturates ($n \sim 0$ in a log–log plot). The relative ΔV_{IT}, ΔV_{HT} and ΔV_{OT} contributions to overall ΔV_T change with changes in T at fixed V_{GSTR} due to differences in the T activation E_A associated with these subcomponents. The relative ΔV_{IT} and ΔV_{OT} contributions increase as T is increased (however, the relative increase in ΔV_{OT} is larger than that of ΔV_{IT} due to higher E_A of ΔV_{OT}), although the overall ΔV_T is always dominated by ΔV_{IT} for the range of T analyzed in these experiments. Similarly, the relative ΔV_{IT}, ΔV_{HT}, and ΔV_{OT} contributions to overall ΔV_T also change with changes V_{GSTR} at fixed T, due to differences in VAF (highest for ΔV_{OT} and lowest for ΔV_{HT}). These aspects are discussed in detail in Chap. 4 through Chap. 7.

The recovery of ΔV_{HT} and ΔV_{OT} subcomponents are fast and slow, respectively. The ΔV_{IT} recovery is observed over an extended time range and is due to two processes: fast recovery (ΔV_{IT_FAST}) by electron capture in the fraction of traps (determined by the parameter f_{FAST}) that go below the Fermi level as the magnitude of V_G is reduced during recovery after stress (this subcomponent shows negligible T

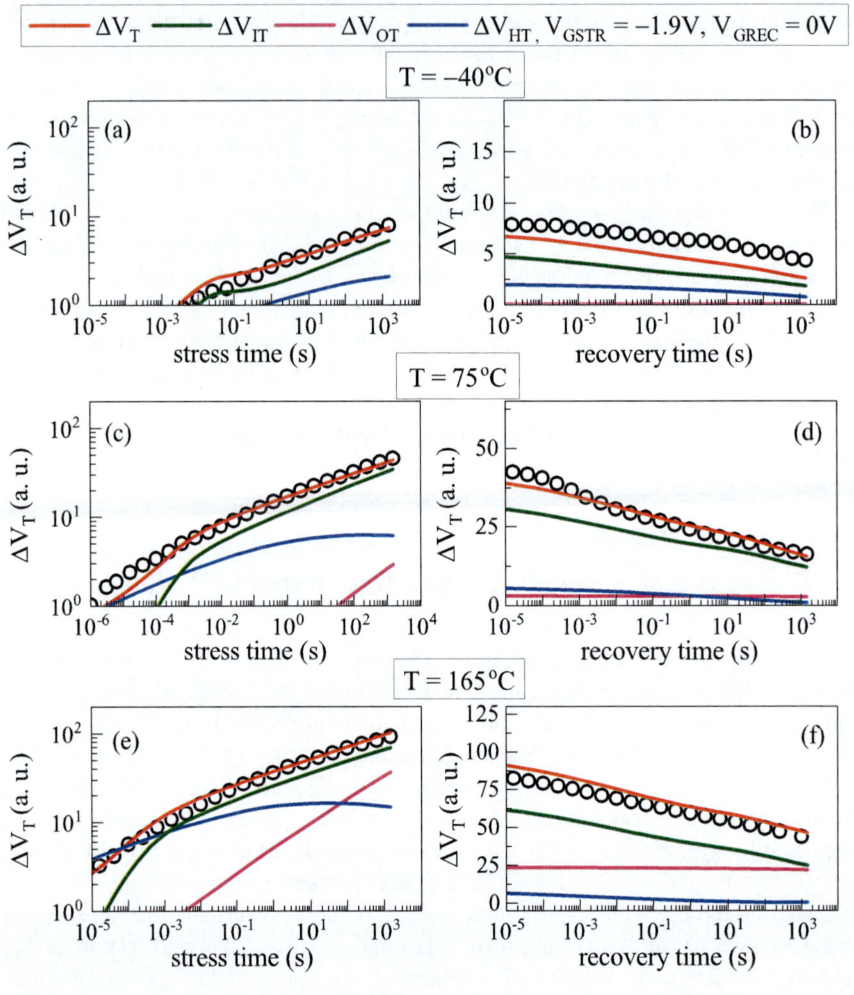

Fig. 9.3 Time evolution of measured and modeled ΔV_T together with the underlying model subcomponents during (left panels) and after (right panels) stress at (top) lowest T, (middle) medium T, and (bottom) highest T but fixed V_{GSTR} in device D0. Symbols: experiment, lines: model calculation. Data from [13]

dependence), and relatively slow recovery (ΔV_{IT_SLOW}) due to the re-passivation of the remaining traps (this subcomponent reduces at higher T), see Chap. 5, Sect. 5.3.

Figure 9.4 shows the time evolution of measured and modeled ΔV_T (a, c) during and (b, d) after DC stress for (a, b) different V_{GSTR} at fixed T, and (c, d) different T at fixed V_{GSTR}. Although the relative ΔV_{OT} contribution increases at higher V_{GSTR} and T, there is no noticeable change in the power-law slope n. It is expected that relatively higher ΔV_{OT} with $n \sim 1/3$ would increase the n of overall ΔV_T, see Chap. 3, Sect. 3.4. However, the stress reduction effect-induced soft saturation (*i.e.*, the reduction in the

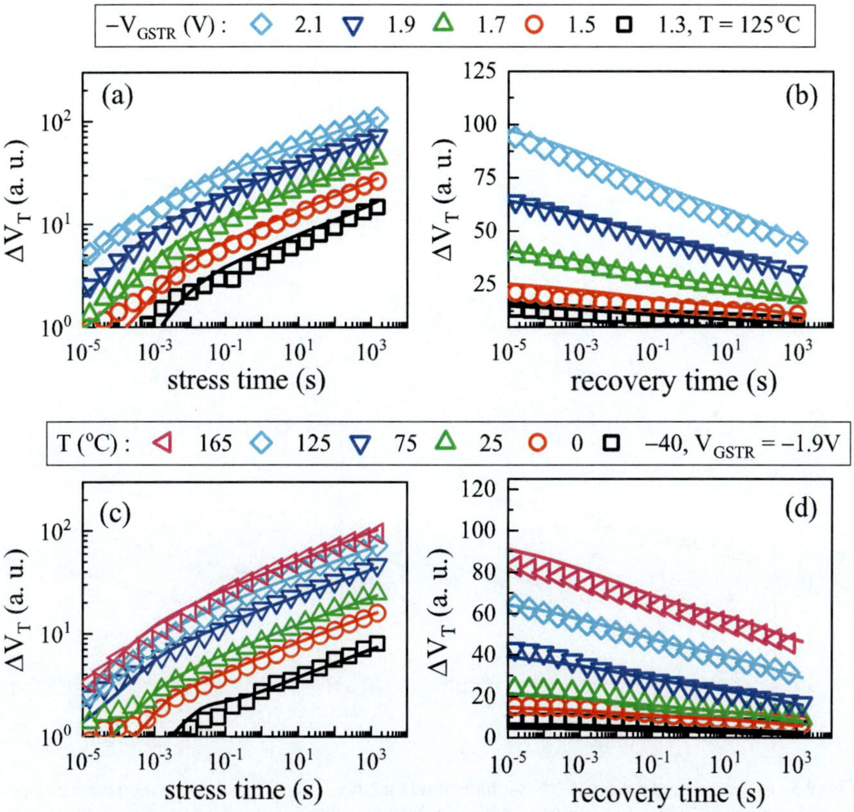

Fig. 9.4 Time evolution of measured and modeled ΔV_T (a, c) during and (b, d) after stress for (a, b) different V_{GSTR} but fixed T and (c, d) different T but fixed V_{GSTR} in device D0. Symbols: experiment, lines: model calculation. Data from [13]

effective electrical stress at higher ΔV_T) would have a counter effect, and therefore, n remains almost invariant across V_{GSTR} and T.

Figure 9.5 shows the V_{GSTR} dependence of measured and modeled ΔV_T at fixed time ($t_{STR} = 1$Ks) at different T and the underlying ΔV_{IT}, ΔV_{HT}, and ΔV_{OT} subcomponents (note that the range of V_{GSTR} is different for different panels). The ΔV_{IT} subcomponent dominates the overall ΔV_T for the range of V_{GSTR} and T used in this experiment. The ΔV_{HT} contribution is always lower than ΔV_{IT}, and their difference increases at higher T, since ΔV_{HT} has lower T activation as compared to ΔV_{IT}, see Chap. 3, Sect. 3.3. The ΔV_{OT} subcomponent is negligible at lower T but increases with increase in T, and it increases beyond ΔV_{HT} at highest T. Note that the T activation is largest for ΔV_{OT}, see Chap. 6, Sect. 6.3, *i.e.*, E_A (ΔV_{OT}) > E_A (ΔV_{IT}) > E_A (ΔV_{HT}). Moreover, the VAF of overall ΔV_T reduces with increase in T due to the reduction of VAF of the underlying ΔV_{IT} and ΔV_{OT} subcomponents. Recall that the bond polarization terms associated with ΔV_{IT} (forward reaction K_{F1}) and ΔV_{OT}

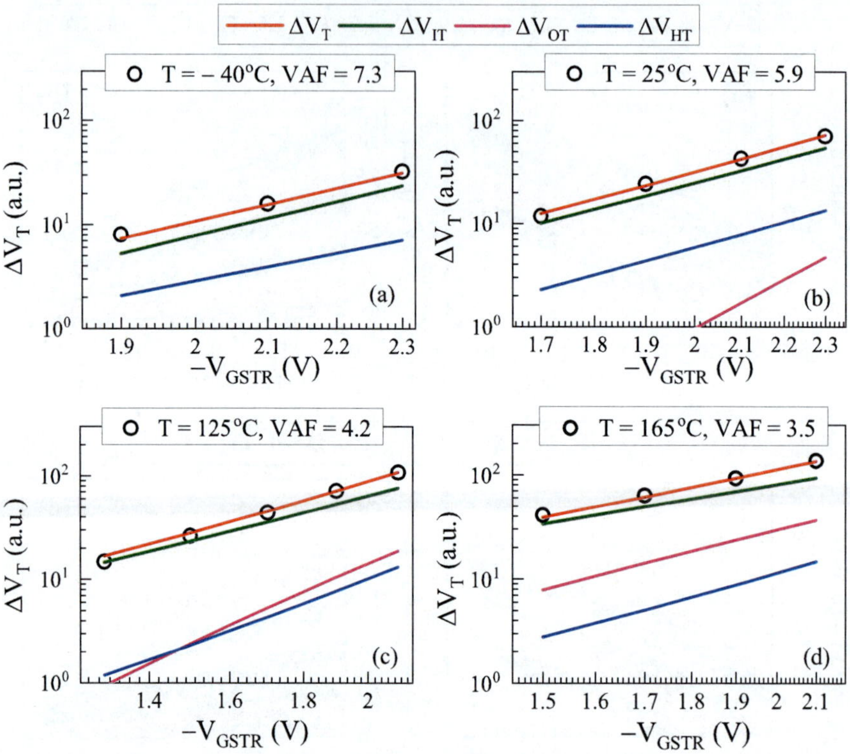

Fig. 9.5 Measured and modeled ΔV_T at fixed time together with the underlying model subcomponents as a function of V_{GSTR} at different stress T in device D0. Symbols: experiment, lines: model calculation. Data from [13]

(forward reaction K_{F3}) result in reducing the field acceleration terms for ΔV_{IT} and ΔV_{OT} subcomponents, see Chap. 4, Sect. 4.3 and Chap. 6, Sect. 6.2 for details. The stress reduction effect also plays a role in reducing the VAF at higher T.

9.3.2 Impact of Ge% and N%

The process-dependent studies presented in this subsection are performed on 14nm FDSOI devices (Table 9.1) having fixed SA. Figure 9.6 shows the time evolution of measured and modeled ΔV_T during DC stress on the reference Si device D1 and SiGe device D6 at different V_{GSTR}. For both cases, the underlying ΔV_{IT}, ΔV_{HT}, and ΔV_{OT} subcomponents for the dataset having lowest ΔV_T are also shown. The SiGe device shows significantly lower ΔV_T compared to the reference Si device at identical stress conditions. Although all the model subcomponents reduce for the SiGe device and ΔV_{IT} dominates the overall ΔV_T, the relative contribution from

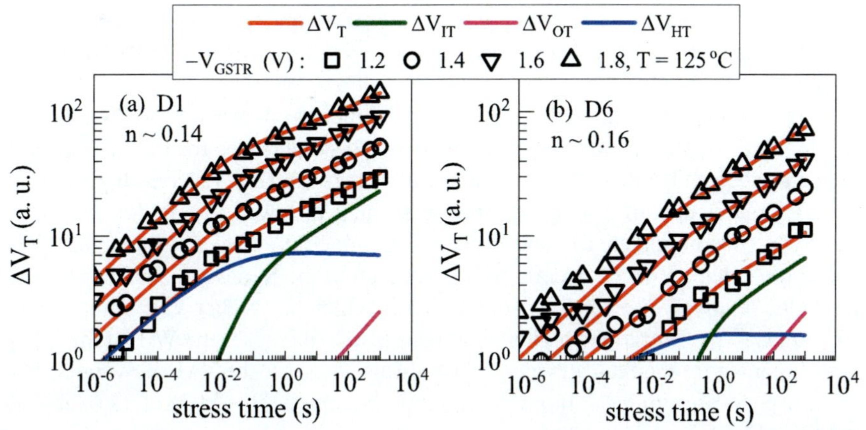

Fig 9.6 Measured and modeled time evolution of ΔV_T under different V_{GSTR} for (a) Si and (b) SiGe channel devices. The underlying model subcomponents are shown for the dataset having lowest V_{GSTR} in both panels. Symbols: experiment, lines: model calculation. Data from [13]

ΔV_{OT} becomes more while that from ΔV_{HT} becomes less, consistent with different technologies modeled in other chapters, see Chap. 8 and Chap. 11. The relatively higher ΔV_{OT} and significantly lower ΔV_{HT} contributions increase the long time slope n of overall ΔV_T for the SiGe device, due to reasons described in Chap. 3, Sects. 3.3 and 3.4.

Figure 9.7 shows the V_{GSTR} dependence of measured and modeled ΔV_T at fixed time ($t_{STR} = 1$Ks) along with the underlying subcomponents for Si (D1) and SiGe (D6) devices. For a particular device, the relative contributions from different

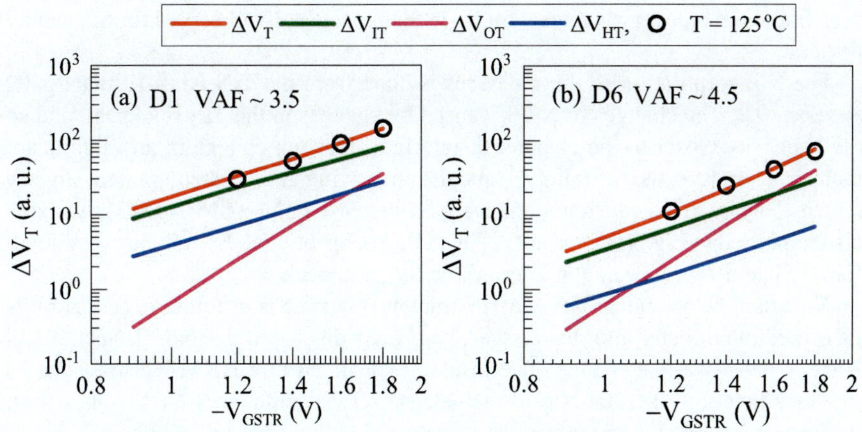

Fig. 9.7 Measured and modeled ΔV_T at fixed time together with the underlying model subcomponents as a function of V_{GSTR} for (a) Si and (b) SiGe channel devices. The VAF values are listed for the overall ΔV_T. Symbols: experiment, lines: model calculation. Data from [13]

subcomponents change as V_{GSTR} is varied, due to differences in VAF associated with the ΔV_{OT} (highest), ΔV_{IT} (moderately high), and ΔV_{HT} (lowest) subcomponents, see Chap. 6, Sect. 6.3.

Note that the contribution from ΔV_{HT} is significantly lower for the SiGe device for all values of V_{GSTR}. Although the relative ΔV_{OT} increases for the SiGe device and both ΔV_{IT} and ΔV_{OT} equally contribute to the overall ΔV_T at very high V_{GSTR}, the ΔV_T under operating bias (when extrapolated to lower V_G) is dominated by the ΔV_{IT} subcomponent for both Si and SiGe devices. Also note that the SiGe device has higher VAF of overall ΔV_T compared to the Si device. This is partially due to the relatively higher ΔV_{OT} contribution, which has higher VAF compared to ΔV_{IT} and ΔV_{HT}. Moreover, higher bond polarization increases the VAF of the ΔV_{IT} subcomponent for the SiGe device, and would also increase the VAF of overall ΔV_T. These aspects are similar to those observed in Si-capped SiGe MOSFETs (Chap. 8) and SiGe FinFETs (Chap. 11).

Figure 9.8 shows the time evolution of measured and modeled ΔV_T during and after DC stress along with the underlying subcomponents for devices D2 through D6 at fixed V_{GSTR} and T. The overall ΔV_T reduces with increase in Ge% (D2–D3–D6), which is re-plotted for clarity in Fig. 9.9. Note that the relatively higher ΔV_{OT} contribution is observed due to SiGe channel and the use of higher V_{GSTR} and T in these experiments. However, due to its higher VAF and E_A, the ΔV_{OT} contribution reduces at lower V_{GSTR} and/or T, and ΔV_{IT} dominates overall ΔV_T for all SiGe devices, see Sect. 9.5. Although the overall ΔV_T and the underlying ΔV_{IT} and ΔV_{OT} subcomponents reduce in absolute terms (ΔV_{HT} is very small for all SiGe devices), the relative contribution from ΔV_{OT} increases with increase in Ge%. The relatively higher ΔV_{OT} and very small contribution from ΔV_{HT} result in higher slope n during stress, and also slow down the long-time recovery after stress, in SiGe channel devices. Note that the recovery tail at long t_{REC} is controlled by the ΔV_{OT} subcomponent (which is semi-permanent), and hence, the FR is higher (*i.e.*, fractional recovery is lower) for SiGe than that usually observed for Si channel devices.

The N incorporation in these devices is done from the TiN High-K metal gate, see Sect. 9.2. The changes in N% is caused by changes in the TiN thickness, and so the N density closer to the channel/IL interface does not change in any significant manner. Therefore, the overall ΔV_T and the underlying ΔV_{IT} subcomponent slightly increase, while ΔV_{OT} does not change with increase in N% (D5-D4-D3), which is also re-plotted in Fig. 9.9 for clarity. The ΔV_{HT} contribution also remains very small for all SiGe devices, even at higher N% in the gate stack.

Note that the change in N% via TiN thickness change is different from that by N incorporation directly into the IL, see Chap. 7. As discussed in Chap. 2, Sect. 2.1, it is the N density at the channel/gate insulator (IL in case for HKMG gate insulator) interface (and not the total N% in the gate stack) that influences NBTI. Therefore, the impact of N% is lower in this case compared to that shown in Chap. 7.

Figure 9.10 shows the time evolution of measured and modeled ΔV_T recovery after DC stress for device D5, in (a) with underlying subcomponents but at a fixed V_{GSTR} and T, and in (b) for multiple V_{GSTR} but fixed T. Note that the FR is higher

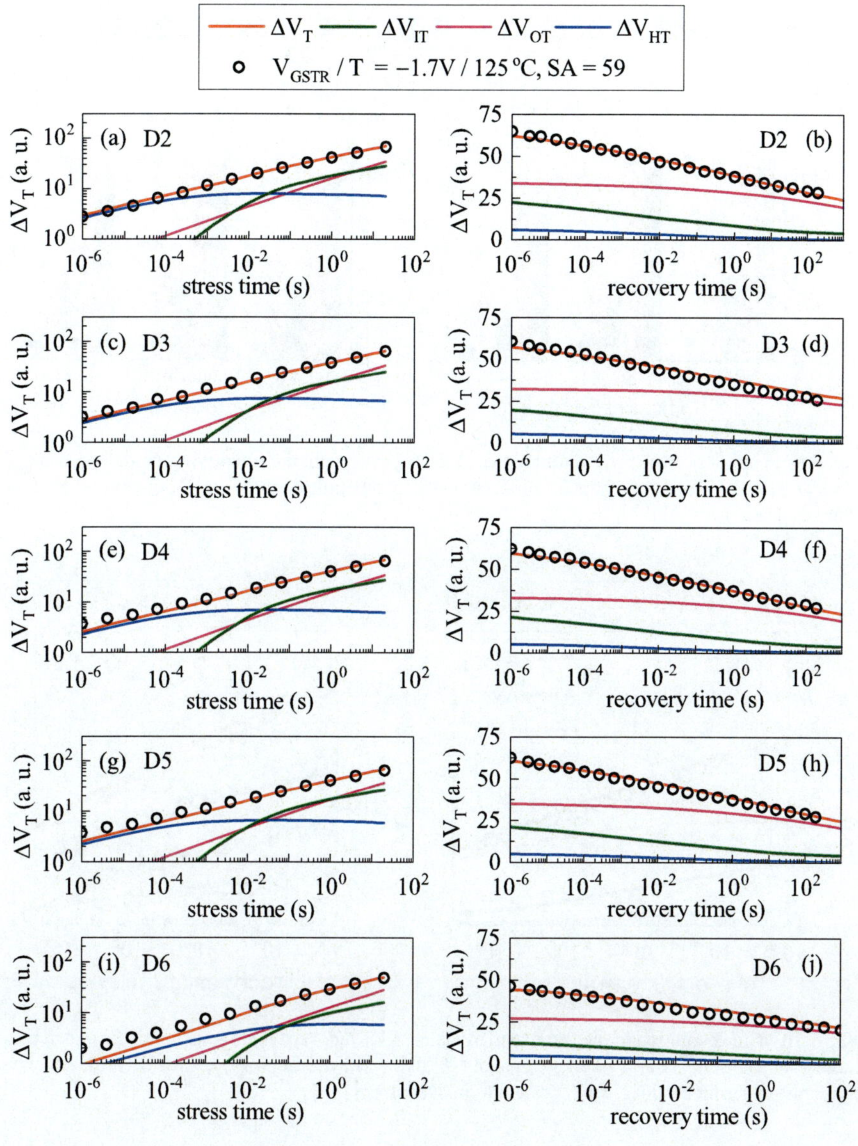

Fig. 9.8 Measured and modeled time evolution of ΔV_{T} together with the underlying model subcomponents (a) during and (b) after stress in different (changes in Ge% and N%) SiGe channel devices. Symbols: experiment, lines: model calculation. Data from [13]

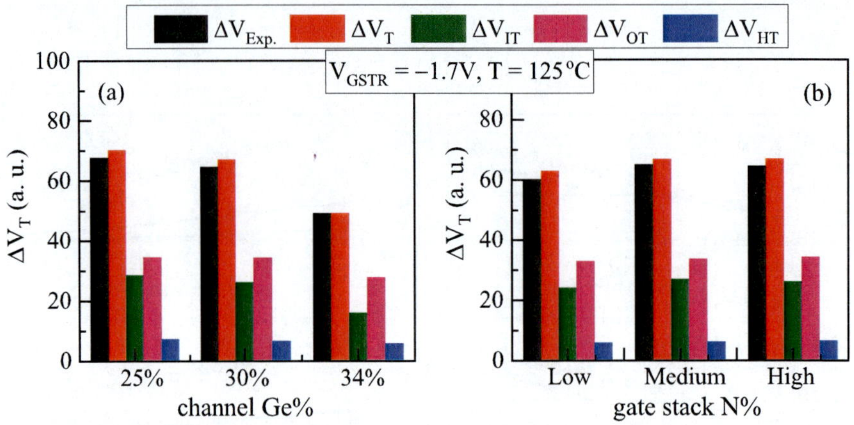

Fig. 9.9 Fixed time measured and modeled ΔV_T together with the underlying model subcomponents at a fixed V_{GSTR} and T for (a) different Ge%, and (b) different N% (TiN thickness) process changes. Data from [13]

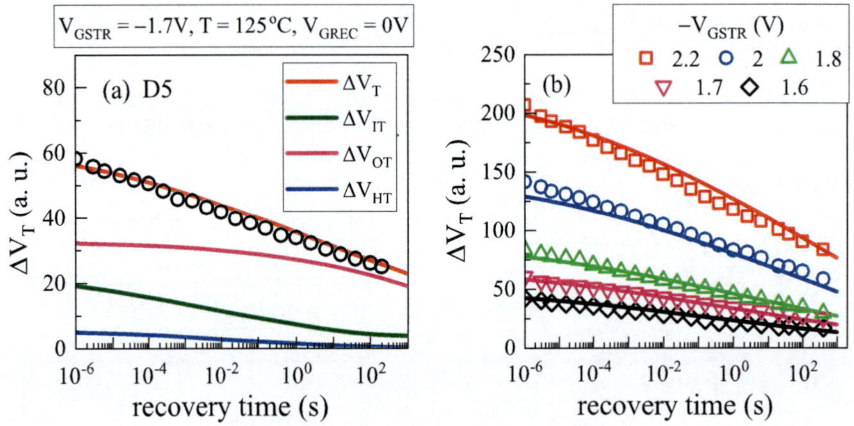

Fig. 9.10 Time evolution of measured and modeled ΔV_T recovery (a) together with the underlying model subcomponents at a fixed V_{GSTR} and T and at (b) different V_{GSTR} for SiGe device D5. Symbols: experiment, lines: model calculation. Data from [13]

(*i.e.*, recovery is lower) at higher V_{GSTR} due to relatively higher contribution from the ΔV_{OT} subcomponent which is semi-permanent.

Figure 9.11 shows the time evolution of measured and modeled ΔV_T for Mode-B AC stress (see Chap. 1, Sect. 1.2) for device D5, in (a) with underlying ΔV_{IT} and ΔV_{OT} subcomponents but at a fixed V_{GSTR} and T, and in (b) for multiple V_{GSTR} but fixed T. As explained in Chap. 5, Sect. 5.3 and Chap. 6, Sect. 6.3, the ΔV_{HT} contribution is negligible for Mode-B AC stress, which is explained later in Chap. 14. Note that similar to DC stress, the relative contribution of ΔV_{OT} on overall ΔV_T is

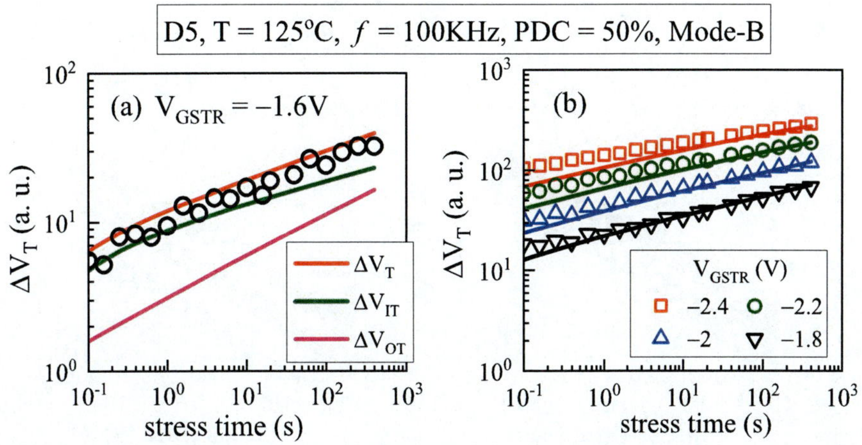

Fig. 9.11 Time evolution of measured and modeled ΔV_T during Mode-B AC stress (a) together with the underlying model subcomponents at a fixed V_{GSTR} and T and at (b) different V_{GSTR} for SiGe device D5. Symbols: experiment, lines: model calculation. Data from [13]

also significant during Mode-B AC stress due to the use of high V_{GSTR} and T in these experiments. However, due to higher VAF and E_A of ΔV_{OT} as compared to ΔV_{IT}, the contribution from ΔV_{OT} would reduce and ΔV_{IT} would be the dominating contributor to overall ΔV_T under use condition (not explicitly shown), same as DC stress, see Fig. 9.7. Note that the K_{F30} pre-factor for Mode-B AC stress is lower than DC stress, all other model parameters remain same. This is due to f dependence of ΔV_{OT}, which is mentioned earlier in Chap. 6, Sect. 6.3, and is analyzed in detail in Chap. 14 [26, 42].

9.3.3 Impact of Layout (Channel to STI/active Edge Spacing)

In this subsection, the impact of different SA (see Fig. 9.2 (a) for the definition of SA) on measured ΔV_T stress and recovery kinetics is modeled in SiGe devices having different processes (changes in Ge% and N%). All devices are DC stressed at fixed V_{GSTR} and T. Note that UCS relaxes as the channel (or transistor active) comes closer to the STI edge when SA is reduced. This is verified using measurements and TCAD simulations, see Fig. 9.2 (b) and (c).

Figure 9.12 shows the time evolution of measured and modeled ΔV_T (a, b) during and (c, d) after stress in SiGe devices with (a, c) lower SA and UCS and (b, d) higher SA and UCS and different processes (Ge% and N%), under identical V_{GSTR} and T. ΔV_T reduces with increase in SA (*i.e.*, increase in UCS), and the reduction is higher in devices having higher Ge% in the channel. However, no noticeable changes are observed in the slope n during stress and in FR during recovery after stress. Note that all model parameters are same as listed in Table 9.2, except the

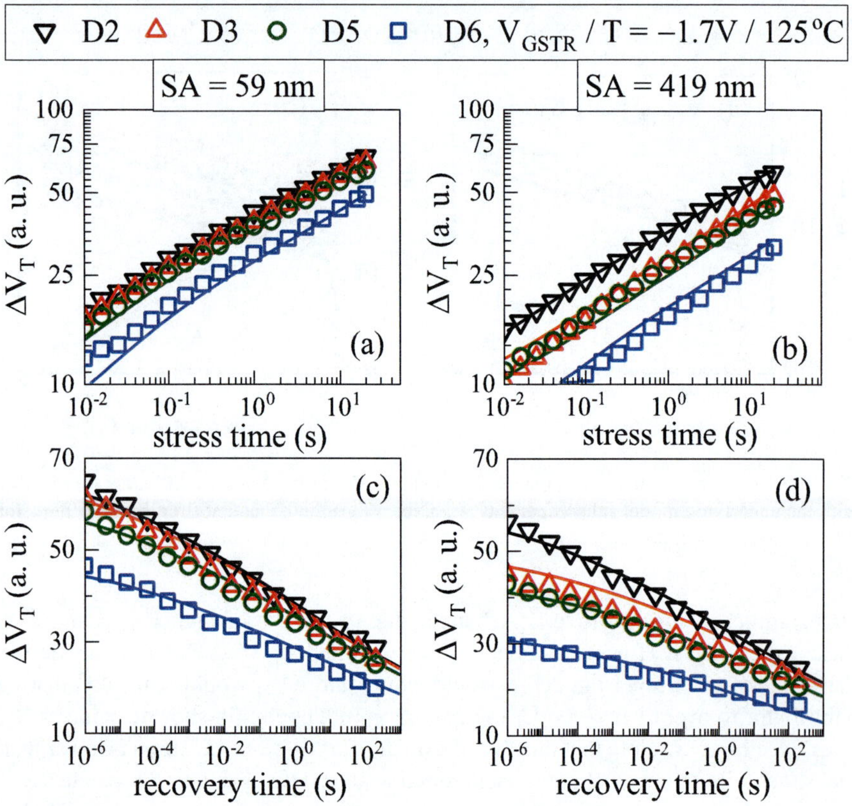

Fig. 9.12 Time evolution of measured and modeled ΔV_T (a, b) during and (c, d) after stress for different SiGe devices (changes in Ge% and N%) having (a, c) low SA and UCS and (b, d) high SA and UCS. Symbols: experiment, lines: model calculation. Data from [13]

pre-factor K_{F10} and T-independent field acceleration factor Γ_0 of the RD model for the ΔV_{IT} subcomponent, these terms change with SA due to the reasons explained in Sect. 9.4, the K_{F30} parameter changes very slightly, that too only at very low SA.

Figure 9.13 shows the (a) measured and modeled ΔV_T at fixed stress time (t_{STR} = 1Ks), V_{GSTR} and T, along with the underlying (b) ΔV_{IT}, as well as (c) ΔV_{HT} and ΔV_{OT} subcomponents, as a function of SA, for various SiGe devices (changes in Ge% and N%). Note that for a particular SA, ΔV_T reduces at higher Ge% and increases slightly at higher N%, which is discussed in the earlier subsection and explained in Sect. 9.4. However, for a particular process (Ge% and N%), ΔV_T and the underlying ΔV_{IT} reduce at larger SA, while the ΔV_{HT} (contribution is very small in any case) and ΔV_{OT} subcomponents show no noticeable change when SA is varied. Hence, the overall ΔV_T is primarily impacted by changes in ΔV_{IT} with changes in SA. This is similar to that observed earlier for changes in Ge% and N%.

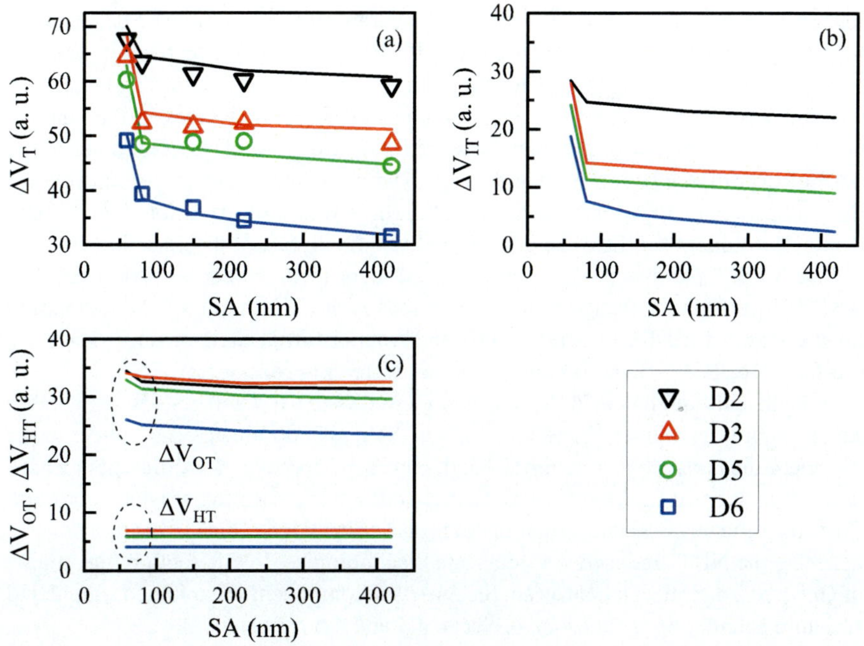

Fig. 9.13 (a) Measured and modeled ΔV_T at fixed time and the underlying (b) ΔV_{IT} and (c) ΔV_{HT} and ΔV_{OT} subcomponents as a function of SA (UCS) for different SiGe (changes in Ge% and N%) devices. Symbols: experiment, lines: model calculation. Data from [13]

As mentioned before (see Fig. 9.7), it is ΔV_{IT} that dominates the overall ΔV_T at use condition. The impact of process (Ge%, N% and SA or UCS) change on the ΔV_{IT} subcomponent is explained next.

9.4 Explanation of Process Dependence

The model parameters are listed in Table 9.2. The contributions from ΔV_{IT}, ΔV_{HT}, and ΔV_{OT} subcomponents and hence from the overall ΔV_T reduce at higher Ge%; however, the relative reduction in ΔV_{HT} is much larger and that of ΔV_{OT} is smaller when compared to the reduction in ΔV_{IT}. On the other hand, both ΔV_{IT} and ΔV_{HT} increase while ΔV_{OT} reduces at higher N%, although in these devices the impact of N% change (caused by change in TiN cap thickness) is not very strong due to less N% change in the IL. Moreover, the ΔV_{HT} subcomponent is very small for all SiGe devices.

As already mentioned, the T activation E_{AKF1} related to the RD model for ΔV_{IT}, see Chap. 4, Sect. 4.3, increases at higher Ge% and reduces at higher N%, the same trends are observed in SiGe p-FinFETs discussed in Chap. 11. Note that the reduction in E_{AKF1} at higher N% has also been reported from atomistic calculations [43], however the same is needed to explain the impact of Ge%, which is not available as of now. The tunneling parameter Γ_0 reduces with increase in Ge% and increases with increase in N%. The parameter α changes only with changes in the channel Ge%. Note that higher α at higher Ge% increases the VAF of ΔV_{IT} and hence the VAF of overall ΔV_{T} for the SiGe devices. The parameter α is also responsible for the T dependence of VAF (*i.e.*, the reduction in VAF at higher T) for ΔV_{IT}, which is more prominent for the SiGe devices and will be discussed in Chap. 11.

The TTOM parameters f_{FAST} and τ_{EC} are kept identical across processes. The ABDWT parameters change in a way that makes the hole trapping and detrapping kinetics faster for D0 compared to D1, and smaller for all SiGe devices. The K_{F30} pre-factor for RDD model is changed across different processes.

Note that although both ΔV_{IT} and ΔV_{OT} contribute to overall ΔV_{T} under harsher stress conditions (higher V_{GSTR} and T), it is ΔV_{IT} that dominates ΔV_{T} under lower V_{G} operating conditions, see Fig. 9.7 and Sect. 9.5. Moreover, for different mechanical strain, only the ΔV_{IT} subcomponent changes, while the others remain constant. Therefore, the physical understanding of process dependence of ΔV_{IT} is necessary to optimize the NBTI-induced ΔV_{T} in scaled technologies. In this section, the impact of Ge%, N% and mechanical strain on ΔV_{IT} subcomponent is explained using band structure calculations (see Chap. 4, Sects. 4.3 and 4.5 for details).

9.4.1 Impact of Ge% and N%

Figure 9.14 illustrates the schematic of hydrogen (H)-passivated bond dissociation process, which causes trap generation at the channel/IL interface by RD model. As explained in Chap. 4, Sect. 4.3, the inversion layer holes under the influence of E_{OX} tunnel to the interfacial X–H bonds get captured and make them weak, and subsequently, the weak bonds get depassivated by thermal activation [37]. The bond dissociation rate depends on the parameters (listed earlier) K_{F10}, Γ_0, α, and E_{AKF1}. The parameter K_{F10} depends on the hole density (p_{H}), hole-tunneling probability

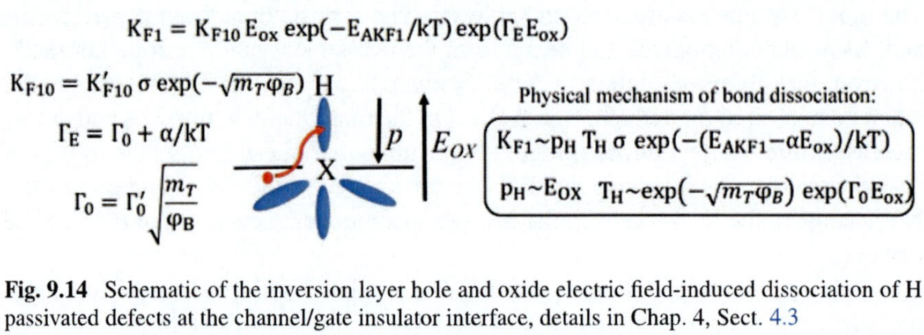

Fig. 9.14 Schematic of the inversion layer hole and oxide electric field-induced dissociation of H passivated defects at the channel/gate insulator interface, details in Chap. 4, Sect. 4.3

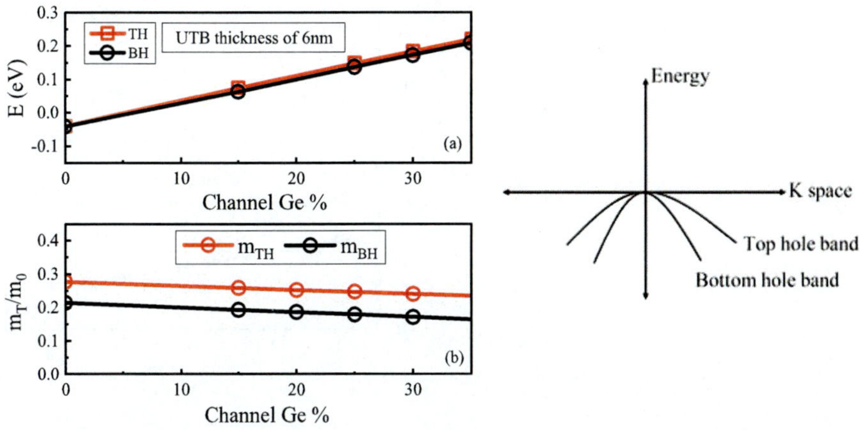

Fig. 9.15 Channel Ge% impact on (a) the degeneracy point (energy maxima) of TH and BH valence bands (see diagram on the right), and (b) tunneling effective mass, from band structure calculations using the tight-binding method

(T_H), and capture cross section (σ). Both T_H and Γ_0 depend on the tunneling barrier height (φ_B) and tunneling effective mass (m_T) of holes.

Figure 9.15 (a) shows the degeneracy point for top hole (TH) and bottom hole (BH) valence bands as a function of Ge%, from calculated energy band dispersion in ultrathin body (UTB) Si and SiGe channels using the tight binding method [44]. For Si channel, the degeneracy point shifts up in energy (shift away from the conduction band) for both TH and BH bands due to the size quantization effect associated with thinner body in FDSOI devices (results are shown for a body thickness of 6nm). However, for SiGe channel at higher Ge%, both TH and BH bands move toward the conduction band. Therefore, the effective barrier (φ_B) for hole tunneling (from channel to X–H bonds at the channel/IL interface) increases with higher Ge%. Note that the tunneling effective mass (m_T) shows negligible change for the range of Ge% analyzed in this chapter, see Fig. 9.15 (b).

The increase in φ_B at higher Ge% exponentially reduces the parameter K_{F10} and therefore reduces the ΔV_{IT} subcomponent. Although higher φ_B would also lower the parameter Γ_0, the overall field acceleration Γ_E (and hence VAF) increases due to increase in α at higher Ge%. The increase in α at higher Ge% can be attributed to the increase in interfacial dielectric constant between the SiGe channel and IL. Higher Ge% also increases E_{AKF1}, although as mentioned before, a first-principles based analysis is needed to understand the impact of Ge% on E_{AKF1}.

Figure 9.16 shows the impact of Ge% on (a) the T-activated parameter K_{F10} (T activation is included as E_{AKF1} changes) and (b) Γ_0. Both K_{F10} (including T activation) and Γ_0 decrease significantly for the range of Ge% used in these devices.

In general, higher N% in the gate stack (particularly at the channel/IL interface) would reduce φ_B (due to lower bandgap in SiON compared to SiO_2) and hence increase K_{F10}, Γ_0, and ΔV_{IT}. However, N incorporation in these devices is from the

Fig. 9.16 Channel Ge% impact on (a) T-activated K_{F10} and (b) Γ_0 parameters for the ΔV_{IT} subcomponent

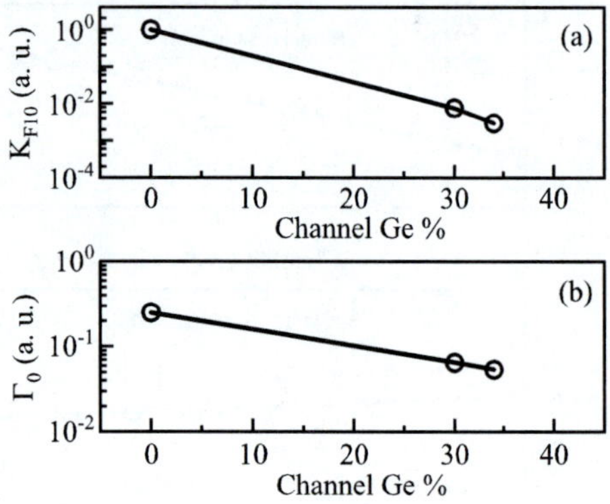

TiN High-K cap. Therefore, not much N diffuses all the way down to IL and most of the N remains confined in High-K, and therefore, only slight impact on ΔV_T is observed. As mentioned before, this is different from the impact of N% discussed in Chap. 7, where the IL itself was nitrided. However, higher N% in the High-K would increase its dielectric constant, and hence, the electric field across the IL would increase. This in turn would increase all the subcomponents, although ΔV_{HT} remains negligible due to SiGe channel. Moreover, the ΔV_{OT} component reduces at higher N%, which is consistent with the time-dependent dielectric breakdown (TDDB) studies [45] (note that the generated bulk gate insulator traps result in stress-induced leakage current (SILC) and TDDB [46], see Chap. 3, Sect. 3.4).

9.4.2 *Impact of Layout*

The impact of UCS (caused due to changes in SA, Fig. 9.2) on NBTI-induced ΔV_T is quantified in Sect. 9.3.3. It is observed that the impact of UCS on ΔV_T is primarily due to changes in ΔV_{IT}, while ΔV_{HT} and ΔV_{OT} remain unaffected. Figure 9.17 shows the impact of UCS on the degeneracy point of the band structure, simulated by the tight-binding method. Calculations are done for a particular Ge $=$ 30% and a body thickness of 6 nm. Note that the TH band shifts down in energy toward the conduction band, while the BH band shifts up in energy away from the conduction band at higher UCS. Therefore, the valence band minimum moves toward the conduction band (hole occupancy would be higher for TH than BH band in this case) and increases φ_B. Furthermore, the TH band has higher m_T compared to the BH band under UCS. The m_T for TH band initially increases with UCS but shows saturation with further increase in stress.

Figure 9.18 shows the impact of UCS on ΔV_{IT} parameters (a) K_{F10} and (b) Γ_0. As expected, K_{F10} reduces with increase in strain (caused by stress) due to changes in φ_B and m_T, while there is negligible impact on Γ_0 (as both φ_B and m_T change with

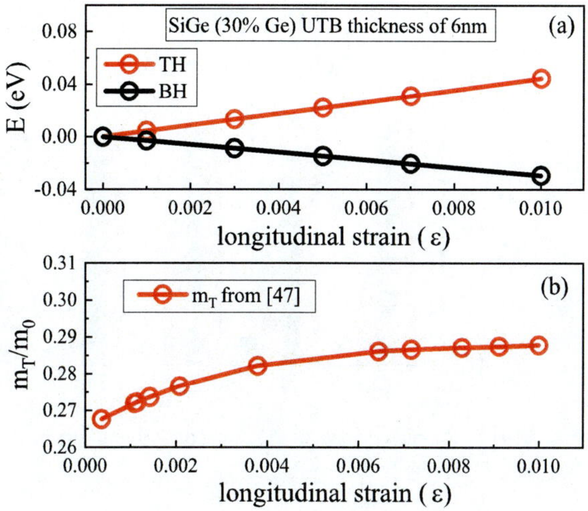

Fig. 9.17 Longitudinal compressive strain (calculated from UCS) impact on (a) the degeneracy point (energy maxima) of TH and BH valence bands and (b) tunneling effective mass, from band structure calculations using the tight-binding method. These results are similar to another published report [47]

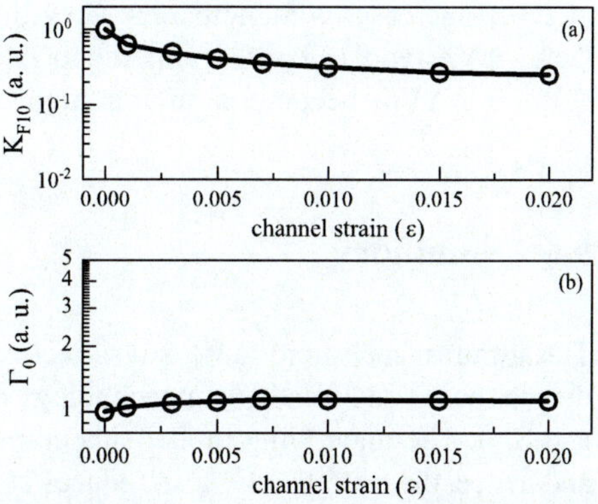

Fig. 9.18 Longitudinal compressive strain (calculated from UCS) impact on (a) K_{F10} and (b) Γ_0 parameters for the ΔV_{IT} subcomponent

UCS). Therefore, similar to higher channel Ge%, higher UCS due to higher SA also reduces ΔV_{IT} and hence overall ΔV_T.

9.5 Estimation of EOL Degradation

The calibrated model is used to project the end of life (EOL) ΔV_T under operating condition for different devices. Figure 9.19 shows the 10 years extrapolated overall ΔV_T and the underlying ΔV_{IT}, ΔV_{HT}, and ΔV_{OT} model subcomponents for different devices. Note, the 14nm technology Si channel device D1 has slightly higher EOL ΔV_T compared to the 28nm technology device D0. However, the EOL ΔV_T reduces for the 14nm technology SiGe channel devices at higher Ge% (D2–D5–D6), although it slightly increases with increase in N% (D5–D4–D3). The EOL ΔV_T is dominated

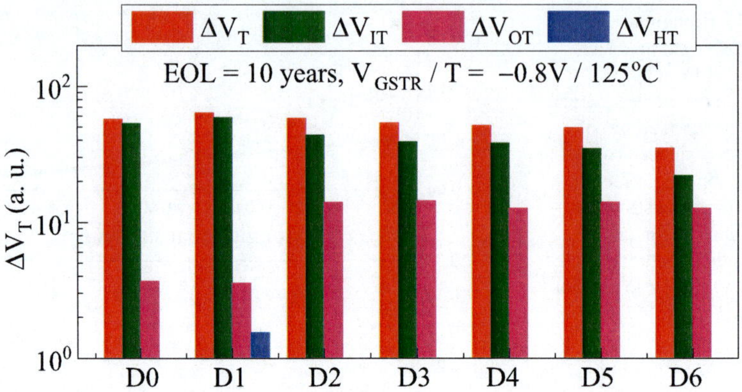

Fig. 9.19 Extrapolated 10 years ΔV_T and underlying subcomponents for DC stress at use condition for different Si and SiGe devices

by ΔV_{IT} for all the processes, while ΔV_{HT} is negligible. The relative contribution of ΔV_{OT} increases with increase in Ge%. It is important to note that with increase in Ge%, ΔV_{IT} reduces faster compared to ΔV_{OT}, and therefore, the contribution from ΔV_{IT} and ΔV_{OT} becomes similar at high Ge% (D6).

9.6 Summary

The ultrafast measured ΔV_T stress-recovery time kinetics is modeled in FDSOI devices based on 28 and 14 nm technology nodes and different processes (Ge%, N%, and SA). The underlying model subcomponents ΔV_{IT}, ΔV_{HT}, and ΔV_{OT} reduce, and hence, the overall ΔV_T also reduces at higher Ge%, although the relative ΔV_{OT} contribution increases and that from ΔV_{HT} reduces at higher Ge%. The relative ΔV_{OT} contribution is found to be significant for SiGe FDSOI devices compared to bulk Si devices discussed earlier in Chap. 7. This is mostly due to the use of higher stress bias in these experiments. However, the EOL ΔV_T under use condition is always dominated by ΔV_{IT}, except for the device with very high Ge% in the gate insulator stack, where both ΔV_{IT} and ΔV_{OT} equally contribute. The ΔV_{IT} and ΔV_{HT} subcomponents increase, and the overall ΔV_T also increases, while ΔV_{OT} reduces, at higher N%. Moreover, ΔV_{IT} and hence ΔV_T reduce with increase in SA due to higher UCS in the channel. Higher Ge% and UCS in the channel direction increase the hole-tunneling barrier and hence lower the ΔV_{IT} subcomponent. These process-dependent features are also observed for other type of devices (*e.g.*, FinFETs) and will be discussed later in Chaps.11 and 13 of this book.

Acknowledgements All experimental data presented in this chapter are re-plotted from previously published reports. The authors would like to acknowledge STMicroelectronics for providing the measurement data used in this chapter and Xavier Federspiel and his team for useful discussion. Karansingh Thakor is acknowledged for help with manuscript preparation.

References

1. N. Kimizuka, K. Yamaguchi, K. Imai, T. Iizuka, C.T. Liu, R.C. Keller, T. Horiuchi, in *Symposium on VLSI Technology Digest of Technical Papers*, 92 (2000).
2. V. Huard, M. Denais, F. Perrier, N. Revil, C. Parthasarathy, A. Bravaix, E. Vincent, Microelectron. Reliab. **45**, 83 (2005)
3. S. Mahapatra, K. Ahmed, D. Varghese, A.E. Islam, G. Gupta, L. Madhav, D. Saha, M.A. Alam, in *IEEE International Reliability Physics Symposium Proceedings*, p. 1 (2007).
4. Y. Mitani, H. Satake, A. Toriumi, IEEE Trans. Device Mater. Reliab. **8**, 6 (2008)
5. S. Pae, M. Agostinelli, M. Brazier, R. Chau, G. Dewey, T. Ghani, M. Hattendorf, J. Hicks, J. Kavalieros, K. Kuhn, M. Kuhn, J. Maiz, M. Metz, K. Mistry, C. Prasad, S. Ramey, A. Roskowski, J. Sandford, C. Thomas, J. Thomas, C. Wiegand, J. Wiedemer, in *IEEE International Reliability Physics Symposium Proceedings*, p. 352 (2008).
6. S. Krishnan, U. Kwon, N. Moumen, M.W. Stoker, E.C.T. Harley, S. Bedell, D. Nair, B. Greene, W. Henson, M. Chowdhury, D.P. Prakash, E. Wu, D. Ioannou, E. Cartier, M.-H. Na, S. Inumiya, K. Mcstay, L. Edge, R. Iijima, J. Cai, M. Frank, M. Hargrove, D. Guo, A. Kerber, H. Jagannathan, T. Ando, J. Shepard, S. Siddiqui, M. Dai, H. Bu, J. Schaeffer, D. Jaeger, K. Barla, T. Wallner, S. Uchimura, Y. Lee, G. Karve, S. Zafar, D. Schepis, Y. Wang, R. Donaton, S. Saroop, P. Montanini, Y. Liang, J. Stathis, R. Carter, R. Pal, V. Paruchuri, H. Yamasaki, J.-H. Lee, M. Ostermayr, J.-P. Han, Y. Hu, M. Gribelyuk, D.-G. Park, X. Chen, S. Samavedam, S. Narasimha, P. Agnello, M. Khare, R. Divakaruni, V. Narayanan, M. Chudzik, in *IEEE International Electron Devices Meeting Technical Digest*, 28.1.1 (2011).
7. K. Joshi, S. Hung, S. Mukhopadhyay, V. Chaudhary, N. Nanaware, B. Rajamohanan, T. Sato, M. Bevan, A. Wei, A. Noori, B. McDougal, C. Ni, G. Saheli, C. Lazik, P. Liu, D. Chu, L. Date, S. Datta, A. Brand, J. Swenberg, S. Mahapatra, in *IEEE International Reliability Physics Symposium Proceedings*, 4C.2.1 (2013).
8. J. Franco, B. Kaczer, P.J. Roussel, J. Mitard, M. Cho, L. Witters, T. Grasser, G. Groeseneken, IEEE Trans. Electron Devices **60**, 396 (2013)
9. P. Srinivasan, J. Fronheiser, K. Akarvardar, A. Kerber, L.F. Edge, R.G. Southwick, E. Cartier, H. Kothari, in *IEEE International Reliability Physics Symposium Proceedings*, 6A.3.1 (2014).
10. N. Parihar, N. Goel, S. Mukhopadhyay, S. Mahapatra, IEEE Trans. Electron Devices **65**, 392 (2018)
11. N. Parihar, S. Mahapatra, in *IEEE International Reliability Physics Symposium Proceedings*, TX.5.1 (2018).
12. V. Huard, C. Ndiaye, M. Arabi, N. Parihar, X. Federspiel, S. Mhira, S. Mahapatra, A. Bravaix, in *IEEE International Reliability Physics Symposium Proceedings*, TX.4.1 (2018).
13. N. Parihar, R. Tiwari, C. Ndiaye, M. Arabi, S. Mhira, H. Wong, S. Motzny, V. Moroz, V. Huard, S. Mahapatra, in *International Conference on Simulation of Semiconductor Processes and Devices*, pp. 167 (2018).
14. S. Ramey, A. Ashutosh, C. Auth, J. Clifford, M. Hattendorf, J. Hicks, R. James, A. Rahman, V. Sharma, A. St. Amour, C. Wiegand, in *IEEE International Reliability Physics Symposium Proceedings*, 4C.5.1 (2013).
15. K. T. Lee, W. Kang, E-A Chung, G. Kim, H. Shim, H. Lee, H. Kim, M. Choe, N.-I. Lee, A. Patel, J. Park, J. Park, in *IEEE International Reliability Physics Symposium Proceedings*, 2D.1.1 (2013).
16. J. Franco, B. Kaczer, A. Chasin, H. Mertens, L.-A. Ragnarsson, R. Ritzenthaler, S. Mukhopadhyay, H. Arimura, P.J. Roussel, E. Bury, N. Horiguchi, D. Linten, G. Groeseneken, A. Thean, in *IEEE International Reliability Physics Symposium Proceedings*, 4B.2.1 (2016).

17. G. Jiao, M. Toledano-Luque, K.-J. Nam, N. Toshiro, S.-H. Lee, J.-S. Kim, T. Kauerauf, E. Chung, D. Bae, G. Bae, D.-W. Kim, K. Hwang, in *IEEE International Electron Devices Meeting Technical Digest*, 31.2.1 (2016).
18. N. Parihar, R.G. Southwick, U. Sharma, M. Wang, J.H. Stathis, S. Mahapatra, in *IEEE International Reliability Physics Symposium Proceedings*, 2D.4.1 (2017).
19. N. Parihar, R. Southwick, M. Wang, J.H. Stathis, S. Mahapatra, in *IEEE International Electron Devices Meeting Technical Digest*, 7.3.1 (2017).
20. N. Parihar, U. Sharma, R.G. Southwick, M. Wang, J.H. Stathis, S. Mahapatra, IEEE Trans. Electron Devices **65**, 23 (2018)
21. N. Parihar, R.G. Southwick, M. Wang, J.H. Stathis, S. Mahapatra, IEEE Trans. Electron Devices **65**, 1699 (2018)
22. N. Parihar, R.G. Southwick, M. Wang, J.H. Stathis, S. Mahapatra, IEEE Trans. Electron Devices **65**, 1707 (2018)
23. N. Parihar, R. Tiwari, S. Mahapatra, in *International Conference on Simulation of Semiconductor Processes and Devices*, p. 176 (2018).
24. R. Tiwari, N. Parihar, K. Thakor, H.Y. Wong, S. Motzny, M. Choi, V. Moroz, S. Mahapatra, IEEE Trans. Electron Devices **66**, 2086 (2019)
25. R. Tiwari, N. Parihar, K. Thakor, H.Y. Wong, S. Motzny, M. Choi, V. Moroz, S. Mahapatra, IEEE Trans. Electron Devices **66**, 2093 (2019)
26. N. Parihar, U. Sharma, R.G. Southwick, M. Wang, J.H. Stathis, S. Mahapatra, in *IEEE International Reliability Physics Symposium Proceedings* (2019).
27. M. Wang, J. Zhang, H. Zhou, R.G. Southwick, R. Hsin, K. Chao, X. Miao, V.S. Basker, T. Yamashita, D. Guo, G. Karve, H. Bu, in *IEEE International Reliability Physics Symposium Proceedings* (2019).
28. N. Choudhury, U. Sharma, H. Zhou, R. G. Southwick, M. Wang, S. Mahapatra, in *IEEE International Reliability Physics Symposium Proceedings* (2020).
29. H. Zhou, M. Wang, J. Zhang, K. Watanabe, C. Durfee, S. Mochizuki, R. Bao, R. Southwick, M. Bhuiyan, B. Veeraraghavan, *IEEE International Reliability Physics Symposium Proceedings* (2020).
30. N. Choudhury, T. Samadder, R. Tiwari, H. Zhou, R.G. Southwick, M. Wang, S. Mahapatra, in *IEEE International Reliability Physics Symposium Proceedings* (2021).
31. O. Weber, E. Josse, F. Andrieu, A. Cros, E. Richard, P. Perreau, E. Baylac, N. Degors, C. Gallon, E. Perrin, S. Chhun, E. Petitprez, S. Delmedico, J. Simon, G. Druais, S. Lasserre, J. Mazurier, N. Guillot, E. Bernard, R. Bianchini, L. Parmigiani, X. Gerard, C. Pribat, O. Gourhant, F. Abbate, C. Gaumer, V. Beugin, P. Gouraud, P. Maury, S. Lagrasta, D. Barge, N. Loubet, R. Beneyton, D. Benoit, S. Zoll, J.-D. Chapon, L. Babaud, M. Bidaud, M. Gregoire, C. Monget, B. Le-Gratiet, P. Brun, M. Mellier, A. Pofelski, L.R. Clement, R. Bingert, S. Puget, J.-F. Kruck, D. Hoguet, P. Scheer, T. Poiroux, J.-P. Manceau, M. Rafik, D. Rideau, M.-A. Jaud, J. Lacord, F. Monsieur, L. Grenouillet, M. Vinet, Q. Liu, B. Doris, M. Celik, S.P. Fetterolf, O. Faynot, M. Haond, in *Symposium on VLSI Technology Digest of Technical Papers* (2014).
32. B. DeSalvo, P. Morin, M. Pala, G. Ghibaudo, O. Rozeau, Q. Liu, A. Pofelski, S. Martini, M. Casse, S. Pilorget, F. Allibert, F. Chafik, T. Poiroux, P. Scheer, R.G. Southwick, D. Chanemougame, L. Grenouillet, K. Cheng, F. Andrieu, S. Barraud, S. Maitrejean, E. Augendre, H. Kothari, N. Loubet, W. Kleemeier, M. Celik, O. Faynot, M. Vinet, R. Sampson, B. Doris, in *IEEE International Electron Devices Meeting Technical Digest*, 7.2.1 (2014).
33. F. Andrieu, M. Casse, E. Baylac, P. Perreau, O. Nier, D. Rideau, R. Berthelon, F. Pourchon, A. Pofelski, B. De Salvo, C. Gallon, V. Mazzocchi, D. Barge, C. Gaumer, O. Gourhant, A. Cros, V. Barral, R. Ranica, N. Planes, W. Schwarzenbach, E. Richard, E. Josse, O. Weber, F. Arnaud, M. Vinet, O. Faynot, M. Haond, in *European Solid State Device Research Conference*, p. 106 (2014).
34. A. Bonnevialle, C. Le Royer, Y. Morand, S. Reboh, C. Plantier, N. Rambal, J.-P. Pedini, S. Kerdiles, P. Besson, J.-M. Hartmann, D. Marseilhan, B. Mathieu, R. Berthelon, M. Casse, F. Andrieu, D. Rouchon, O. Weber, F. Boeuf, M. Haond, A. Claverie, M. Vinet, in *Symposium on VLSI Technology Digest of Technical Papers* (2016).

35. R. Berthelon, F. Andrieu, P. Perreau, E. Baylac, A. Pofelski, E. Josse, D. Dutartre, A. Claverie, M. Haond, in *European Solid-State Device Research Conference*, p. 127 (2016).
36. S. Mahapatra, N. Goel, S. Desai, S. Gupta, B. Jose, S. Mukhopadhyay, K. Joshi, A. Jain, A.E. Islam, M.A. Alam, IEEE Trans. Electron Devices **60**, 901 (2013)
37. A.E. Islam, N. Goel, S. Mahapatra, M.A. Alam, in *Fundamentals of Bias Temperature Instability in MOS Transistors* (Springer India, 2015), pp. 181–207.
38. S. Mukhopadhyay, K. Joshi, V. Chaudhary, N. Goel, S. De, R.K. Pandey, K.V.R.M. Murali, S. Mahapatra, in *IEEE International Reliability Physics Symposium Proceedings*, GD 3.1 (2014).
39. N. Choudhury, N. Parihar, N. Goel, A. Thirunavukkarasu, S. Mahapatra, IEEE J. Electron Devices Soc. **8**, 1281 (2020)
40. T. Samadder, N. Choudhury, S. Kumar, D. Kochar, N. Parihar, S. Mahapatra, IEEE Trans. Electron Devices **68**, 485 (2021)
41. A. Béché, J.L. Rouvière, L. Clément, J.M. Hartmann, Appl. Phys. Lett. **95**, 123114 (2009).
42. L. Zhou, Q. Zhang, H. Yang, Z. Ji, Z. Zhang, Q. Liu, H. Xu, B. Tang, E. Simoen, X. Ma, X. Wang, Y. Li, H. Yin, J. Luo, C. Zhao, W. Wang, IEEE Electron Device Lett. **41**, 965 (2020)
43. S.S. Tan, T.P. Chen, C.H. Ang, L. Chan, Atomic modeling of nitrogen neighboring effect on negative bias temperature instability of pMOSFETs. IEEE Electron Device Lett. **25**, 504 (2004)
44. https://nanohub.org/resources/bandstrlab.
45. L.K. Han, S. Crowder, M. Hargrove, E. Wu, S.H. Lo, F. Guarin, E. Crabbe, L. Su, in *IEEE International Electron Devices Meeting Technical Digest*, p. 643 (1997).
46. M.A. Alam, IEEE Trans. Electron Devices **49**, 226 (2002)
47. G. Sun, Ph.D. Dissertation, University of Florida, 2007.

Chapter 10
BAT Framework Modeling of RMG HKMG SOI FinFETs

Narendra Parihar, Nilotpal Choudhury, Tarun Samadder, Uma Sharma, Richard Southwick, Miaomiao Wang, James H. Stathis, and Souvik Mahapatra

10.1 Introduction

As described in the earlier chapters, Negative Bias Temperature Instability (NBTI) emerged as an important reliability issue with the migration from Silicon Dioxide (SiO_2) to Silicon Oxynitride (SiON) MOSFETs [1–4]. It continues to impact the dual-layer (SiO_2 or SiON based Interlayer (IL) and Hafnium Dioxide (HfO_2) High-K layer) High-K Metal Gate (HKMG) bulk [5–10] and Fully Depleted Silicon On Insulator (FDSOI) [11, 12] planar MOSFETs, bulk and SOI FinFETs [13–25], and Gate All Around Stacked Nanosheet (GAA-SNS) FETs [26–29], with either Silicon (Si) or Silicon Germanium (SiGe) channel. Several key NBTI features are summarized below (reproduced from Chap. 3, Sect. 3.1):

As described in Chap. 1, Sect. 1.3, NBTI results in a gradual buildup of positive charges in the gate insulator of the device when the gate is held at a negative bias ($V_G < 0$ V), leading to a shift of various device parameters, *e.g.*, threshold voltage shift (ΔV_T), over time. NBTI accelerates at higher stress gate bias ($V_G = V_{GSTR}$) and at higher temperature (T), respectively, governed by the voltage acceleration factor (VAF) and Arrhenius T activation energy (E_A). However, the accrued gate insulator charges during stress partially reduce after the stress V_G is reduced or removed ($V_G = V_{GREC}$ or 0 V), resulting in a partial recovery of ΔV_T over time. NBTI recovery results in lower ΔV_T during AC compared to DC stress. The ratio of AC to DC NBTI ΔV_T depends on the Pulse Duty Cycle (PDC), pulse low bias (V_{GLOW}), AC

N. Parihar · N. Choudhury · T. Samadder · U. Sharma · S. Mahapatra (✉)
Department of Electrical Engineering, Indian Institute of Technology Bombay, Powai, Mumbai 400076, India
e-mail: souvik@ee.iitb.ac.in

R. Southwick · M. Wang
IBM Research Division, Albany Nanotech, Albany, NY 12203, USA

J. H. Stathis
IBM Research Division, T. J. Watson Research Center, Yorktown Heights, NY 10598, USA

S. Mahapatra (ed.), *Recent Advances in PMOS Negative Bias Temperature Instability*,
https://doi.org/10.1007/978-981-16-6120-4_10

stress mode, and may or may not depend on the pulse frequency (f). Furthermore, the recovery of NBTI necessitates the use of ultra-fast methods to measure the device parametric shift without any recovery artifacts, see Chap. 1, Sect. 1.2. Therefore, the ultra-fast Measure Stress Measure (MSM) method is used throughout this book.

FinFETs were introduced to alleviate the short channel effects of conventional planar MOSFETs at scaled technology nodes. However, FinFETs feature (110) fin sidewall-dominated channel, which is different from the (100) channel orientation of planar MOSFETs. Although early reports have shown higher NBTI in FinFETs compared to planar devices due to difference in channel orientation, process optimization has helped reduce the degradation in matured nodes [13, 14]. However, note that all the past reports of NBTI in FinFETs were based on conventional (not ultra-fast) measurements and that too only for DC stress [13–16]. Moreover, detailed DC and AC stress and recovery data at different V_{GSTR}, T, PDC, and f are not shown. Furthermore, none has attempted to model the stress and recovery time kinetics of ΔV_T during and after DC and AC stress at various V_{GSTR}, T, PDC, and f.

Recently, the present authors have reported detail characterization and modeling of NBTI time kinetics during and after DC and AC stress in the Replacement Metal Gate (RMG) process-based HKMG bulk and SOI FinFETs [17–25]. The modeling is done using the physics-based BTI Analysis Tool (BAT) framework, which is explained in Chaps. 4–6 and is briefly described below (reproduced from Chap. 6, Sect. 6.3). The BAT framework is also recently used to model the GAA-SNS FETs, refer to [27, 29] for details.

Figure 10.1 illustrates the BAT framework [9] used to model the time kinetics of ΔV_T during and after stress. It uses uncorrelated contributions from generated interface (ΔV_{IT}) and bulk (ΔV_{OT}) gate insulator traps, and hole trapping in process-related preexisting bulk gate insulator traps (ΔV_{HT}). The Reaction Diffusion (RD) model is used to calculate the time kinetics of interface traps (density ΔN_{IT}) [9, 30, 31]. In Chap. 4, the RD model is described and independently validated by measured data from Direct Current IV (DCIV) method [32]. Transient Trap Occupancy Model

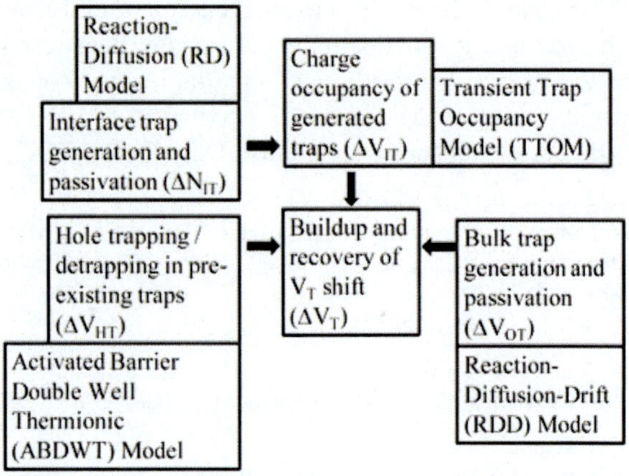

Fig. 10.1 Schematic of the BTI Analysis Tool (BAT) framework used in this book to model measured ΔV_T kinetics during and after DC and AC NBTI stress, reproduced from Chap. 4

(TTOM) is used to calculate the occupancy of generated interface traps and their contribution (ΔV_{IT}) [9], which is described and validated in Chap. 5, Sect. 5.3. The ΔV_{HT} and ΔV_{OT} kinetics are modeled by the Activated Barrier Double Well Thermionic (ABDWT) model [33] and Reaction Diffusion Drift (RDD) model [34], and these are described and validated in Chap. 5, Sect. 5.4 and Chap. 6, Sect. 6.2.

The framework has been used in the earlier chapters of this book to model the measured NBTI stress-recovery kinetics in planar bulk MOSFETs, having Silicon (Si) channel and differently processed HKMG gate insulators (Chap. 7), having Si-capped SiGe channel with differently processed Si cap and SiGe quantum well (Chap. 8), as well as in FDSOI MOSFETs having Si and SiGe channels (Chap. 9). The impact of stress conditions (V_{GSTR}, T) and processes (channel material, gate insulator stack, and layout) on the measured ΔV_T magnitude and its time kinetics during stress (i.e., the power-law time slope (n) at longer stress time (t_{STR})) and after stress (i.e., the Fraction Remaining (FR) at a given recovery time (t_{REC}) after stress, FR is defined as the ratio of ΔV_T at $t = t_{REC}$ to that at the end of stress at $t = t_{STR}$) have been modeled. In this chapter, the BAT framework is used to model the NBTI kinetics in SOI FinFETs. In the later chapters, bulk Si and SiGe FinFETs having different channel and gate insulator processes (Chap. 11), as well as bulk and SOI FinFETs and GAA-SNS FETs having different device (channel) dimensions (Chaps. 12 and 13) are modeled. Finally, detailed analysis and modeling of the AC stress (PDC, f, stress modes, see Chap. 1, Sect. 1.2) are done in Chap. 14.

As shown in Chap. 2, Sect. 2.5, stochastic variation of transistor entities such as doping, gate work function (for HKMG process), and device dimensions, as well as process variation across the wafer, results in variation of time-zero parameters, e.g., threshold voltage (V_{T0}), in small area devices [35–41]. Measured V_{T0} is normally distributed, whose variance is inversely proportional to the square root of the area of the transistor (Pelgrom law) [35, 41]. Due to V_{T0} variation, the electric field (E_{OX}) in the gate insulator would vary across devices during NBTI stress at fixed V_{GSTR} [37, 40]. Furthermore, the generation and passivation of interface and bulk gate insulator traps, occupancy of the generated interface traps, and hole trapping in and detrapping from the preexisting bulk gate insulator traps during and after NBTI stress are stochastic in nature [42–49]. This implies that from a microscopic perspective, these processes occur randomly in time and the traps are at random locations in the gate insulator over the channel of a device. Therefore, the measured ΔV_T in small area devices also shows variation [35–40, 48], although the exact nature of the distribution is debated [35–38, 45, 48, 50, 51]. There is very little evidence of any meaningful correlation between V_{T0} and ΔV_T [35, 40], and also see Chap. 2, Fig. 2.12, although some has reported a weak correlation [37]. The post-stress V_T distribution is also normal. It is important to note that both the mean and variance of ΔV_T distribution, and only the mean (and not the variance) of post-stress V_T distribution shift during and after stress, when the stress is done at constant V_{GSTR} and not at constant V_{OV} ($= V_{GSTR} - V_{T0}$) [51, 52], see Chap. 2, Figs. 2.10 and 2.11. This is because the variance due to the time-zero process variation usually dominates over the variance due to stochastic NBTI.

The temporal and spatial averaged out effects of charges associated with interface and bulk gate insulator traps are observed in large area devices, as shown and modeled in Chaps. 7–14. The same holds when the averaging is done on measured data from multiple small area devices (~ few-fin FinFETs). Therefore, the same (macroscopic) BAT framework can be used to model the measured stress-recovery time kinetics, which is shown in this chapter.

10.2 Device Details and Model Parameters

Measured data from RMG HKMG Si channel-based SOI p-FinFETs fabricated using proprietary IBM processes are modeled in this chapter. The FinFETs have FL of 20nm, FW of 10nm, and 24 fins. Moreover, two-fin devices are used for variability study. The gate insulators for these devices are based on standard chemical oxide-based IL, HfO_2 based High-K, and have low nitrogen content (N%) in the IL, resulting in an equivalent oxide thickness of approximately 1.1nm.

All data are obtained by the One Point Drop Down (OPDD) MSM method with 10μs measurement delay, see Chap. 1, Sect. 1.2. The ΔV_T kinetics is measured during and after DC stress at different V_{GSTR}, V_{GREC}, T (extended range from –40 °C to 150 °C), and t_{STR}. The AC ΔV_T kinetics at different V_{GSTR} and T is modeled for the Mode-B stress only, which is more realistic from a practical standpoint of circuit operation (see Chap. 1, Sect. 1.2).

As mentioned before, the measured ΔV_T time kinetics is modeled using uncorrelated contributions from ΔV_{IT}, ΔV_{HT}, and ΔV_{OT} subcomponents using the framework illustrated in Fig. 10.1. The model parameters are same as listed in Table 6.3 of Chap. 6, and are also re-listed in Table 10.1; their description is as follows (reproduced from Chap. 6, Sect. 6.3).

As described in Chap. 4, Sect. 4.3, the process-dependent RD model parameters are related to the pre-factor (K_{F10}), T-independent field acceleration (Γ_0), bond polarization (α), and T activation energy (E_{AKF1}) of the inversion layer hole-assisted bond dissociation process, see Chap. 4, Fig. 4.5. The process-dependent TTOM parameters are related to fast fraction of traps that undergo electron capture (f_{FAST}) and the electron capture time constant (τ_{EC}), see Chap. 5, Sect. 5.3. As described in Chap. 5, Sect. 5.4, the process-dependent ABDWT model parameters are related to the density of pre-existing defects (N_{0HT}), the energy barrier (E_{BM}), and the factors associated with E_{OX} dependence of the barrier (γ_B) and trap energy level ($\gamma_{E2}= m\gamma_B$), see Chap. 5, Fig. 5.10. The process-dependent RDD model parameters are related to the pre-factor (K_{F30}), T-independent field acceleration (Γ_{0OT}), and T activation energy (E_{AOT}) of the Anode Hole Injection (AHI) process-assisted bond dissociation (note, the bond polarization factor (α_{OT}) is not varied across devices but listed for completeness), and the forward reaction rate for ions (K_{F50}), see Chap. 6, Sect. 6.2. Other model parameters are process agnostic and are listed in the respective sections of Chaps. 4–6 (Tables 4.1, 5.1, 5.2 and 6.1).

Table. 10.1. Process-dependent RD, TTOM, ABDWT and RDD parameters (as per classification done in Chaps. 4–6) used in this chapter

Parameter	Unit	Value
K_{F10}	cm/Vs	–
E_{AKF1}	eV	0.3
Γ_0	cm/MV	0.19
α	qÅ	1.8
f_{FAST}	–	0.67
τ_{EC}	s	0.03
N_{0HT}	1/cm^2	–
E_{BM}	eV	1.3
γ_B	C.cm	5.7×10^{-9}
m	–	2.4
K_{F30}	1/s	–
E_{AOT}	eV	0.90
Γ_{0OT}	cm/MV	0.80
α_{OT}	qÅ	3.6
K_{F50}	cm^3/s	80

The parameters K_{F10}, N_{0HT}, and K_{F30} are not shown to maintain confidentiality. Device details in [19]

10.3 DC Stress Kinetics, Voltage, and Temperature Dependence

Figure 10.2 shows the time evolution of measured and modeled ΔV_T and the underlying ΔV_{IT}, ΔV_{HT}, and ΔV_{OT} subcomponents during stress at different V_{GSTR} and T. Figure 10.3 (a) through (g) show the time evolution of measured and modeled ΔV_T during stress for extended range of T and for different V_{GSTR} at each T. The modeling of the measured time kinetics of ΔV_T at fixed $V_{GSTR,}$ but different T is shown in Fig. 10.3 (h). Note that the time evolution of ΔV_T, when plotted in a log–log scale, evolves rapidly at the initiation of stress at very short time (this part may or may not be visible in the measurement time window) and asymptotically settles into a power-law dependence having slope n at longer time. The power-law time slope n usually varies if the stress T is varied over an extended range.

As shown in Fig. 10.2, the time kinetics of ΔV_{HT} increases rapidly at very short time but saturates at longer time, while ΔV_{IT} and ΔV_{OT} evolve relatively slowly at very short time and show power-law time dependence with n of ~ 1/6 and ~ 1/3 respectively at longer time. The ΔV_T kinetics at short time is dominated by ΔV_{HT} for lowest V_{GSTR} and T, since ΔV_{IT} is low and ΔV_{OT} is negligible. However, the relative ΔV_{IT} and ΔV_{OT} contributions increase at higher V_{GSTR} and T. This is due to the relatively higher VAF and T activation E_A for the ΔV_{IT} and ΔV_{OT} subcomponents

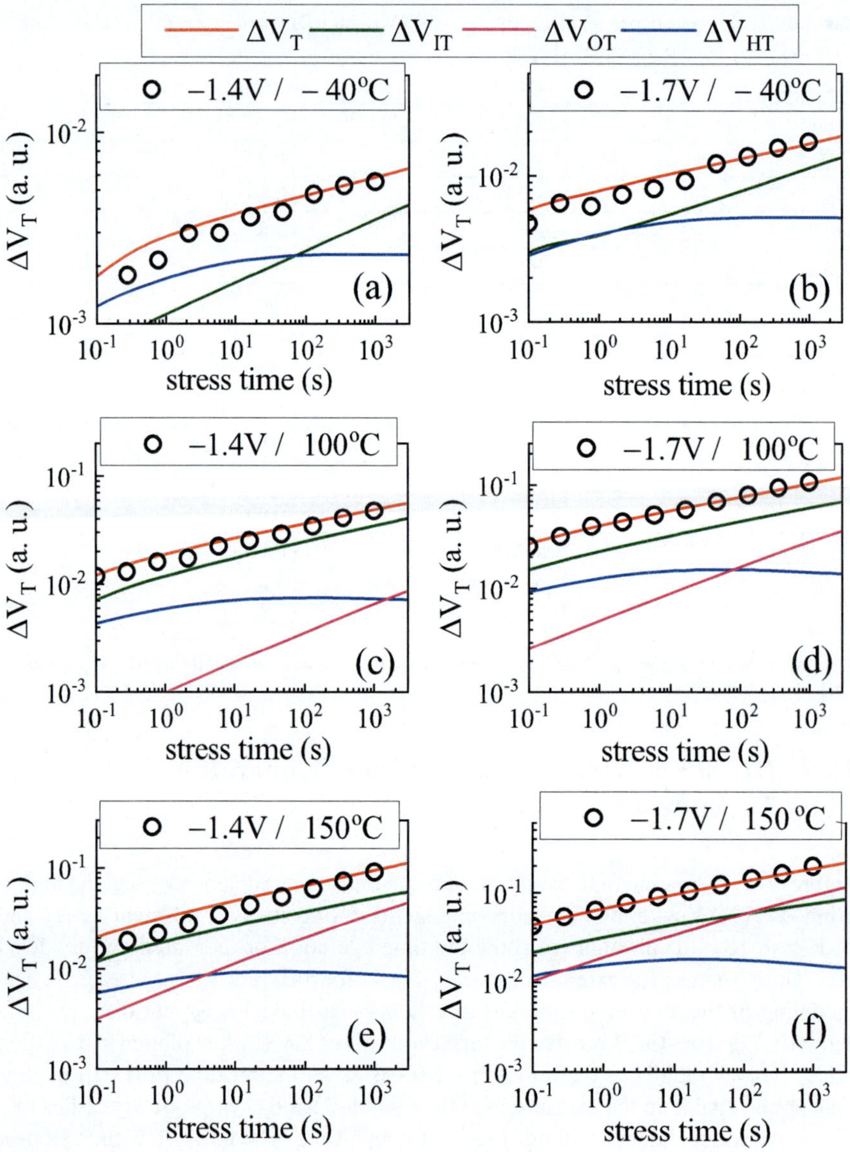

Fig. 10.2 Time evolution of measured and modeled ΔV_T together with the underlying model subcomponents for (a, b) low, (c, d) moderate, and (e, f) high T stress. Symbols: experiment, lines: model calculation. Data from [19]

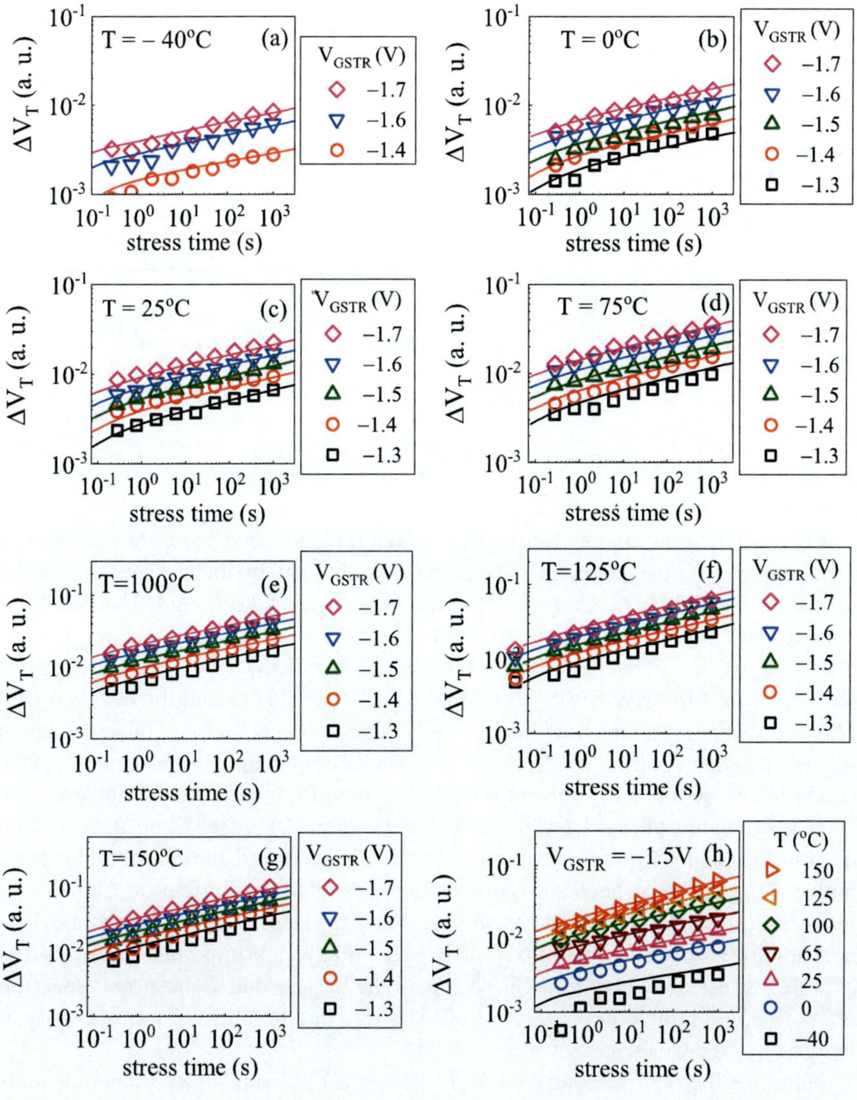

Fig. 10.3 Time evolution of measured and modeled ΔV_T during stress under (a–g) different V_{GSTR} at fixed T in each panel and increase in T from panels (a) through (g), and (h) different T at fixed V_{GSTR}. Symbols: experiment, lines: model calculation. Data from [19]

compared to that of ΔV_{HT} (note, the VAF and E_A are largest for ΔV_{OT}), which is discussed later in this chapter. The overall ΔV_T is dominated by ΔV_{IT} at moderate V_{GSTR} and T, and by both ΔV_{IT} and ΔV_{OT} at very high V_{GSTR} and T, especially at very long time. The capability of modeling the measured ΔV_T stress time kinetics at wide range of T and at different V_{GSTR} at each T (in a $V_{GSTR} \times T$ matrix) ascertain

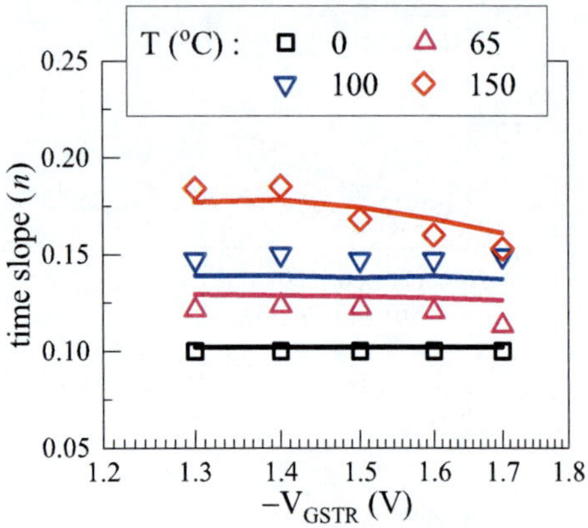

Fig. 10.4 Measured and modeled power-law slope n (from linear regression of ΔV_T time kinetics in t_{STR} range of 1s to 1Ks) as a function of V_{GSTR} at different T. Symbols: experiment, lines: model calculation. Data from [19]

the validity of the relative dominance of the different model subcomponents under different experimental conditions.

Figure 10.4 shows the measured and modeled longer time power-law slope n as a function of V_{GSTR} at different T. Note that n is obtained by linear regression of the time kinetics of ΔV_T (see Fig. 10.3) in the t_{STR} range of 1s through 1Ks, which is done throughout the book (unless mentioned otherwise), see Chap. 1, Sect. 1.3. As discussed in Chap. 3, Sects. 3.3 and 3.4, the addition of ΔV_{HT} ($n \sim 0$ in a log–log plot) with ΔV_{IT} ($n \sim 1/6$) reduces the time slope of ΔV_T, while the addition of ΔV_{OT} ($n \sim 1/3$) to ΔV_{IT} increases the time slope of ΔV_T. The slope n is lowest at lowest T due to relatively higher contribution from the ΔV_{HT} subcomponent. For moderate T, ΔV_T is dominated by the ΔV_{IT} subcomponent, and therefore, n increases. However, the slope is still below the $n \sim 1/6$ value as dictated by pure ΔV_{IT} (see Chap. 4, Sect. 4.4), due to additional ΔV_{HT} contribution. At further increase in T, the contributions from both ΔV_{IT} and ΔV_{OT} become significant and the relative contribution from ΔV_{HT} reduces, resulting in further increase in n. For very large T, higher relative ΔV_{OT} contribution increases the slope beyond $n \sim 1/6$. Finally, it is important to remark that at high T stress, a reduction in n is seen at higher V_{GSTR} due to the stress reduction effect (i.e., the reduction in effective NBTI stress due to higher ΔV_T especially at longer t_{STR}, see Chap. 4, Sect. 4.3).

Note that the relative contributions of ΔV_{IT}, ΔV_{HT}, and ΔV_{OT} subcomponents change with changes in V_{GSTR} and T due to differences in their VAF and T activation E_A, which is discussed next. This is consistent with the uncorrelated nature of these subcomponents, see Chaps. 4–6, and is necessary to explain the measured time kinetics of ΔV_T across different V_{GSTR} and over an extended range of T (in a $V_{GSTR} \times T$ matrix), as done in Fig. 10.3.

As discussed in Chap. 1, Sect. 1.2, the increase in n at higher T can also be triggered by the measurement delay artifacts. However, this is not the case in Fig. 10.4 due to the use of ultra-fast MSM method. Nevertheless, note that the change in n with V_{GSTR} and T would make the VAF and E_A a function of t_{STR} at which the parameters are determined. Furthermore, the VAF and E_A corresponding to a fixed ΔV_T level

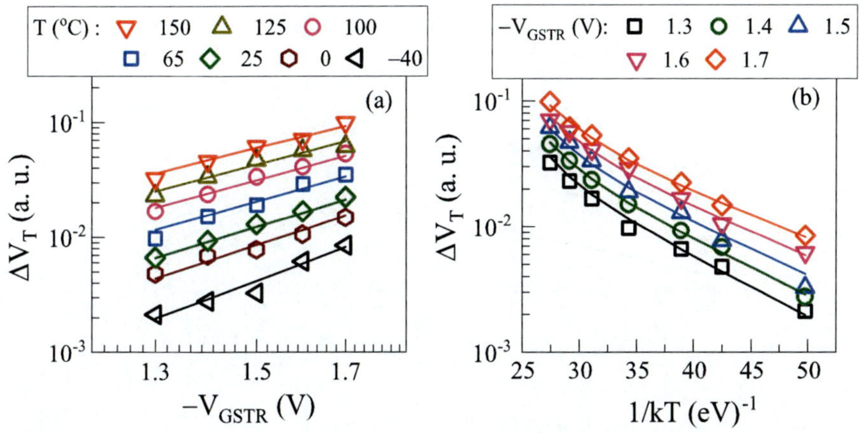

Fig. 10.5 Measured and modeled fixed time ΔV_T as a function of (a) V_{GSTR} at different T and (b) T at different V_{GSTR}. Symbols: experiment, lines: model calculation. Data from [19]

(or lifetime) would depend on the choice of ΔV_T for parameter extraction. This is unlike the data shown in in Chap. 1, Sect. 1.3 and Chap. 7, and therefore, the fixed t_{STR} and fixed ΔV_T level parameters cannot be related by the time slope n, unlike the example shown in Chap. 1, Fig. 1.12.

Figure 10.5 shows the measured and modeled fixed time ($t_{STR} = 1$Ks) ΔV_T as a function of (a) V_{GSTR} at different T and (b) T at different V_{GSTR}. Note, the power-law VAF reduces slightly at higher T, because of the polarization factors associated with the bond dissociation processes for the ΔV_{IT} and ΔV_{OT} subcomponents, as discussed in Chap. 4, Sect. 4.3 and Chap. 6, Sect. 6.2. Moreover, the stress reduction effect due to higher ΔV_T at higher V_{GSTR} and T would also lower the VAF at higher T. Of particular interest is the non-Arrhenius nature of the T activation of ΔV_T when measured over an extended range of T (also see Chap. 3, Fig. 3.15 (a)). Such non-Arrhenius T dependence is explained by uncorrelated contributions of the underlying ΔV_{IT}, ΔV_{HT}, and ΔV_{OT} subcomponents having different E_A. Note that the ΔV_{HT} subcomponent with lower E_A dominates ΔV_T at lower T, while the ΔV_{IT} and ΔV_{OT} subcomponents with higher E_A dominate ΔV_T at higher T.

Figure 10.6 shows the simulated ΔV_{IT}, ΔV_{HT}, and ΔV_{OT} subcomponents at fixed stress time ($t_{STR} = 1$Ks) as a function of V_{GSTR} at different T (left panels) and stress T at different V_{GSTR} (right panels), using the parameters listed in Table 10.1. The power-law VAF for ΔV_{IT} is similar to that of ΔV_{HT}, while ΔV_{OT} shows very high VAF. The T activation of ΔV_{IT} is higher than ΔV_{HT}, while that for ΔV_{OT} is significantly higher than both ΔV_{IT} and ΔV_{HT}. Note that the stress reduction effect (the reduction in the effective electrical stress at higher ΔV_T, see Chap. 4, Sect. 4.3) is prominent at higher V_{GSTR} and T, and results in a soft saturation of ΔV_{IT} and ΔV_{OT} (and a slight reduction of saturated ΔV_{HT}) time kinetics at longer t_{STR}, and apparently reduces the actually obtained VAF at higher T and E_A at higher V_{GSTR}.

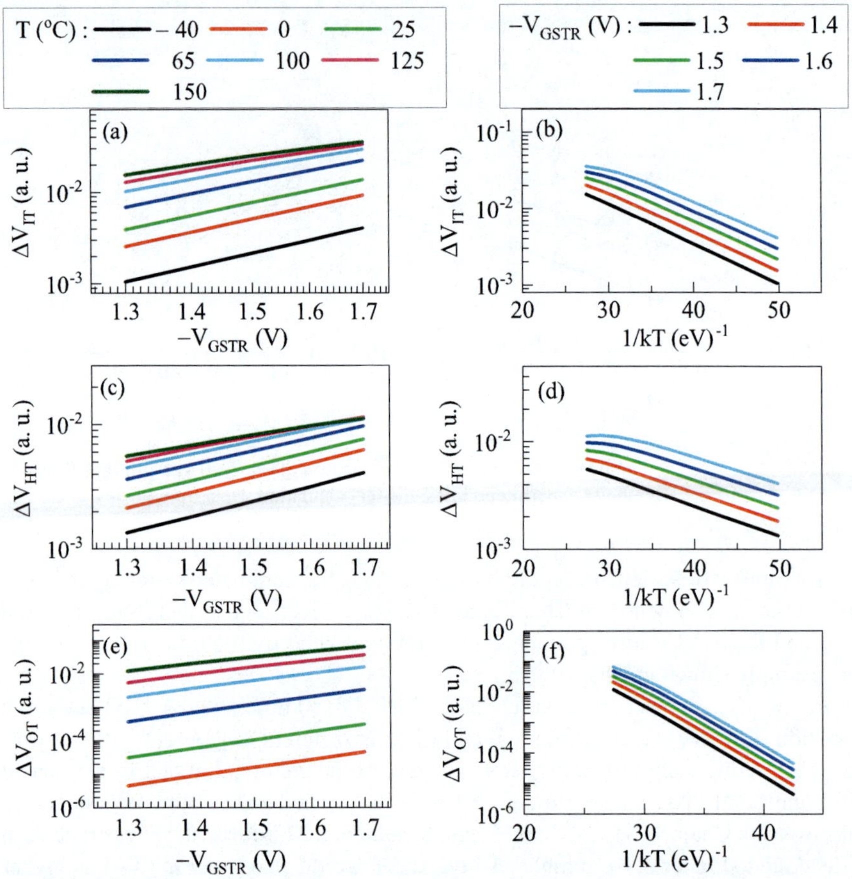

Fig. 10.6 Fixed time (a, b) ΔV_{IT}, (c, d) ΔV_{HT}, and (e, f) ΔV_{OT} subcomponents as a function of (a, c, e) V_{GSTR} at different T and (b, d, f) T at different V_{GSTR}, calculated using the model parameters given in Table 10.1

Therefore, the differences in VAF and E_{A} of the ΔV_{IT}, ΔV_{HT}, and ΔV_{OT} subcomponents result in differences in their relative contributions to overall ΔV_{T} as V_{GSTR} and T are changed. This consequently results in the differences in the time kinetics of ΔV_{T} at different V_{GSTR} and T, see Fig. 10.3, and different slope n, see Fig. 10.4, since the subcomponents have very unique and different stress kinetics. These subcomponents also have different and unique recovery kinetics and therefore impact overall ΔV_{T} recovery at different V_{GSTR} and T, as discussed next.

10.4 Recovery Kinetics After DC Stress

Figure 10.7 shows the time evolution of measured and modeled ΔV_T recovery (at $V_{GREC} = 0$ V) together with the underlying ΔV_{IT}, ΔV_{HT}, and ΔV_{OT} subcomponents after DC stress at different V_{GSTR} and T (see Fig. 10.2 for the corresponding stress data). Figure 10.8 (a) through (g) show the time evolution of measured and modeled ΔV_T recovery after DC stress over an extended T range, and for a wide range of V_{GSTR} at each T, while the T dependence of measured and modeled ΔV_T recovery

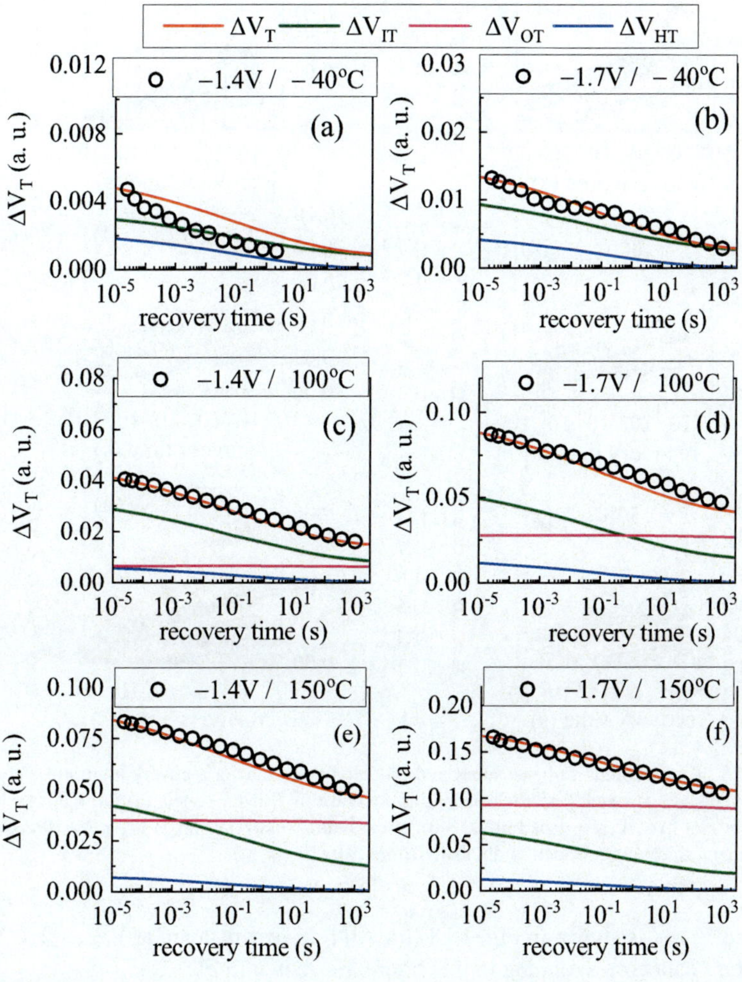

Fig. 10.7 Time evolution of measured and modeled ΔV_T together with the underlying model subcomponents during recovery after stress at (a, b) low, (c, d) moderate, and (e, f) high T (see Fig. 10.2 for stress data). Recovery bias is $V_{GREC} = 0$ V. Symbols: experiment, lines: model calculation. Data from [19]

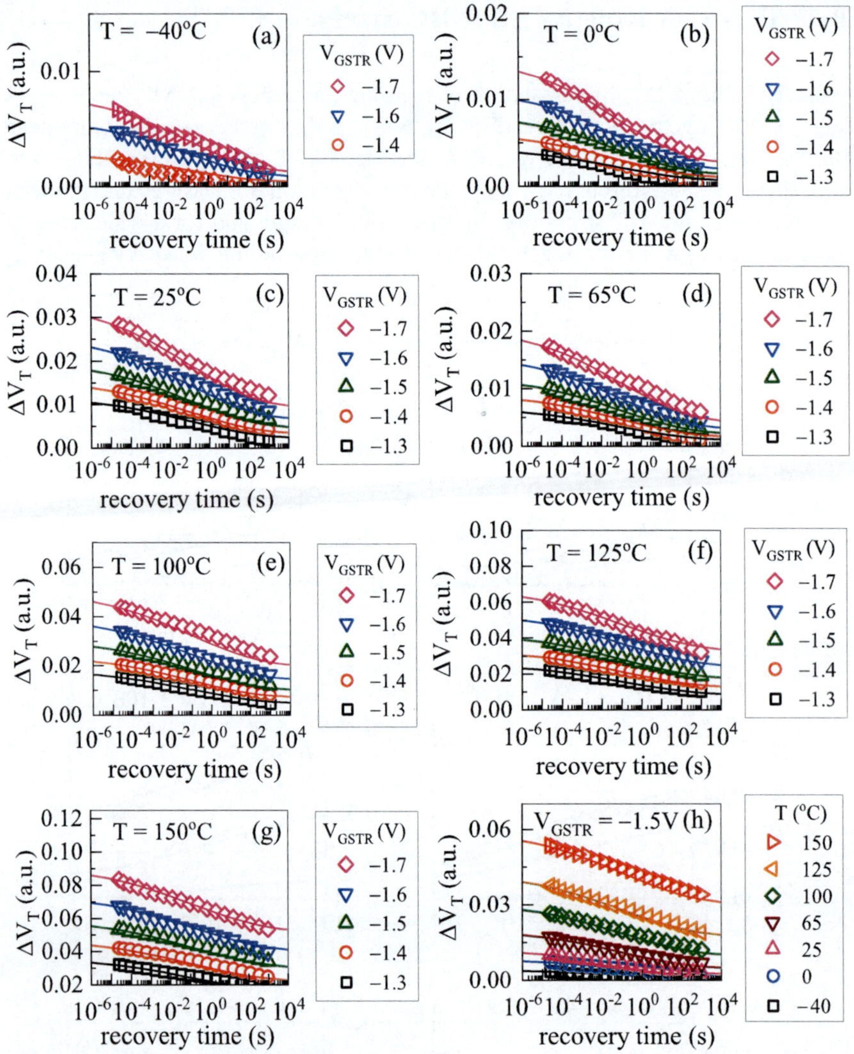

Fig. 10.8 Time evolution of measured and modeled ΔV_T during recovery after stress for (a–g) different V_{GSTR} at fixed T in each panel and increase in T from panels (a) through (g), and (h) different T at fixed V_{GSTR} (see Fig. 10.3 for stress data). Recovery bias is $V_{GREC} = 0V$. Symbols: experiment, lines: model calculation. Data from [19]

at fixed V_{GSTR} is shown in Fig. 10.8 (h). All recovery data are at $V_{GREC} = 0$ V, and the stress data corresponding to this figure are shown in Fig. 10.3.

As shown in Fig. 10.7, the ΔV_{HT} subcomponent recovers fast, and ΔV_{OT} is semi-permanent and recovers slowly. The ΔV_{IT} recovery is over an extended timescale and is a sum of two processes: ΔV_{IT_FAST} related to fast electron capture in traps that go below the Fermi level as the magnitude of gate bias is lowered from V_{GSTR} to

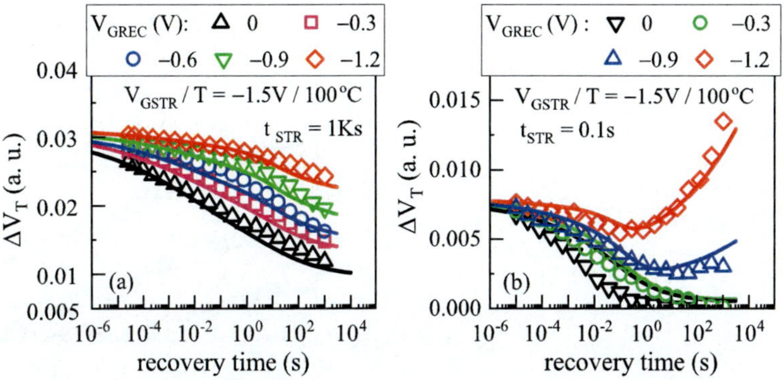

Fig. 10.9 Time evolution of measured and modeled ΔV_T recovery at different V_{GREC}, after stress for (a) long and (b) short t_{STR} at fixed V_{GSTR} and T. Symbols: experiment, lines: model calculation. Data from [19]

V_{GREC}, and ΔV_{IT_SLOW} due to re-passivation of generated interface traps by hydrogen (H/H_2) back diffusion, see Chap. 5 for details. Note, the stochastic hopping and lock-in related H_2 diffusivity reduction factor (A) for the ΔV_{IT_SLOW} component (see Chap. 4, Sect. 4.3) is higher ($A = 35$), due to the three-dimensional radial diffusion in FinFETs, as compared to the planar devices ($A = 7$, see Chap. 7). However, the parameter A is only dependent on device architecture and is independent of other process changes; e.g., $A = 35$ is also used for different SiGe p-FinFETs and GAA-SNS FETs modeled, respectively, in Chaps. 11 and 12.

The ΔV_{HT} domination of ΔV_T at lower T causes faster recovery in Fig. 10.7 (a) and (b). However, the increased contributions from ΔV_{IT} and ΔV_{OT} subcomponents slow down the recovery time kinetics at higher T, see Fig. 10.7 (c) through (f). Therefore, the relative dominance of different uncorrelated subcomponents at different T impacts the overall recovery rate at different V_{GSTR} and T.

Figure 10.9 shows the time evolution of measured and modeled ΔV_T recovery at different V_{GREC} after stress for (a) long and (b) short t_{STR} at fixed V_{GSTR} and T. The rate of recovery slows down at higher magnitude of V_{GREC}, and re-stressing related increase in ΔV_T is observed when shorter time stress is followed by long recovery and the difference between V_{GSTR} and V_{GREC} is small (similar results are also analyzed in Chap. 7, Sect. 7.5). The time kinetics of ΔV_T recovery is primarily governed by that of ΔV_{IT} and ΔV_{HT} subcomponents, as ΔV_{OT} is semi-permanent. The slowing down of ΔV_T recovery at higher magnitude of V_{GREC} is due to lower f_{FAST} (fraction of traps that go below the Fermi level during recovery after stress, see Chap. 5, Sect. 5.3) and the corresponding ΔV_{IT_FAST} subcomponent as handled by TTOM, resulting in slowing down of the ΔV_{IT} recovery kinetics, as well as the lower move back of the barrier E_B and trap level E_2 of the ABDWT model, resulting in lower ΔV_{HT} recovery, see Chap. 5, Sect. 5.4. Furthermore, the re-stressing in Fig. 10.9b is attributed to the generation of new interface and bulk traps, and the model can capture this feature without any parameter adjustment. Note that only

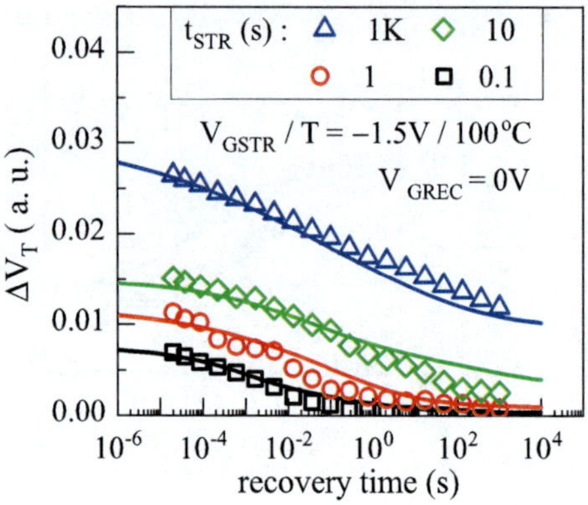

Fig. 10.10 Time evolution of measured and modeled ΔV_T recovery after stress at fixed V_{GSTR} and T but different t_{STR}. Recovery bias is $V_{GREC} = 0$ V. Symbols: experiment, lines: model calculation. Data from [19]

the TTOM parameters (see Chap. 5, Sect. 5.3) are adjusted to model the V_{GREC} dependence of ΔV_T recovery. All the other parameters related to the RD, ABDWT, and RDD models of Table 10.1 are kept fixed across V_{GREC}.

Figure 10.10 shows the time evolution of measured and modeled ΔV_T recovery at $V_{GREC} = 0$ V, after stress at fixed V_{GSTR} and T but different t_{STR}. Note that during stress, ΔV_T at shorter t_{STR} is dominated by ΔV_{HT} (see Fig. 10.2), and hence, the corresponding recovery is faster (see Fig. 10.7 (a) when recovery is dominated by hole detrapping). Slower recovery after longer t_{STR} is due to the higher relative contributions from ΔV_{IT} and ΔV_{OT} during stress, since ΔV_{IT} recovers slowly compared to ΔV_{HT} and ΔV_{OT} is semi-permanent.

10.5 AC Stress and Recovery Kinetics

As mentioned in Chap. 6, Sect. 6.3, the contribution from ΔV_{OT} subcomponent is much smaller compared to the corresponding DC stress (after the effective stress duration is computed by accounting for PDC of the AC pulse) [25, 53]. This is also reported for AC Time-Dependent Dielectric Breakdown (TDDB) experiments (note that generated bulk gate insulator traps are responsible for TDDB) [54, 55]. However, the exact mechanism responsible for the f dependence of ΔV_{OT} is not understood at this time, and this aspect is analyzed in detail in Chap. 14. Therefore, only the K_{F30} parameter in Table 10.1 is re-adjusted to model AC stress (note that the parameter is only dependent on f and on nothing else, and reduces at higher f), all other parameters are kept same as DC stress.

Figure 10.11 shows the time evolution of measured and modeled ΔV_T (a) during and (b) after Mode-B AC stress at different V_{GSTR} (which is V_{GHIGH} for AC stress) and T (see Chap. 1, Sect. 1.2 for AC stress modes). Note that the ΔV_{IT} subcomponent dominates the overall ΔV_T for all cases. This is because ΔV_{HT} is negligible for Mode-B AC stress, as explained in Chap. 5, Sect. 5.3 and also in Chap. 14. Furthermore, the ΔV_{OT} contribution is smaller than that of ΔV_{IT} for the range of V_{GSTR} and T

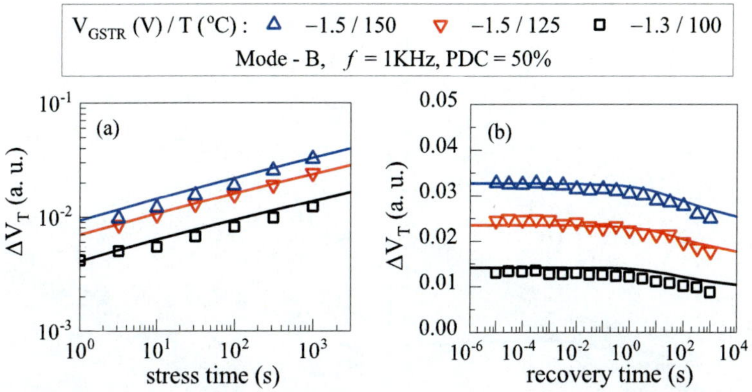

Fig. 10.11 Time evolution of measured and modeled ΔV_{T} (a) during and (b) after Mode-B AC stress at different V_{GSTR} and T. Recovery bias is $V_{\mathrm{GREC}} = 0$ V. Symbols: experiment, lines: model calculation. Data from [19]

used in these experiments. The recovery is delayed after Mode-B AC stress due to the absence of ΔV_{HT} (i.e., hole detrapping) and fast electron capture processes, as explained in detail in Chap. 14.

10.6 Variation in Few-Fin FinFETs

As mentioned before, measured ΔV_{T} in small area devices shows variability due to the stochastic effects as only a handful of defects (generated and pre-existing) are involved. However, the mean of ΔV_{T} stress and recovery kinetics obtained from the measurements of several small area devices behaves as large area device data, since the stochastic effects get averaged out. Therefore, the mean kinetics can be modeled using the macroscopic BAT framework of Chaps. 4–6.

Figure 10.12 shows the measured individual traces and their mean ΔV_{T} (a) during and (c) after DC stress at fixed V_{GSTR} and T from multiple fin# = 2 devices. The model calculated mean ΔV_{T} kinetics together with the underlying ΔV_{IT}, ΔV_{HT} and ΔV_{OT} subcomponents are also shown. Note that all features of the mean ΔV_{T} stress and recovery kinetics for the fin# = 2 devices are similar to that of the fin# = 24 devices discussed earlier in Sects. 10.3 and 10.4. The mean of the measured ΔV_{T} kinetics shows power-law time dependence at longer time during stress, with similar slope n as observed for multi-fin devices. The relative contribution from ΔV_{HT} is higher only at short t_{STR}, while the contributions from ΔV_{IT} and ΔV_{OT} are higher at longer t_{STR}. Note that ΔV_{IT} evolves faster than ΔV_{OT} at the initiation of stress, and these subcomponents show power-law time dependence with $n \sim 1/6$ and $\sim 1/3$, respectively, at longer t_{STR}. The early kinetics of ΔV_{HT} is even faster, but it saturates ($n \sim 0$) at longer t_{STR}. The recovery of ΔV_{HT} is fast, that of ΔV_{IT} is distributed over

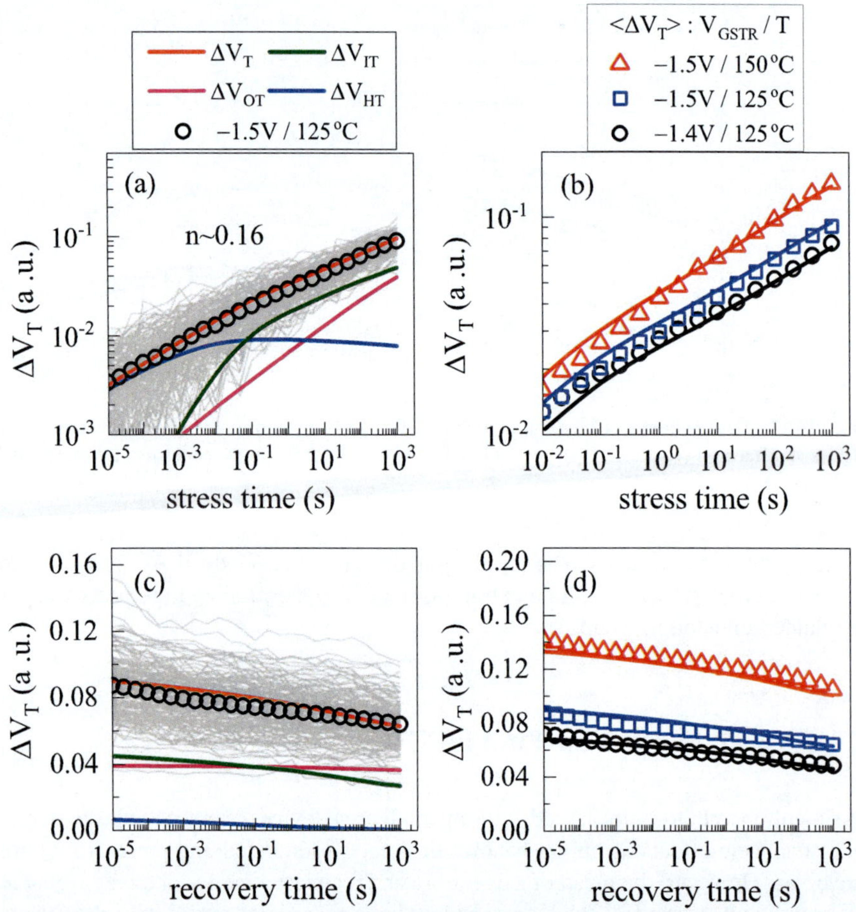

Fig. 10.12 Time evolution of measured (a, c) individual ΔV_T transients from multiple few-fin devices and their mean at fixed V_{GSTR} and T, and (b, d) mean of measured ΔV_T transients from multiple few-fin devices at different V_{GSTR} and T, (a, b) during and (c, d) after DC stress. The overall ΔV_T modeling is shown in all panels, and the underlying subcomponents in panels (a, c). Recovery bias is $V_{GREC} = 0$ V. Symbols: experiment, lines: model calculation. Data from [48]

several decades in time, while ΔV_{OT} shows semi-permanent behavior and it slows down the overall ΔV_T recovery at longer t_{REC}.

Figure 10.12 also shows the time evolution of measured and modeled mean ΔV_T from multiple fin# = 2 devices (b) during and (d) after DC stress at different V_{GSTR} and T. The deterministic framework can model the mean kinetics in few-fin devices with exactly the same model parameters as multi-fin devices as listed in Table 10.1, and therefore, verifies the universality of NBTI modeling framework in large (fin# = 24) and small (fin# = 2) area devices.

Figure 10.13 shows the measured and modeled mean ΔV_T from multiple fin#

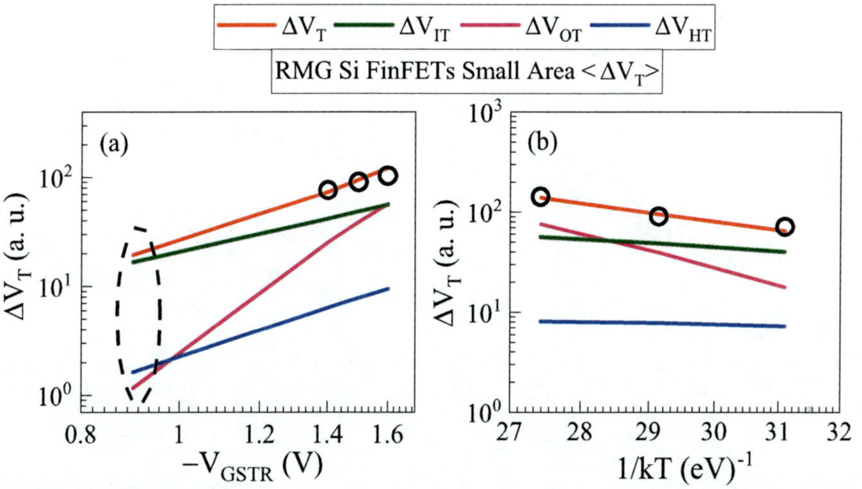

Fig. 10.13 Mean of measured ΔV_T at fixed time (t_{STR} = 1Ks) from multiple few-fin devices together with the overall ΔV_T modeling and underlying model subcomponents versus (a) V_{GSTR} at fixed T and (b) T at fixed V_{GSTR}. Symbols: experiment, lines: model calculation. Data from [48]

= 2 devices at a fixed stress time (t_{STR} = 1Ks) and the underlying ΔV_{IT}, ΔV_{HT}, and ΔV_{OT} subcomponents versus (a) V_{GSTR} at fixed T and (b) T at fixed V_{GSTR}. Note that the VAF and E_A values of overall ΔV_T are identical to that of multi-fin devices shown earlier in Sect. 10.3. The overall mean ΔV_T is dominated by ΔV_{IT} unless the stress is performed at very high V_{GSTR} and T, where both the ΔV_{IT} and ΔV_{OT} subcomponents equally contribute. The VAF and E_A values for ΔV_{OT} are much larger when compared to the corresponding values for the ΔV_{IT} and ΔV_{HT} subcomponents. Due to high VAF of the ΔV_{OT} subcomponent, the mean ΔV_T at low V_{GSTR} close to use (~ operating) condition is dominated by the ΔV_{IT} subcomponent, similar to the multi-fin devices.

Therefore, it is demonstrated that the deterministic BAT framework can model the mean ΔV_T stress and recovery kinetics from multiple small area devices, and the model subcomponents have identical parameters as large area devices.

10.7 Estimation of EOL Degradation

Figure 10.14 shows the time evolution of measured and modeled ΔV_T for (a) DC and (b) Mode-B AC stress at different V_{GSTR} (V_{GHIGH} for AC) but constant T. The model uses ΔV_{IT}, ΔV_{HT}, and ΔV_{OT} for DC but only ΔV_{IT} and ΔV_{OT} for Mode-B AC stress, and is capable of explaining the short-time kinetics at higher V_{GSTR} and the long-time kinetics at lower V_{GSTR}, with identical set of parameters as listed in Table 10.1 (note, only the RDD model K_{F30} pre-factor for ΔV_{OT} is different between

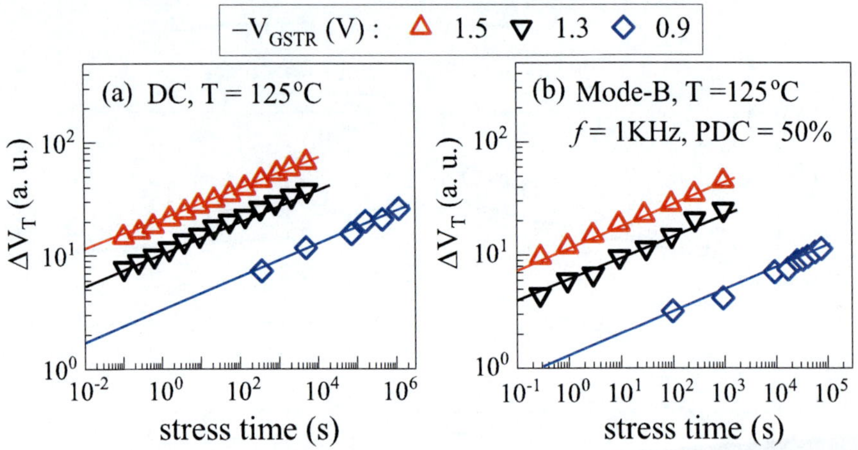

Fig. 10.14 Time evolution of measured and modeled ΔV_T during (a) DC and (b) Mode-B AC stress at different V_{GSTR} but fixed T. Symbols: experiment, lines: model calculation. Data from [19]

DC and AC stress due to the reason mentioned before). The calibrated model is used to determine the extrapolated ΔV_T at end of life (EOL) under use condition.

As discussed in Chap. 1, Sect. 1.4, the conventional lifetime method fits the time kinetics of ΔV_T measured over short time and under high V_{GSTR} using a power-law time dependence and extrapolate to EOL. The extrapolated values at EOL, usually obtained at several V_{GSTR}, is then extrapolated to operating bias (V_{DD}) either using power-law or exponential V_G dependence. Note that the accuracy of the fit-based method depends on the range of V_{GSTR} (better if closer to V_{DD}) and t_{STR} (better if closer to EOL) used in this exercise, see Chap. 6, Sect. 6.4. The accuracy of the fit-based method for the data measured in SOI p-FinFET is analyzed below.

Figure 10.15 (a) compares the measured and calculated ΔV_T at t_{STR} = 100Ks, with calculated ΔV_T obtained using the model based and empirical regression (or fit) based methods (the empirical methods are explained in Chap. 1, Sect. 1.4). Note that the model calculated ΔV_T values are close to the experimental ones for DC and Mode-B AC stress. However, the fit-based method (fitting of measured data is done over t_{STR} range of 1s-1Ks) underestimates DC but overestimates Mode-B AC ΔV_T when compared to measured data. Moreover, note that lower error is obtained for exponential V_G dependence based extrapolation for DC but for power-law V_G dependence based extrapolation for Mode-B AC stress.

Figure 10.15 (b) compares the extrapolated ΔV_T at an EOL value of 10 years, with ΔV_T calculated using the model based and fit-based methods. The power-law V_G dependence extrapolation yields lower ΔV_T while the exponential V_G dependence extrapolation yields higher ΔV_T compared to the model calculated ΔV_T for DC stress. However, both power-law and exponential methods yield higher ΔV_T than model for Mode-B AC stress.

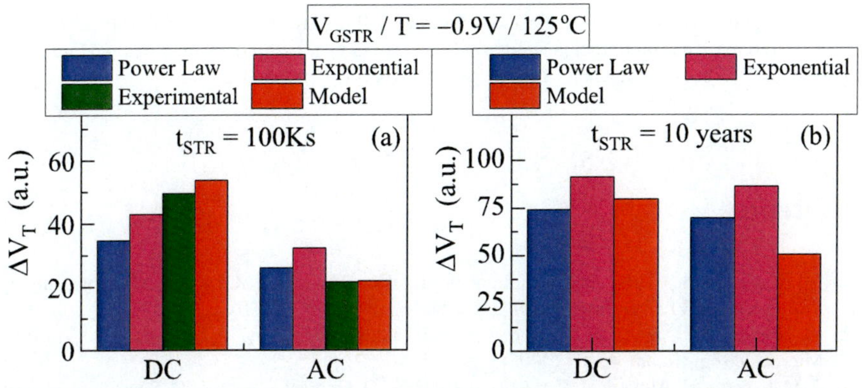

Fig. 10.15 Comparison of the (a) measured, modeled, and regression (fit)-based calculated ΔV_T at $t_{STR} = 100$Ks, and (b) modeled and fit-based calculated ΔV_T at EOL under use condition, for DC and Mode-B AC conditions. Fit-based calculation use both exponential and power-law V_G dependence. Data from [19]

The difference between the model-based and regression-based methods is due to the different underlying subcomponents that impact the overall ΔV_T during the short time stress at elevated V_G and long time at use V_G. Therefore, it is preferable to use the model-based extrapolation, since the model can accurately estimate the relative contributions of different subcomponents for such different situations.

10.8 Summary

DC and AC NBTI stress and recovery time kinetics are measured in RMG HKMG SOI p-FinFETs by an ultra-fast method and modeled by the BAT framework that uses uncorrelated contributions from the ΔV_{IT}, ΔV_{HT}, and ΔV_{OT} subcomponents. A wide range of experimental data measured at different V_{GSTR} (V_{GHIGH} for AC), T, and V_{GREC} are modeled. The subcomponents have different V_{GSTR} and T dependencies and different time kinetics during stress and recovery. The relative contributions of these subcomponents vary with stress conditions. ΔV_{IT} dominates DC and AC stress at longer t_{STR} for moderate V_{GSTR} and T, while ΔV_{HT} dominates DC stress at very short t_{STR} and very low T. Furthermore, ΔV_{OT} dominates DC stress at longer t_{STR} and very high V_{GSTR} and T. Identical BAT parameters are used to model the mean of measured data from several few-fin devices and multi-fin devices. The extrapolated EOL ΔV_T at use condition is calculated by the calibrated model and compared to regression-based empirical methods. The empirical methods provide inaccurate estimation of ΔV_T at EOL for both DC and AC stress. This necessitates proper modeling for reliable estimation of NBTI lifetime.

Acknowledgements All experimental data presented in this chapter are re-plotted from previously published reports. The authors acknowledge IBM for providing measurement facilities and FinFETs. Karansingh Thakor is acknowledged for help with manuscript preparation.

References

1. N. Kimizuka, K. Yamaguchi, K. Imai, T. Iizuka, C.T. Liu, R.C. Keller, T. Horiuchi, in *Symposium on VLSI Technology Digest of Technical Papers*, (2000), p. 92
2. V. Huard, M. Denais, F. Perrier, N. Revil, C. Parthasarathy, A. Bravaix, E. Vincent, Microelectron. Reliab. **45**, 83 (2005)
3. S. Mahapatra, K. Ahmed, D. Varghese, A.E. Islam, G. Gupta, L. Madhav, D. Saha, M.A. Alam, in *IEEE International Reliability Physics Symposium Proceedings* (2007), p. 1
4. Y. Mitani, H. Satake, A. Toriumi, IEEE Trans. Device Mater. Reliab. **8**, 6 (2008)
5. S. Pae, M. Agostinelli, M. Brazier, R. Chau, G. Dewey, T. Ghani, M. Hattendorf, J. Hicks, J. Kavalieros, K. Kuhn, M. Kuhn, J. Maiz, M. Metz, K. Mistry, C. Prasad, S. Ramey, A. Roskowski, J. Sandford, C. Thomas, J. Thomas, C. Wiegand, J. Wiedemer, in *IEEE International Reliability Physics Symposium Proceedings*, (2008), p. 352
6. K. Joshi, S. Hung, S. Mukhopadhyay, V. Chaudhary, N. Nanaware, B. Rajamohanan, T. Sato, M. Bevan, A. Wei, A. Noori, B. McDougal, C. Ni, G. Saheli, C. Lazik, P. Liu, D. Chu, L. Date, S. Datta, A. Brand, J. Swenberg, S. Mahapatra, in *IEEE International Reliability Physics Symposium Proceedings* (2013), p. 4C.2.1
7. J. Franco, B. Kaczer, P.J. Roussel, J. Mitard, M. Cho, L. Witters, T. Grasser, G. Groeseneken, IEEE Trans. Electron. Devices **60**, 396 (2013)
8. P. Srinivasan, J. Fronheiser, K. Akarvardar, A. Kerber, L.F. Edge, R.G. Southwick, E. Cartier, H. Kothari, in *IEEE International Reliability Physics Symposium Proceedings* (2014), p. 6A.3.1
9. N. Parihar, N. Goel, S. Mukhopadhyay, S. Mahapatra, IEEE Trans. Electron. Devices **65**, 392 (2018)
10. N. Parihar, S. Mahapatra, in *IEEE International Reliability Physics Symposium Proceedings* (2018), p. TX.5.1
11. V. Huard, C. Ndiaye, M. Arabi, N. Parihar, X. Federspiel, S. Mhira, S. Mahapatra, A. Bravaix, in *IEEE International Reliability Physics Symposium Proceedings* (2018), p. TX.4.1
12. N. Parihar, R. Tiwari, C. Ndiaye, M. Arabi, S. Mhira, H. Wong, S. Motzny, V. Moroz, V. Huard, S. Mahapatra, in *International Conference on Simulation of Semiconductor Processes and Devices* (2018), p. 167
13. S. Ramey, A. Ashutosh, C. Auth, J. Clifford, M. Hattendorf, J. Hicks, R. James, A. Rahman, V. Sharma, A. St. Amour, C. Wiegand, in *IEEE International Reliability Physics Symposium Proceedings* (2013), p. 4C.5.1
14. K.T. Lee, W. Kang, E.-A. Chung, G. Kim, H. Shim, H. Lee, H. Kim, M. Choe, N.-I. Lee, A. Patel, J. Park, J. Park, in *IEEE International Reliability Physics Symposium Proceedings* (2013), p. 2D.1.1
15. J. Franco, B. Kaczer, A. Chasin, H. Mertens, L.-A. Ragnarsson, R. Ritzenthaler, S. Mukhopadhyay, H. Arimura, P.J. Roussel, E. Bury, N. Horiguchi, D. Linten, G. Groeseneken, A. Thean, in *IEEE International Reliability Physics Symposium Proceedings* (2016), p. 4B.2.1
16. G. Jiao, M. Toledano-Luque, K.-J. Nam, N. Toshiro, S.-H. Lee, J.-S. Kim, T. Kauerauf, E. Chung, D. Bae, G. Bae, D.-W. Kim, K. Hwang, in *IEEE International Electron Devices Meeting Technical Digest* (2016), p. 31.2.1
17. N. Parihar, R.G. Southwick, U. Sharma, M. Wang, J.H. Stathis, S. Mahapatra, in *IEEE International Reliability Physics Symposium Proceedings* (2017), p. 2D.4.1
18. N. Parihar, R. Southwick, M. Wang, J.H. Stathis, S. Mahapatra, in *IEEE International Electron Devices Meeting Technical Digest*, (2017), p. 7.3.1

19. N. Parihar, U. Sharma, R.G. Southwick, M. Wang, J.H. Stathis, S. Mahapatra, IEEE Trans. Electron. Devices **65**, 23 (2018)
20. N. Parihar, R.G. Southwick, M. Wang, J.H. Stathis, S. Mahapatra, IEEE Trans. Electron. Devices **65**, 1699 (2018)
21. N. Parihar, R.G. Southwick, M. Wang, J.H. Stathis, S. Mahapatra, IEEE Trans. Electron. Devices **65**, 1707 (2018)
22. N. Parihar, R. Tiwari, S. Mahapatra, in *International Conference on Simulation of Semiconductor Processes and Devices*, (2018), p. 176
23. R. Tiwari, N. Parihar, K. Thakor, H.Y. Wong, S. Motzny, M. Choi, V. Moroz, S. Mahapatra, IEEE Trans. Electron. Devices **66**, 2086 (2019)
24. R. Tiwari, N. Parihar, K. Thakor, H.Y. Wong, S. Motzny, M. Choi, V. Moroz, S. Mahapatra, IEEE Trans. Electron. Devices **66**, 2093 (2019)
25. N. Parihar, U. Sharma, R.G. Southwick, M. Wang, J.H. Stathis, S. Mahapatra, in *IEEE International Reliability Physics Symposium Proceedings* (2019)
26. M. Wang, J. Zhang, H. Zhou, R.G. Southwick, R. Hsin, K. Chao, X. Miao, V.S. Basker, T. Yamashita, D. Guo, G. Karve, H. Bu, in *IEEE International Reliability Physics Symposium Proceedings* (2019)
27. N. Choudhury, U. Sharma, H. Zhou, R.G. Southwick, M. Wang, S. Mahapatra, in *IEEE International Reliability Physics Symposium Proceedings* (2020)
28. H. Zhou, M. Wang, J. Zhang, K. Watanabe, C. Durfee, S. Mochizuki, R. Bao, R. Southwick, M. Bhuiyan, B. Veeraraghavan, *IEEE International Reliability Physics Symposium Proceedings* (2020)
29. N. Choudhury, T. Samadder, R. Tiwari, H. Zhou, R.G. Southwick, M. Wang, S. Mahapatra, in *IEEE International Reliability Physics Symposium Proceedings* (2021)
30. S. Mahapatra, N. Goel, S. Desai, S. Gupta, B. Jose, S. Mukhopadhyay, K. Joshi, A. Jain, A.E. Islam, M.A. Alam, IEEE Trans. Electron. Devices **60**, 901 (2013)
31. A.E. Islam, N. Goel, S. Mahapatra, M.A. Alam, in *Fundamentals of Bias Temperature Instability in MOS Transistors* (Springer India, 2015), pp. 181–207.
32. S. Mukhopadhyay, K. Joshi, V. Chaudhary, N. Goel, S. De, R. K. Pandey, K.V.R.M. Murali, S. Mahapatra, in *IEEE International Reliability Physics Symposium Proceedings* (2014), p. GD 3.1
33. N. Choudhury, N. Parihar, N. Goel, A. Thirunavukkarasu, S. Mahapatra, IEEE J Electron. Devices Soc. **8**, 1281 (2020)
34. T. Samadder, N. Choudhury, S. Kumar, D. Kochar, N. Parihar, S. Mahapatra, IEEE Trans. Electron. Devices **68**, 485 (2021)
35. D. Angot, V. Huard, L. Rahhal, A. Cros, X. Federspiel, A. Bajolet, Y. Carminati, M. Saliva, E. Pion, F. Cacho, A. Bravaix, in *IEEE International Electron Devices Meeting Technical Digest* (2013), p. 15.4.1
36. A. Kerber, IEEE Electron. Device Lett. **35**, 294 (2014)
37. A. Kerber, T. Nigam, in *IEEE International Reliability Physics Symposium Proceedings* (2014), p. 6A.6.1
38. C. Prasad, M. Agostinelli, J. Hicks, S. Ramey, C. Auth, K. Mistry, S. Natarajan, P. Packan, I. Post, S. Bodapati, M. Giles, S. Gupta, S. Mudanai, K. Kuhn, in *IEEE International Reliability Physics Symposium Proceedings* (2014), p. 6A.5.1
39. C. Liu, H. Nam, K. Kim, S. Choo, H. Kim, H. Kim, Y. Kim, S. Lee, S. Yoon, J. Kim, J.J. Kim, L. Hwang, S. Ha, M.-J. Jin, H.C. Sagong, J.-K. Park, S. Pae, J. Park, in *IEEE International Electron Devices Meeting Technical Digest* (2015), p. 11.3.1
40. S. Ramey, M. Chahal, P. Nayak, S. Novak, C. Prasad, J. Hicks, in *IEEE International Reliability Physics Symposium Proceedings* (2015), p. 3B.2.1
41. M.D. Giles, N. Arkali Radhakrishna, D. Becher, A. Kornfeld, K. Maurice, S. Mudanai, S. Natarajan, P. Newman, P. Packan, T. Rakshit, in *Symposium on VLSI Technology Digest of Technical Papers* (2015), p. T150
42. T. Grasser, Microelectron. Reliab. **52**, 39 (2012)

43. F. Scanovsky, T. Grasser, in *IEEE International Reliability Physics Symposium Proceedings* (2012), p. XT.10.1
44. T. Naphade, N. Goel, P.R. Nair, S. Mahapatra, in *IEEE International Reliability Physics Symposium Proceedings* (2013), p. XT.5.1
45. T. Naphade, K. Roy, S. Mahapatra, in *IEEE International Electron Devices Meeting Technical Digest* (2013), p. 33.6.1
46. A. Chaudhary, B. Fernandez, N. Parihar, S. Mahapatra, IEEE Trans. Electron. Devices **64**, 256 (2017)
47. R. Anandkrishnan, S. Bhagdikar, N. Choudhury, R. Rao, B. Fernandez, A. Chaudhury, N. Parihar, S. Mahapatra, in *International Conference on Simulation of Semiconductor Processes and Devices* (2018), p. 181
48. N. Parihar, R. Anandkrishnan, A. Chaudhary, S. Mahapatra, IEEE Trans. Electron. Devices, **66**, 3273 (2019)
49. S. Bhagdikar, S. Mahapatra, in *International Conference on Simulation of Semiconductor Processes and Devices* (2018)
50. S.E. Rauch, IEEE Trans. Device Mater. Reliab. **7**, 524 (2007)
51. B. Kaczer, T. Grasser, P.J. Roussel, J. Franco, R. Degraeve, L.-A. Ragnarsson, E. Simoen, G. Groeseneken, H. Reisinger, in *IEEE International Reliability Physics Symposium Proceedings* (2010), p. 26
52. A. Kerber, P. Srinivasan, IEEE Electron. Device Lett. **35**, 431 (2014)
53. L. Zhou, Q. Zhang, H. Yang, Z. Ji, Z. Zhang, Q. Liu, H. Xu, B. Tang, E. Simoen, X. Ma, X. Wang, Y. Li, H. Yin, J. Luo, C. Zhao, W. Wang, IEEE Electron. Device Lett. **41**, 965 (2020)
54. R. Ranjan, Y. Liu, T. Nigam, A. Kerber, B. Parameshwaran, in *IEEE International Reliability Physics Symposium Proceedings* (2017), p. DG.10.1
55. M. Rafik, A.P. Nguyen, X. Garros, M. Arabi, X. Federspiel, C. Diouf, in *IEEE International Reliability Physics Symposium Proceedings* (2018), p. 4A.3.1

Chapter 11
BAT Framework Modeling of RMG HKMG Si and SiGe Channel FinFETs

Narendra Parihar, Nilotpal Choudhury, Tarun Samadder, Richard Southwick, Miaomiao Wang, James H. Stathis, and Souvik Mahapatra

11.1 Introduction

As shown in the previous chapters, Negative Bias Temperature Instability (NBTI) became an important reliability concern during the migration from Silicon Dioxide (SiO_2) to Silicon Oxynitride (SiON) MOSFETs [1–4]. It continues to impact the dual layer (SiO_2 or SiON based interlayer (IL) and Hafnium Dioxide (HfO_2) High-K layer) High-K Metal Gate (HKMG) bulk [5–11] and Fully Depleted Silicon On Insulator (FDSOI) [12, 13] planar MOSFETs, bulk and SOI FinFETs [13–26], and Gate All Around Stacked Nanosheet (GAA-SNS) FETs [27–30], with either Silicon (Si) or Silicon Germanium (SiGe) channel. The key features of NBTI are summarized below (reproduced from Chap. 3, Sect. 3.1).

As described in Chap. 1, Sect. 1.3, NBTI results in a gradual buildup of positive charges in the gate insulator of the device and shifts various transistor parameters, e.g., threshold voltage shift (ΔV_T), over time. It gets accelerated at more negative gate bias (V_G) during stress ($V_G = V_{GSTR}$) and at higher temperature (T), governed by the Voltage Acceleration Factor (VAF) and Arrhenius T activation energy (E_A) respectively. However, the positive charges accrued during stress reduce if the magnitude of V_G is reduced or removed ($V_G = V_{GREC}$ or 0 V), which reduces ΔV_T. Therefore, AC stress results in lower ΔV_T than DC stress. The ratio of AC to DC NBTI ΔV_T depends on the pulse duty cycle (PDC), pulse low bias (V_{GLOW}), AC stress mode and may or may not depend on the pulse frequency (f). Moreover, the recovery of NBTI

N. Parihar · N. Choudhury · T. Samadder · S. Mahapatra (✉)
Department of Electrical Engineering, Indian Institute of Technology Bombay, Powai, Mumbai 400076, India
e-mail: souvik@ee.iitb.ac.in

R. Southwick · M. Wang
IBM Research Division, Albany Nanotech, Albany, NY 12203, USA

J. H. Stathis
IBM Research Division, T. J. Watson Research Center, Yorktown Heights, NY 10598, USA

S. Mahapatra (ed.), *Recent Advances in PMOS Negative Bias Temperature Instability*,
https://doi.org/10.1007/978-981-16-6120-4_11

necessitates the use of various ultra-fast methods to measure the device parametric shift without any recovery artifacts, see Chap. 1, Sect. 1.2. Therefore, the ultra-fast Measure-Stress-Measure (MSM) method is used in this book.

As mentioned earlier in Chap. 10, Sect. 10.1, FinFETs were introduced to reduce the short channel effects associated with conventional planar MOSFETs at scaled technology nodes. However, FinFETs feature (110) fin sidewall-dominated channel, which is different from the (100) channel orientation of planar MOSFETs. It should be noted that early reports have shown higher NBTI in FinFETs compared to planar devices due to the difference in channel orientation. However, process optimization was used to reduce the degradation in matured nodes [14, 15].

Note that all the past reports of NBTI in FinFETs used conventional (not ultra-fast) measurements and only used DC stress [14–17]. Detailed DC and AC stress and recovery kinetics at different V_{GSTR}, T, PDC, and f are not measured. Furthermore, none has attempted to model the ΔV_T stress and recovery time kinetics during and after both DC and AC stress at various V_{GSTR}, T, PDC, and f. These aspects have been recently addressed by the present authors [18–26], and are described in this book.

Moreover, SiGe channel MOSFETs and FinFETs are shown to have lower ΔV_T than their Si channel counterparts [6, 8, 9, 16, 17]. As mentioned in Chap. 8, Sect. 8.1, the mechanism responsible for the impact of Ge% on ΔV_T is debated. Lower NBTI is suggested to be due to higher pre-stress interface trap density that results in lower stress induced interface trap generation [6], unfavorable valence band alignment resulting in either lower hole trapping [8, 16], or lower contribution from generated interface traps [9], and lowering of the oxide electric field (E_{OX}) near the channel/gate insulator interface due to large negative charge density in the gate insulator at the valence band edge [17]. However, none of the above reports have modeled the measured ΔV_T time kinetics across V_{GSTR} and T, which is a prerequisite for the validation of any physical mechanism.

As discussed in Chap. 2, the reduced NBTI for SiGe channel devices compared to their Si channel counterparts is universally observed in bulk [6, 8, 9, 11] and FDSOI [12, 13] planar MOSFETs and FinFETs [18, 19, 21, 22]. On the other hand, the NBTI magnitude increases with Nitrogen content (N%) in the gate insulator stack for both SiON [1–4] and HKMG [7, 10, 13, 18, 19, 21, 22] devices. Note that besides the ΔV_T magnitude, Germanium content (Ge%) in the channel and N% in the gate insulator impact the time kinetics of ΔV_T during and after stress, i.e., the power-law time slope n at longer stress time (t_{STR}) during stress and fraction remaining (FR) at a particular recovery time (t_{REC}) after stress (FR is defined as the ratio of ΔV_T at $t = t_{REC}$ to that at the end of stress at $t = t_{STR}$). These processes also impact the VAF, E_A and T dependence of VAF (i.e., the reduction in VAF at higher T). Increase in Ge% increases n at long t_{STR}, FR for a particular ratio of t_{REC} to t_{STR} (i.e., the recovery becomes slower), VAF, E_A and the T dependence of VAF (higher VAF reduction at higher T). However, increase in N% reduces n at long t_{STR}, FR for a particular ratio of t_{REC} to t_{STR} (i.e., the recovery becomes faster), VAF, $E_{A,}$ and the T dependence of VAF (i.e., lower VAF reduction at higher T). Moreover, changes in the compressive mechanical stress in the channel due to changes in the layout or device dimensions

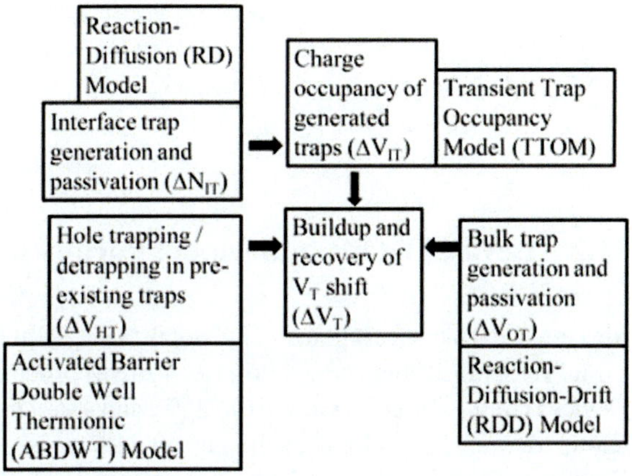

Fig. 11.1 Schematic of the BTI Analysis Tool (BAT) framework used in this book to model measured ΔV_T kinetics during and after DC and AC NBTI stress, reproduced from Chap. 4

also impact NBTI, which is reported for FDSOI planar MOSFETs [12, 13], FinFETs [23, 25], and GAA-SNS FETs [29, 30].

In Chap. 7 thorough Chap. 9, the BTI Analysis Tool (BAT) framework is used to model the impact of N% and Ge% on NBTI in bulk and FDSOI planar MOSFETs, the layout impact is also modeled in Chap. 9. The framework is used in Chaps. 6 and 10 to model the ultra-fast measured stress-recovery kinetics in Si bulk and SOI FinFETs. It is used in this chapter to model the N% and Ge% impact on ultra-fast measured stress-recovery kinetics in SiGe channel bulk FinFETs, as well as in Chaps. 12 and 13 to model the dimensions dependence of FinFETs and GAA-SNS FETs. Detailed modeling of AC NBTI is done in Chap. 14. The BAT framework is explained in Chap. 4 through Chap. 6 and is briefly described below (reproduced from Chap. 6, Sect. 6.3).

Figure 11.1 illustrates the BAT framework [10] used to model the time kinetics of ΔV_T during and after stress. It uses uncorrelated contributions from generated interface (ΔV_{IT}) and bulk (ΔV_{OT}) gate insulator traps, and hole trapping in process related pre-existing bulk gate insulator traps (ΔV_{HT}). The Reaction Diffusion (RD) model is used to calculate the time kinetics of interface traps (density ΔN_{IT}) [10, 31, 32]. In Chap. 4, the RD model is described and independently validated by measured data from Direct Current IV (DCIV) method [33]. The Transient Trap Occupancy Model (TTOM) is used to calculate the occupancy of generated interface traps and their contribution (ΔV_{IT}) [10], which is described and validated in Chap. 5, Sect. 5.3. The ΔV_{HT} and ΔV_{OT} kinetics are modeled by the Activated Barrier Double Well Thermionic (ABDWT) model [34] and Reaction Diffusion Drift (RDD) model [35] throughout this book, these models are described and validated in Chap. 5, Sect. 5.4 and Chap. 6, Sect. 6.2.

In Chaps. 7 and 9, the impact of N% on ΔV_T is shown to be due to increase in ΔV_{IT} and ΔV_{HT} (with relatively higher increase in ΔV_{HT}) and reduction in ΔV_{OT} subcomponents, although ΔV_{IT} dominates the measured ΔV_T across V_{GSTR} and T. In Chaps. 8 and 9, the impact of Ge% on ΔV_T is shown to be due to reduction in ΔV_{IT}, ΔV_{HT} and ΔV_{OT} subcomponents (with relatively higher reduction in ΔV_{HT}

and lower reduction in ΔV_{OT}), although once again, ΔV_{IT} dominates the measured ΔV_{T} across V_{GSTR} and T (for moderate to low values of V_{GSTR} and T). The impact of N% and Ge% changes on the measured ΔV_{T} in SiGe FinFETs is also explained by the above mechanisms in this chapter.

11.2 Device Details and Model Parameters

Measured data from Replacement Metal Gate (RMG) process based HKMG bulk p-FinFETs, fabricated using a proprietary IBM process, are modeled in this chapter. Devices having different Ge% (0%, 25%, and 45%) in the channel and different N% (Low, Medium, and High) in the gate stack are used to study the process impact. Devices having fin length (FL) of 200 nm, fin width (FW) of 10 nm, and 24 fins are used to minimize any impact of process variation. The RMG gate insulator stacks have standard, low temperature Chemical Oxide based IL and HfO_2 based High-K, resulting in an equivalent oxide thickness of approximately 1.1 nm. Note that a specialized treatment has been used to remove the Ge suboxide from the IL during gate stack formation. All data are obtained by the One Point Drop Down (OPDD) MSM method with 10 μs measurement delay, see Chap. 1, Sect. 1.2.

As mentioned before, the measured ΔV_{T} time kinetics is modeled using uncorrelated contributions from ΔV_{IT}, ΔV_{HT}, and ΔV_{OT} subcomponents using the framework illustrated in Fig. 11.1. The model parameters are listed in Table 11.1, and described below (reproduced from Chap. 6, Sect. 6.3). D1 is the reference Si channel and D2 through D6 are SiGe channel FinFETs, with 25% (D2, D3, D4) and 45% (D5, D6) Ge in the channel, and low (D2, D5), medium (D3, D6) and high (D4) N% in the gate insulator. Note that the D1 device is the same as the D3 device of Chap. 4, where DCIV measured ΔN_{IT} data have been quantified by the RD model. However, the D2 device has different IL process, and so the RD model parameters are also different, compared to the D4 device of Chap. 4.

As described in Chap. 4, Sect. 4.3, the process dependent RD model parameters are related to the pre-factor (K_{F10}), T independent field acceleration (Γ_0), bond polarization (α) and T activation energy (E_{AKF1}) of the inversion layer hole assisted bond dissociation process, see Chap. 4, Fig. 4.5, reproduced below as Fig. 11.2. The process dependent TTOM parameters are related to fast fraction of traps that undergo electron capture (f_{FAST}) and the electron capture time constant (τ_{EC}), see Chap. 5, Sect. 5.3. As shown in Chap. 5, Sect. 5.4, the process dependent ABDWT model parameters are related to the density of pre-existing defects (N_{0HT}), the energy barrier (E_{BM}), and the factors associated with E_{OX} dependence of the barrier (γ_B) and trap energy level ($\gamma_{E2} = m\gamma_B$), see Chap. 5, Fig. 5.10. Note that the ΔV_{HT} magnitude is generally smaller (than in Chap. 7) in these devices. The process dependent RDD model parameters are related to the pre-factor (K_{F30}), T independent field acceleration (Γ_{0OT}), and T activation energy (E_{AOT}) of the Anode Hole Injection (AHI) mechanism assisted bond dissociation (note, the bond polarization factor (α_{OT}) is not varied across devices but listed for completeness), and the forward reaction rate for ions

Table 11.1 Process dependent RD, TTOM, ABDWT, and RDD parameters (as per classification done in Chap. 4 through Chap. 6) used in this chapter

Parameter	Unit	D1	D2	D3
K_{F10}	cm/Vs	1	3.67×10^3	2.5×10^2
E_{AKF1}	eV	0.29	0.67	0.55
Γ_0	cm/MV	0.29	0.05	0.1
α	qÅ	1.8	2.3	2.3
f_{FAST}	–	0.42	0.45	0.44
τ_{EC}	s	0.03	0.03	0.03
N_{0HT}	1/cm^2	1	–	1.4×10^{-2}
E_{BM}	eV	1.3	–	1.3
τ_B	C.cm	4.5×10^{-9}	–	8.0×10^{-9}
K_{F30}	1/s	1	1.1×10^{-2}	1.1×10^{-2}
Parameter	Unit	D4	D5	D6
K_{F10}	cm/Vs	1.67×10^3	2.0×10^{11}	1.0×10^4
E_{AKF1}	eV	0.54	1.35	0.75
Γ_0	cm/MV	0.12	0.001	0.008
α	qÅ	2.3	3.0	3.0
f_{FAST}	–	0.43	0.60	0.40
τ_{EC}	s	0.03	0.008	0.008
N_{0HT}	1/cm^2	2.9×10^{-2}	–	–
E_{BM}	eV	0.9	–	–
γ_B	C cm	7.5×10^{-9}	–	–
K_{F30}	1/s	9.1×10^{-3}	9.8×10^{-3}	9.1×10^{-3}

As per Chap. 4, Sect. 4.4, the channel/IL precursor density for the RD model is suitably adjusted according to Ge% in the channel. The parameters K_{F10}, N_{0HT}, and K_{F30} are normalized to those for the D1 device to maintain confidentiality. The product of K_{F10} and exp ($-E_{AKF1}$/kT) reduces at higher Ge% (fixed N%) and increases at higher N% (fixed Ge%). Other adjustable model parameters are $m = 2.4$ for ABDWT, as well as $E_{AOT} = 1.14$ eV, $\Gamma_{0OT} = 0.13$ cm/MV, $\alpha_{OT} = 3.6$ qÅ and $K_{F50} = 23$ cm^3/s for RDD, for all devices. ΔV_{HT} is negligible in devices D2, D5, and D6

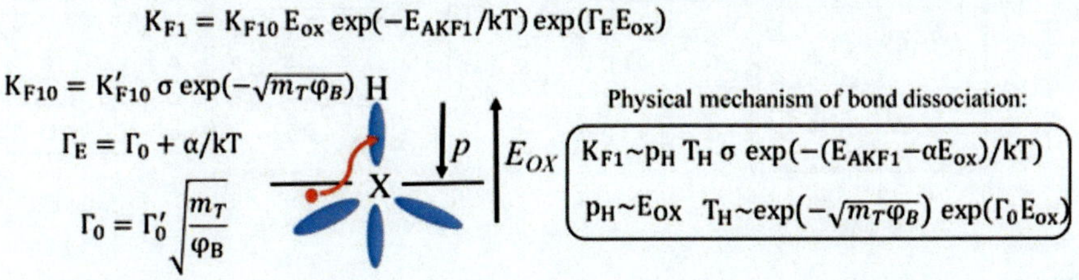

Fig. 11.2 Schematic of the inversion layer hole and oxide electric field induced dissociation of H passivated defects at the channel/gate insulator interface, details in Chap. 4, Sect. 4.3

(K_{F50}), see Chap. 6, Sect. 6.2. Other model parameters are process agnostic and are listed in the respective sections of Chap. 4 through Chap. 6 (Table 4.1, Table 5.1, Table 5.2, and Table 6.1).

Note, different devices have different E_{OX} at a particular V_{GSTR}, due to differences in their flat band voltages related to the differences in Ge% and N% between them (more so for the Ge = 45% devices). Moreover, the interface trap precursor density at the channel/IL interface also varies across devices due to the differences in Ge%, as discussed in Chap. 4, Sect. 4.4. The bulk trap precursor density, however, is kept same across all devices, as discussed in Chap. 6, Sect. 6.2.

11.3 Stress and Recovery Time Kinetics—DC Stress

Figure 11.3 shows the time evolution of measured and modeled ΔV_T from Si and

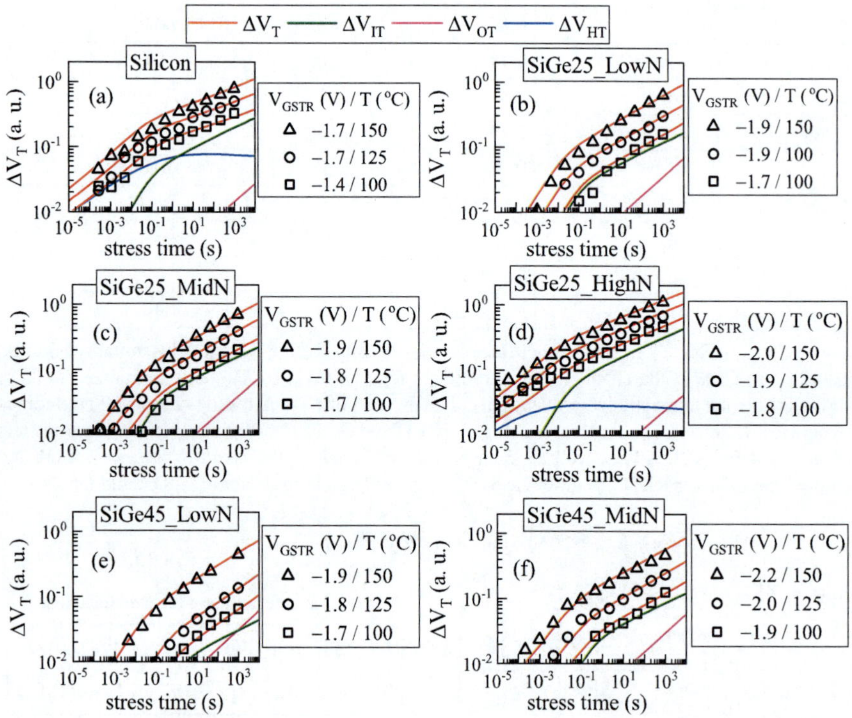

Fig. 11.3 Measured and modeled ΔV_T time kinetics for Si and SiGe (different Ge% and N%) p-FinFETs during DC stress. The underlying ΔV_{IT}, $\Delta V_{HT,}$ and ΔV_{OT} subcomponents are shown for the lowest dataset. All measured and modeled data including subcomponents are normalized to the Si device modeled ΔV_T at −1.7 V, 150 °C, 10 Ks. Symbols: experiment, lines: model calculation. Data from [21]

SiGe (different channel Ge% and different gate stack N%) devices, for different V_{GSTR} and T conditions. The model subcomponents (ΔV_{IT}, ΔV_{HT}, and ΔV_{OT}) for the dataset having lowest ΔV_T magnitude are also shown for all devices. Note that all data for all devices are normalized to the maximum modeled ΔV_T at −1.7 V/150 °C and 10 Ks of the reference Si device.

The time kinetics of ΔV_T evolves rapidly at the initiation of stress and asymptotically settles into a power-law dependence at long stress time ($t_{STR} > 1$ s) when plotted in the usual log–log scale. The ΔV_{HT} subcomponent is only seen for the Si and SiGe25 (Ge = 25%) moderate and high N% devices (see Sect. 11.5). It evolves rapidly at the initiation of stress, however, it saturates ($n \sim 0$ in a log–log plot) for a longer time. The ΔV_{HT} contribution is found to be negligible for all the other SiGe devices. The ΔV_{IT} and ΔV_{OT} subcomponents show a gradual buildup at the initiation of stress, and power-law time kinetics at long stress time with slope (n) of ~1/6 and ~1/3 respectively. The ΔV_{IT} subcomponent is present in all devices and often dominates. The relative ΔV_{OT} contribution is significantly lower for the Si compared to all SiGe devices, while it is highest for the SiGe45_LowN device.

All the subcomponents and hence the overall ΔV_T reduces with increase in Ge% (Fig. 11.3(a), (b), (e)). However, the reduction in ΔV_{IT} and ΔV_{HT} is larger than the reduction in ΔV_{OT}, so the relative contribution from ΔV_{OT} increases with increase in Ge%. Moreover, with increase in N% (Fig. 11.3(b)–(f)), the ΔV_{IT} and ΔV_{HT} contributions increase while that of ΔV_{OT} reduces (ΔV_{HT} is negligible in (e), (f)). The increase in ΔV_{IT} and ΔV_{HT} is considerably higher than the reduction in ΔV_{OT}, hence the overall ΔV_T increases with increase in N%. Note, the reduction of ΔV_{OT} at higher N% is consistent with the Time Dependent Dielectric Breakdown (TDDB) data shown in [36]; it is well known that the generated bulk gate insulator traps are responsible for TDDB (refer to Chaps. 3 and 6 for details).

Note that eight processes (variation with Ge%, N%) dependent model parameters are needed to explain the stress time kinetics for diverse V_{GSTR} and T conditions as listed in Table 11.1: four for RD model (K_{F10}, E_{AKF1}, Γ_0, and α), three for ABDWT model (N_{0HT}, E_{BM} and γ_B), and one for RDD model (K_{F30}). Among these, the relative variations in the parameters K_{F10} and Γ_0 can be obtained using band structure calculations, which are discussed in Chap. 4, Sect. 4.3.

For each device, the stress time kinetics is measured and modeled at four different T (only three T points for the SiGe45 devices as ΔV_T becomes negligible at lower T), and for four different V_{GSTR} at each T, only a few are shown in Fig. 11.3 for brevity. It should be noted that two additional parameters for TTOM (f_{FAST} and τ_{EC}) are adjusted to model the recovery time kinetics after DC stress as discussed later. Only one parameter (K_{F30} for RDD) is re-adjusted to model the Mode-B AC stress and recovery kinetics across V_{GSTR} and T ($V_{GSTR} \times T$ matrix, same as DC stress), see Sect. 11.4, due to the reasons mentioned in Chap. 6, Sect. 6.3. Therefore, a total of 64 datasets (48 for SiGe45 devices) are modeled with ten adjustable parameters for every device.

Note that the T activated pre-factor (K_{F10}*exp(–E_{AKF1}/kT) of the ΔV_{IT} subcomponent reduces at higher Ge% and increases at higher N%. The T activation energy E_{AKF1} increases at higher Ge% and reduces at higher N%, the impact of N% on E_{AKF1}

is same as reported in [37]. The field acceleration factor Γ_0 reduces at higher Ge% and slightly increases at higher N%. The process impact on K_{F10} and Γ_0 is discussed in detail in Chap. 4, Sect. 4.3. Note that the impact of N% on Γ_0 is different for the SiGe FinFETs (Γ_0 slightly increase with N%) than the Si MOSFETs (Γ_0 reduces with N%) modeled in Chap. 7. The polarization term α increases with increase in Ge% but does not vary for different N%.

The trap density N_{0HT} related to ΔV_{HT} reduces with increase in Ge% while it increases with increase in N%. Note that the reduction in N_{0HT} with increase in Ge% is presumably a reflection of the unfavorable defect band alignment as discussed in [8]. The increase in N_{0HT} at higher N% is a universal feature and is verified by flicker noise measurements [7], also see Chap. 3, Sect. 3.3. The other ΔV_{HT} parameters indicate faster saturation with increase in N%. The parameter K_{F30} for ΔV_{OT} reduces with increase in Ge% and N%. As mentioned above, the reduction in bulk trap generation with N% has also been reported in TDDB experiments [36].

Figure 11.4 shows the Ge% and N% dependence of the measured and modeled

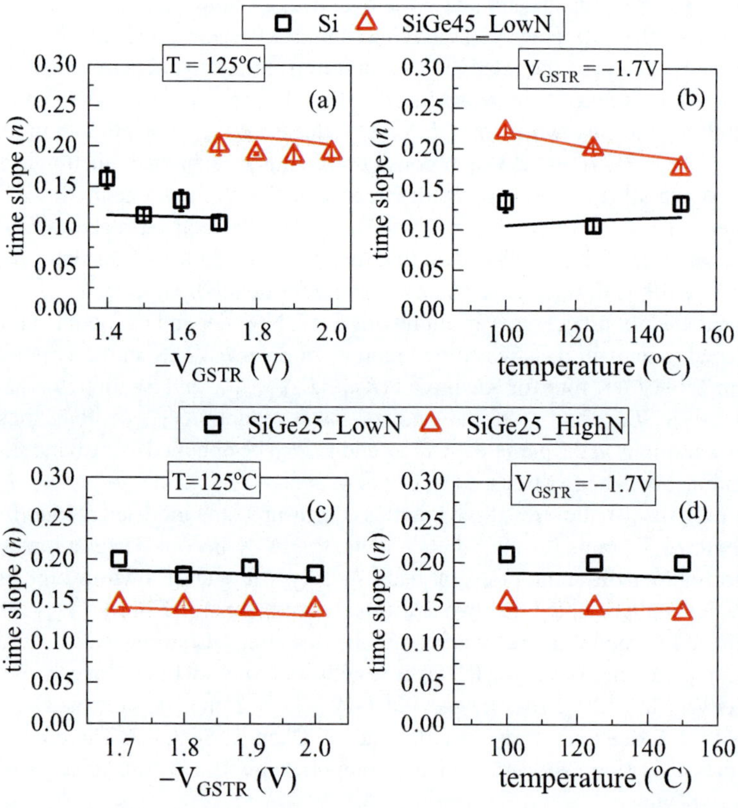

Fig. 11.4 Measured and modeled power-law time slope (n) as a function (a, c) V_{GSTR} and (b, d) T for different (a, b) Ge% and (c, d) N% p-FinFETs during DC stress. Symbols: experiment, lines: model calculation. Data from [21]

long-time power-law slope n at different (a, c) V_{GSTR} and (b, d) stress T. The n is extracted using linear regression of the ΔV_T time kinetics (from a log–log plot) in the t_{STR} interval of 10 s to 1 Ks. Note that different subcomponents (ΔV_{IT}, ΔV_{HT}, and ΔV_{OT}) have different time kinetics as described above, and they also have different VAF and E_A (both values are highest for ΔV_{OT}, VAF is somewhat similar for ΔV_{IT} and ΔV_{HT}, while E_A is lowest for ΔV_{HT}, see Sect. 11.5). The subcomponents have different relative dominance (e.g., higher relative ΔV_{OT} contribution at higher V_{GSTR} and/or T), and therefore, they differently impact the long-time ΔV_T kinetics as V_{GSTR} and/or T are varied.

The ΔV_{IT} subcomponent shows power-law time slope of $n \sim 1/6$ at a longer time, which is independently verified using (delay corrected) DCIV measurements in Chap. 4, Sect. 4.4. Higher relative ΔV_{HT} contribution reduces the slope n of overall ΔV_T at long time (since n is ~ 0 for ΔV_{HT} in a log–log plot), see Chap. 3, Sect. 3.3 Likewise, n of overall ΔV_T increases with relative increase in ΔV_{OT} (since n is ~ 1/3 for ΔV_{OT}), see Chap. 3, Sect. 3.4.

Although ΔV_{IT} dominates, the relatively higher ΔV_{HT} for the Si device reduces the longer-time n of overall ΔV_T below ~ 1/6 (the ΔV_{OT} contribution is smaller in this device), see Fig. 11.4(a, b). As discussed above, the relative contribution of ΔV_{OT} increases with the increase in Ge%, while the contribution from ΔV_{HT} decreases. Therefore, n increases (beyond ~ 1/6) with increase in Ge%. However, the relative ΔV_{HT} contribution increases while that of ΔV_{OT} decreases with increase in N%, and hence the longer-time n reduces, see Fig. 11.4(c, d). Moreover, for Si and SiGe devices with high N%, the long-time n can further reduce due to the stress reduction effect, since the resulting higher ΔV_T in these devices would result in a reduction in effective stress (see Chap. 4, Sect. 4.3). The SiGe45_LowN device (highest Ge% and lowest N%) shows strong T dependence of n due to significant contribution from ΔV_{OT}, since the T activation of ΔV_{OT} is very high compared to that of ΔV_{IT} and ΔV_{HT}, refer to Chap. 4 through Chap. 6 for further details.

Figure 11.5 shows the time evolution of measured and modeled ΔV_T recovery after stress under different V_{GSTR} and T conditions in Si and SiGe devices having different Ge% and N%. As mentioned before, all data are normalized to the modeled ΔV_T at −1.7 V/150 °C/10 Ks for the Si reference device during stress (Fig. 11.3(a)). Identical recovery bias (V_{GREC}) of 0 V is used for all devices and experimental conditions. The underlying ΔV_{IT}, $\Delta V_{HT,}$ and ΔV_{OT} model subcomponents for the dataset having lowest ΔV_T magnitude are also shown for all cases.

The recovery kinetics of different subcomponents has been discussed earlier in Chap. 4 through Chap. 6. Note that the ΔV_{HT} and ΔV_{OT} subcomponents recover fast and slow respectively (ΔV_{OT} is semi-permanent). The recovery time kinetics for ΔV_{HT} remain almost similar for different processes discussed here, and the same holds for that of and ΔV_{OT}. However, it should be noted that the ΔV_{HT} contribution is negligible for most SiGe devices, and so it is not possible to accurately determine the process dependence of hole detrapping. The ΔV_{IT} recovery is distributed over several decades in time and is due to two processes: recovery by fast electron capture (ΔV_{IT_FAST}) for traps that go below the Fermi level as the magnitude of V_G is reduced

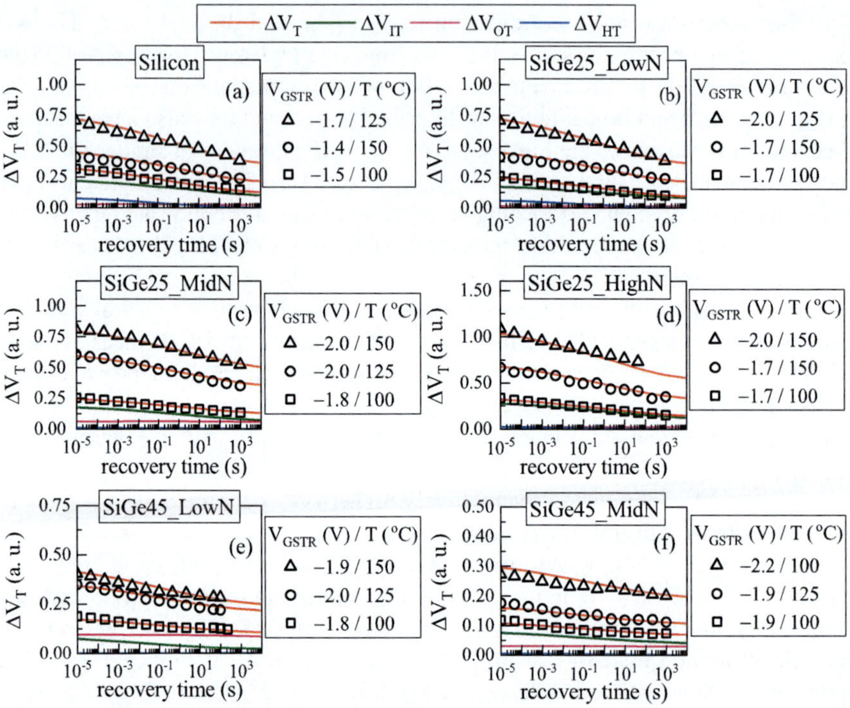

Fig. 11.5 Measured and modeled ΔV_{T} kinetics for Si and SiGe (different Ge% and N%) p-FinFETs after DC stress. The underlying ΔV_{IT}, ΔV_{HT}, and ΔV_{OT} subcomponents are shown for the lowest dataset. All measured and modeled data including subcomponents are normalized to the Si device modeled ΔV_{T} during stress at −1.7 V, 150°C, 10 Ks. Symbols: experiment, lines: model calculation. Data from [21]

for recovery and the recovery by passivation of the generated interface traps by Hydrogen (H/H_2) back diffusion ($\Delta V_{\mathrm{IT\_SLOW}}$).

The recovery due to $\Delta V_{\mathrm{IT\_FAST}}$ depends on the time constant for electron capture (τ_{EC}) and the fraction of traps (f_{FAST}) that go below the Fermi level as the gate bias is reduced in magnitude from V_{GSTR} to V_{GREC}, and is handled by TTOM. The parameter f_{FAST} slightly increases at higher Ge% but reduces with increase in N%. The electron capture time (τ_{EC}) reduces with higher Ge% but does not change with variation in N%. Note that increase in f_{FAST} and reduction in τ_{EC} for higher Ge% channels have also been seen for Si-capped SiGe devices, discussed in Chap. 8.

The recovery of $\Delta V_{\mathrm{IT\_SLOW}}$ depends on the stochastic hopping and lock-in of the H_2 molecules, which is modeled by slowing down the H_2 diffusivity in time only during recovery in the continuum framework, see Chap. 4, Sect. 4.3. Note that the diffusivity reduction factor used for FinFETs ($A = 35$) is somewhat higher as compared to that used for planar devices ($A = 7$) in earlier chapters, which is presumably due to the 3-dimensional nature of FinFETs that enhances radial diffusion. The parameter only depends on the device architecture and is found to be independent of any other process changes like variation in Ge% or N% ($A = 35$ is also used in Chaps. 10 and 12 for other FinFETs and GAA-SNS FETs). These aspects are verified using a stochastic implementation of the RD model in [38].

The recovery kinetics slows down for higher Ge% devices due to the relatively higher contribution from ΔV_{OT} and lower contribution from ΔV_{HT} subcomponents. However, higher N% increases the recovery because of the relatively higher contribution from ΔV_{HT} and lower contribution from ΔV_{OT}. As mentioned before, two additional parameters (τ_{EC} and f_{FAST}), apart from the eight used earlier to model the stress kinetics (the stress parameters are unchanged during recovery), are used to model the recovery kinetics for a wide range of V_{GSTR} and T conditions shown in Fig. 11.5.

11.4 Stress and Recovery Time Kinetics—AC Stress

Figure 11.6 shows the time evolution of measured and modeled ΔV_T for Mode-B AC stress at different V_{GSTR} ($=V_{GHIGH}$) and T in Si and SiGe (different Ge%, N%) devices. As explained earlier in Chap. 1, Sect. 1.2, measurement is done at the end of pulse off phase during Mode-B AC stress; identical delay is used for DC and AC stress. The PDC and f are kept identical for all the devices shown in Fig. 11.6.

Note, all data are normalized to the modeled ΔV_T at −1.7 V/150 °C/10 Ks for the Si reference device in Fig. 11.3(a) for DC stress. The ΔV_{IT} and ΔV_{OT} subcomponents for the dataset having the lowest magnitude of ΔV_T are also shown for all devices (note that ΔV_{HT} is negligible for Mode-B AC stress, see Chap. 5, Sect. 5.3 and Chap. 14). As mentioned before, except for the RDD model pre-factor K_{F30}, all other parameters that are used to model the AC time kinetics under different V_{GSTR} and T conditions are identical to DC stress, for all Si and SiGe (different Ge%, N%) devices studied in this chapter.

As mentioned before, four T conditions (only three for the SiGe45 devices) are used for stress, and four V_{GHIGH} values are used at each T. The ΔV_{OT} is f dependent on these devices (similar to those shown in Chaps. 6, 7, 9 and 10), and the pre-factor K_{F30} reduces at higher f AC stress (but it does not depend on any other stress conditions). Although the exact reason is not well understood as yet, the f dependence of ΔV_{OT} is analyzed in detail in [26, 39] and described later in Chap. 14, and is consistent with the f dependence of AC TDDB experiments as shown in [40, 41].

The impact of Ge% and N% on AC stress kinetics is similar to that observed for DC stress (i.e., ΔV_T reduces at higher Ge% but increases at higher N%), except that the contribution of ΔV_{HT} is negligible for Mode-B AC stress in all the devices (including in Si channel reference), especially so at PDC = 50% as used in these experiments. This is because any holes trapped during the pulse on phase get detrapped during the pulse off phase before measurement for the Mode-B AC stress, which is explained in Chap. 5, Sect. 5.3 and also Chap. 14. Hence, the overall ΔV_T time kinetics is only governed by ΔV_{IT} and ΔV_{OT} subcomponents.

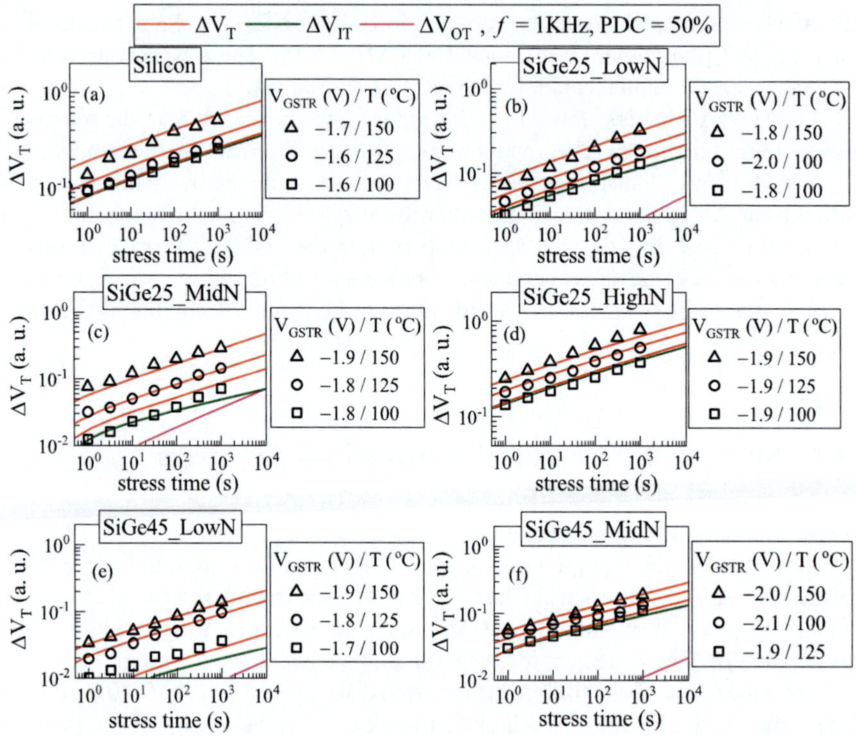

Fig. 11.6 Measured and modeled ΔV_T kinetics for Si and SiGe (different Ge% and N%) p-FinFETs during Mode-B AC stress. The underlying ΔV_{IT} and ΔV_{OT} subcomponents are shown for the lowest dataset, and ΔV_{HT} is found to be negligible for all devices. All measured and modeled data including subcomponents are normalized to the Si device modeled ΔV_T during DC stress at -1.7 V, 150 °C, 10 Ks. Symbols: experiment, lines: model calculation. Data from [22]

The process dependence of the ΔV_{IT} and ΔV_{OT} subcomponents for AC stress is identical to that for DC stress. Increase in Ge% (Fig. 11.6(a), (b), (e)) results in reduction for both ΔV_{IT} and ΔV_{OT}, although the reduction in ΔV_{IT} is larger than that of ΔV_{OT}. This results in larger relative ΔV_{OT} contribution at higher Ge%. However, ΔV_{IT} increases while ΔV_{OT} reduces with an increase in N% (Fig. 11.6(b)–(f)), and hence, the relative ΔV_{OT} contribution reduces at higher N%. Note that the longer-time power-law slope n for Mode-B AC stress is always equal to or higher than the corresponding value under DC stress. This is because ΔV_{HT} contribution is negligible for Mode-B AC stress (ΔV_{HT} would saturate at longer stress time and reduce n of overall ΔV_T if present, see Chap. 3, Sect. 3.3). Likewise, for higher Ge%, the long-time n increases due to higher ΔV_{OT} as the n for ΔV_{OT} is higher than n for ΔV_{IT} as mentioned earlier in this chapter (see Chap. 3, Sect. 3.4).

Figure 11.7 shows the time evolution of measured and modeled ΔV_T recovery after Mode-B AC stress at different V_{GSTR} (=V_{GHIGH}) and T in Si and SiGe (different Ge%, N%) devices. As before, all data are normalized to the modeled ΔV_T at $-$

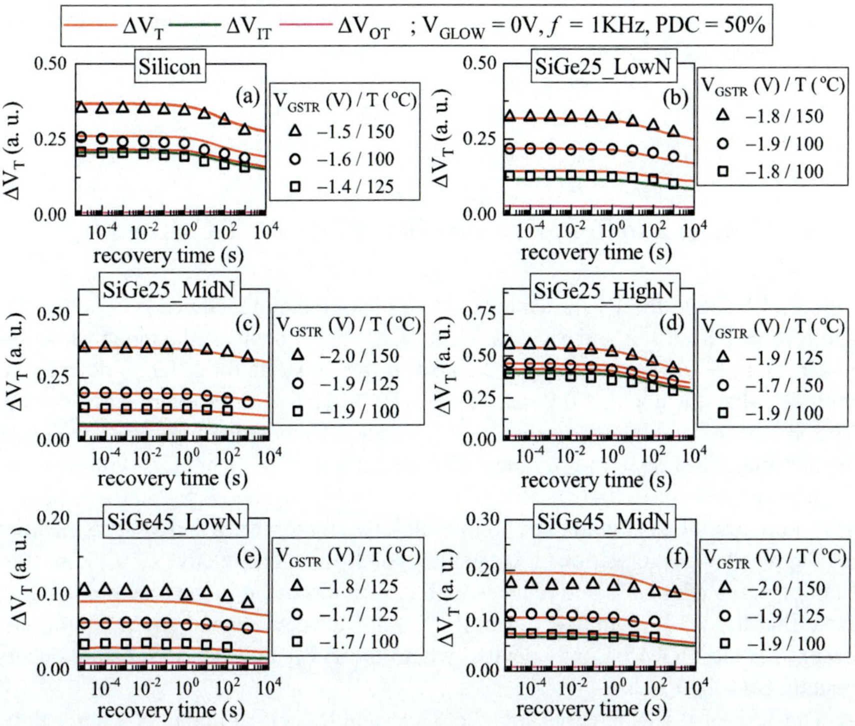

Fig. 11.7 Measured and modeled ΔV_T kinetics for Si and SiGe (different Ge% and N%) p-FinFETs after Mode-B AC stress. The underlying ΔV_{IT} and ΔV_{OT} subcomponents are shown for the lowest dataset, and ΔV_{HT} is found to be negligible for all devices. All measured and modeled data including subcomponents are normalized to the Si device modeled ΔV_T during DC stress at −1.7 V, 150 °C, 10 Ks. Symbols: experiment, lines: model calculation. Data from [22]

1.7 V/150 °C/10 Ks for the Si reference device in Fig. 11.3(a) for DC stress, and the model subcomponents for the dataset having lowest ΔV_T magnitude are shown for all devices. Note that as ΔV_{HT} is negligible, the recovery of ΔV_T after Mode-B AC stress is due to that of ΔV_{IT} and ΔV_{OT} subcomponents. The recovery of ΔV_{OT} is negligible, and the recovery of ΔV_{IT} is only due to the ΔV_{IT_SLOW} fraction for Mode-B AC stress. Note that the contribution from ΔV_{IT_FAST} fraction is negligible during recovery, as the electron capture process occurs during the pulse off phase of stress before the onset of measurement. This is discussed in further detail in Chap. 14. So, the recovery after the end of Mode-B AC stress is delayed due to the absence of ΔV_{IT_FAST} and ΔV_{HT} contributions. For Si and SiGe devices with high N%, the recovery kinetics is governed only by ΔV_{IT} due to negligible ΔV_{OT}. The recovery is further delayed at higher Ge% due to increased relative contribution from the ΔV_{OT} subcomponent.

Note that the fractional recovery after Mode-B AC stress for low to moderately high t_{REC} values is always smaller (FR is higher) compared to DC stress, due to the absence of the ΔV_{HT} and ΔV_{IT_FAST} subcomponents as mentioned above. However, DC and Mode-B AC stress show similar FR at very long t_{REC}, as this is controlled by ΔV_{IT_SLOW} and ΔV_{OT} even for DC stress. As mentioned before, no parameter is adjusted to model recovery after Mode-B AC stress.

11.5 Voltage and Temperature Dependence—DC Stress

Figure 11.8 shows the V_{GSTR} dependence of measured and modeled ΔV_T at fixed T and $t_{STR} = 1$ Ks for Si and SiGe (different Ge%, N%) devices; the model subcomponents (ΔV_{IT}, $\Delta V_{HT,}$ and ΔV_{OT}) are also shown. All data for different devices are normalized to the modeled ΔV_T at −1.7 V/150 °C/10 Ks for the reference Si device. The Si device has highest magnitude of overall ΔV_T when compared to the other devices under a similar V_{GSTR} range. The magnitude of ΔV_T reduces with increase in Ge% (Fig. 11.8(a)–(b)–(e)) and increases with higher N% (Fig. 11.8(b)–(f)). The ΔV_{HT} contribution is found to be non-negligible only for the Si and SiGe25 medium and high N devices for various V_{GSTR} and T conditions. The relative ΔV_{OT} contribution increases with Ge% and reduces with N% as discussed in the previous section. Note that the ΔV_{IT} dominates overall ΔV_T for all devices and stress conditions, except for the SiGe45_LowN device, where the ΔV_{IT} and ΔV_{OT} subcomponents equally contribute.

The VAF of ΔV_T increases at higher Ge% and reduces at higher N% and can be explained using the process dependence of the VAF of the underlying subcomponents. Higher VAF at higher Ge% is due to increase in VAF of the ΔV_{IT} subcomponent. Moreover, higher relative ΔV_{OT} contribution at higher Ge% also contributes to increasing the VAF of overall ΔV_T, since it has highest VAF among all the subcomponents. On the other hand, reduction in the VAF of ΔV_{IT} at higher N% and lower relative ΔV_{OT} contribution reduce the VAF of overall ΔV_T. Note that although ΔV_{HT} increases with increase in N%, its relative contribution to overall ΔV_T remains small even for the SiGe25_HighN device. The VAF of the overall modeled ΔV_T and the underlying subcomponents are shown for all devices.

The field dependence (hence voltage dependence or VAF) of the ΔV_{IT} subcomponent is process dependent and changes with change in Ge% and N%. Note that the field acceleration is given by $\Gamma_E = \Gamma_0 + \alpha/kT$, where Γ_0 is associated with hole tunneling and reduces, while the polarization factor α and hence Γ_E increases at higher Ge% (kT is thermal energy), see Table 11.1. Hence, the VAF of ΔV_{IT} increases at higher Ge%. On the other hand, although Γ_0 increases slightly at higher N%, the presence of higher N% in the IL increases the effective dielectric constant (ε_{IL}) and reduces the VAF of V_{IT} due to the reason discussed in Chap. 4, Sect. 4.3. The field dependence of ΔV_{HT} is process independent, but VAF (related to the voltage dependence) changes between devices due to the same reason. The VAF of ΔV_{OT} is independent of processes as well, but it reduces at higher T due to the bond polarization factor, see Chap. 6, Sect. 6.2 for further details.

Figure 11.9 shows the T dependence of measured and modeled ΔV_T at fixed

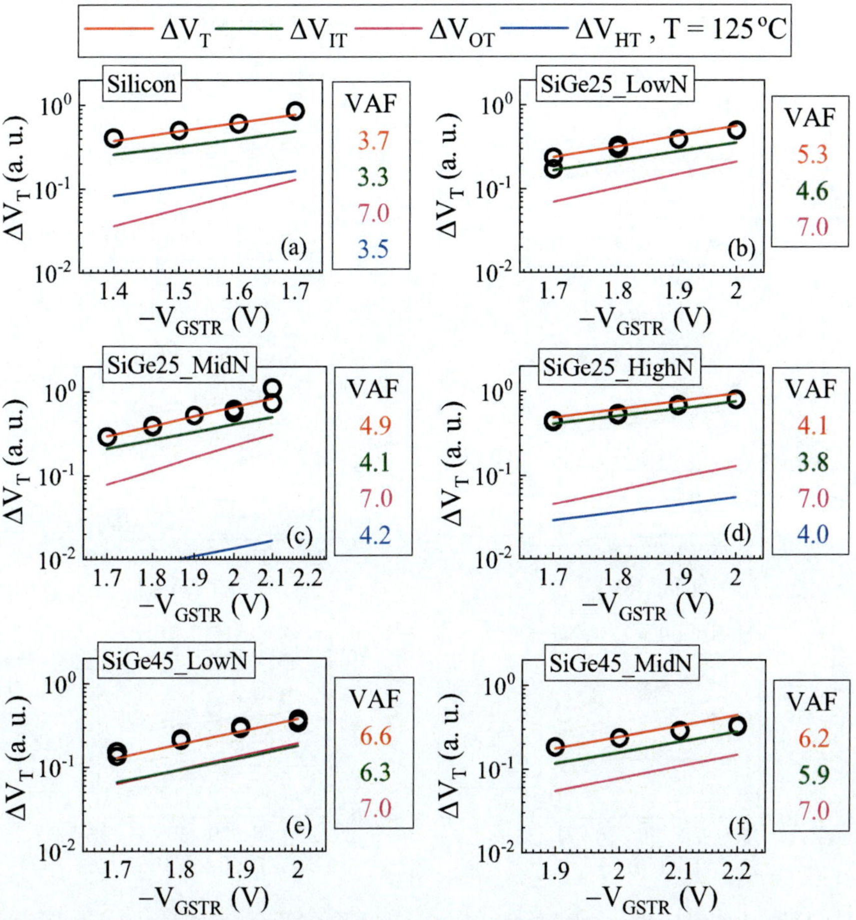

Fig. 11.8 Measured and modeled fixed time ($t_{STR} = 1$ Ks) ΔV_T as a function of V_{GSTR} for Si and SiGe (different Ge% and N%) p-FinFETs during DC stress. The underlying ΔV_{IT}, ΔV_{HT}, and ΔV_{OT} subcomponents are shown. All measured and modeled data including subcomponents are normalized to the Si device modeled ΔV_T at -1.7 V, 150 °C, 10 Ks. The VAF values are mentioned for the overall model and underlying subcomponents. Symbols: experiment, lines: model calculation. Data from [21]

V_{GSTR} and $t_{STR} = 1$ Ks for Si and SiGe (different Ge%, N%) devices; the model subcomponents are also shown. Note, all data for different devices are normalized to the modeled ΔV_T at -1.7 V/150 °C/10 Ks for the reference Si device. Note that the T dependence is obtained at different V_{GSTR} for Si and SiGe devices. The T activation E_A of modeled ΔV_T and the underlying subcomponents are shown for all devices. Although the parameters for the T activation of ΔV_{HT} and ΔV_{OT} subcomponents are process independent, the obtained E_A values changes with a change in E_{OX} (and V_{GSTR}) due to the soft saturation related stress reduction effect.

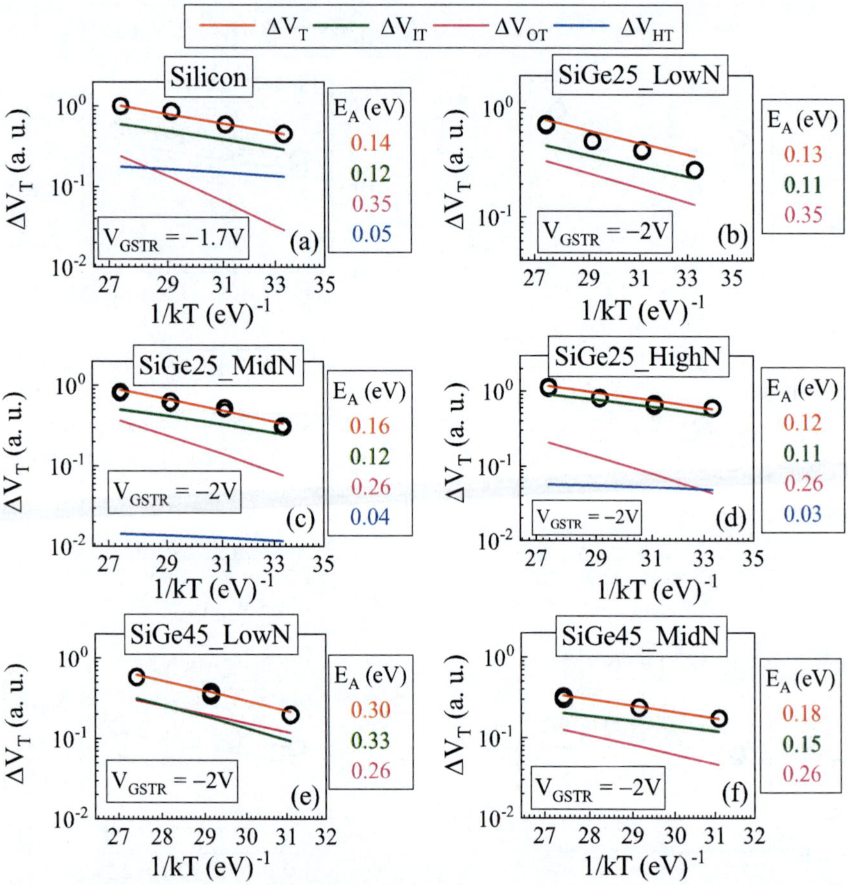

Fig. 11.9 Measured and modeled fixed time ($t_{STR} = 1$ Ks) ΔV_T as a function of T for Si and SiGe (different Ge% and N%) p-FinFETs during DC stress. The underlying ΔV_{IT}, ΔV_{HT}, and ΔV_{OT} subcomponents are shown. All measured and modeled data including subcomponents are normalized to the Si device modeled ΔV_T at −1.7 V, 150 °C, 10Ks. The E_A values are mentioned for the overall model and underlying subcomponents. Symbols: experiment, lines: model calculation. Data from [21]

The T activation of ΔV_{IT} (E_{AKF1}) is process dependent and it increases at higher Ge% and reduces at higher N%. First principles calculation is needed to explain the Ge% impact on E_{AKF1}, which is beyond the scope of the present analysis. The reduction in dissociation energy for X–H bonds in the presence of Nitrogen is shown to be responsible for the reduction in E_{AKF1} [37].

Note that the T activation of ΔV_{OT} is higher than that for ΔV_{IT} and ΔV_{HT} while that of ΔV_{HT} is very small (experimental evidences are provided in Chap. 3), which means E_A (ΔV_{OT}) > E_A (ΔV_{IT}) > E_A (ΔV_{HT}). The increase in E_A of overall ΔV_T at higher Ge% is due to increase in E_{AKF1} for ΔV_{IT} and also partly due to the relative increase in ΔV_{OT} contribution. It is remarkable that the SiGe45_LowN device has similar contributions from ΔV_{OT} and ΔV_{IT} subcomponents and the T activation of ΔV_{IT} is also highest. Hence, highest E_A of overall ΔV_T is observed for this device. The reduction in E_A of overall ΔV_T at higher N% is primarily due to the reduction in E_A of the ΔV_{IT} subcomponent, and relatively lower ΔV_{OT} contribution.

Figure 11.10 shows the measured and modeled fixed time ΔV_T as a function of V_{GSTR} for different T in Si and SiGe (different Ge%, N%) devices. All data for different devices are normalized to the modeled ΔV_T at −1.7 V/150 °C/10 Ks for the reference Si device. Note that VAF reduces with increase in T, and the extent of reduction is higher for higher Ge% but is lower for higher N% devices.

Figure 11.11 shows the corresponding measured and modeled T dependence of fixed time ΔV_T at different V_{GSTR}. Once again, all data for different devices are normalized to the modeled ΔV_T at −1.7 V/150 °C/10 Ks for the reference Si device. Note that the E_A reduces with increase in V_{GSTR} (consistent with VAF reduction at

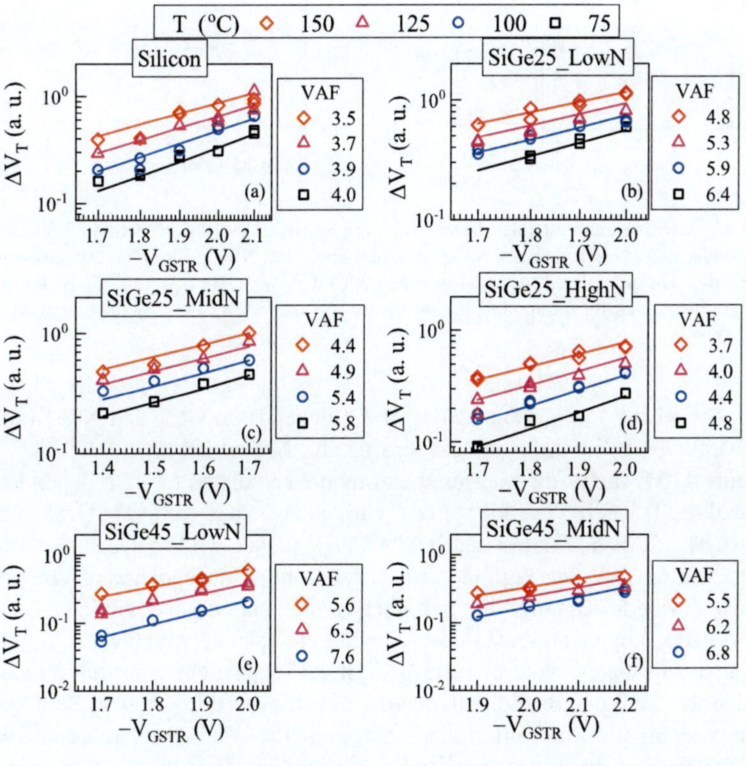

Fig. 11.10 Measured and modeled fixed time (t_{STR} = 1 Ks) ΔV_T as a function of V_{GSTR} at different T under DC stress for Si and SiGe (different Ge% and N%) p-FinFETs. All measured and modeled data are normalized to the Si device modeled ΔV_T at −1.7 V, 150 °C, 10 Ks. The VAF values are mentioned for the modeled data. Symbols: experiment, lines: model calculation. Data from [21]

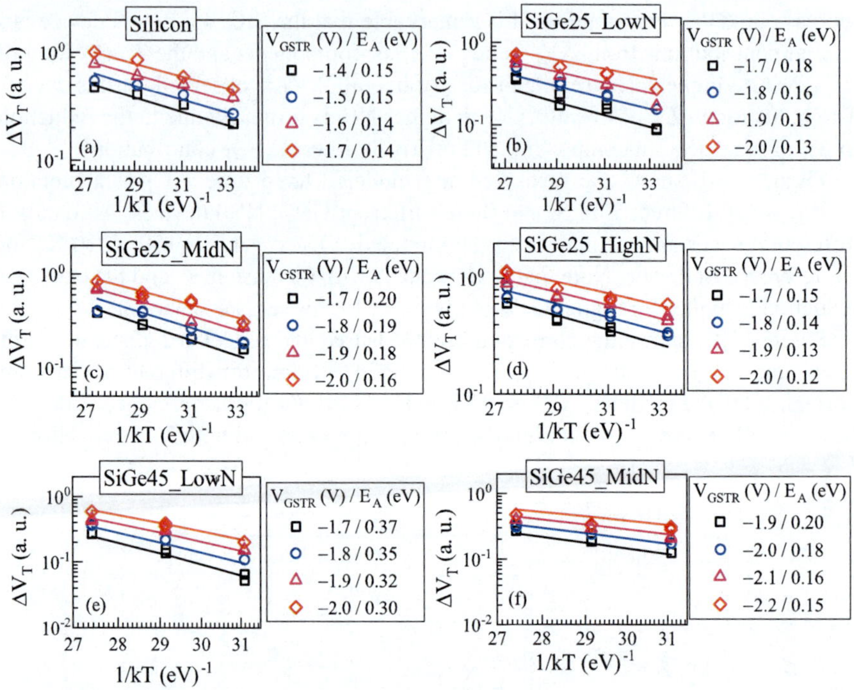

Fig. 11.11 Measured and modeled fixed time ($t_{STR} = 1$ Ks) ΔV_T as a function of T at different V_{GSTR} under DC stress for Si and SiGe (different Ge% and N%) p-FinFETs. All measured and modeled data are normalized to the Si device modeled ΔV_T at −1.7 V, 150 °C, 10 Ks. The E_A values are mentioned for the modeled data. Symbols: experiment, lines: model calculation. Data from [21]

higher T) and the V_{GSTR} dependence of E_A depends on Ge% and N% (it is more sensitive for Ge% variation and less sensitive for N% variation).

Figure 11.12 shows the measured and model calculated (a, c) T dependence of VAF and (b, d) V_{GSTR} dependence of E_A for Si and SiGe (different Ge% and N%) devices. The Si device shows negligible T dependence while SiGe shows strong T dependence of VAF, see Fig. 11.12(a). The strong T dependence of the VAF for the SiGe device is primarily due to higher effective bond polarization factor (α) of the ΔV_{IT} subcomponent, see Table 11.1, while the relatively higher ΔV_{OT} at higher Ge% also contributes. Note that the VAF of ΔV_{OT} also changes with a change in T due to the bond polarization effect, which is discussed in Chap. 6, Sect. 6.2. For devices with high N% in the IL, higher ΔV_{HT} and lower ΔV_{OT} relative contributions make the VAF relatively less sensitive to T, see Fig. 11.12(c). As a consequence, the Si device has weak V_{GSTR} dependence while the SiGe device has strong V_{GSTR} dependence of E_A, see Fig. 11.12(b). The magnitude of E_A becomes smaller and

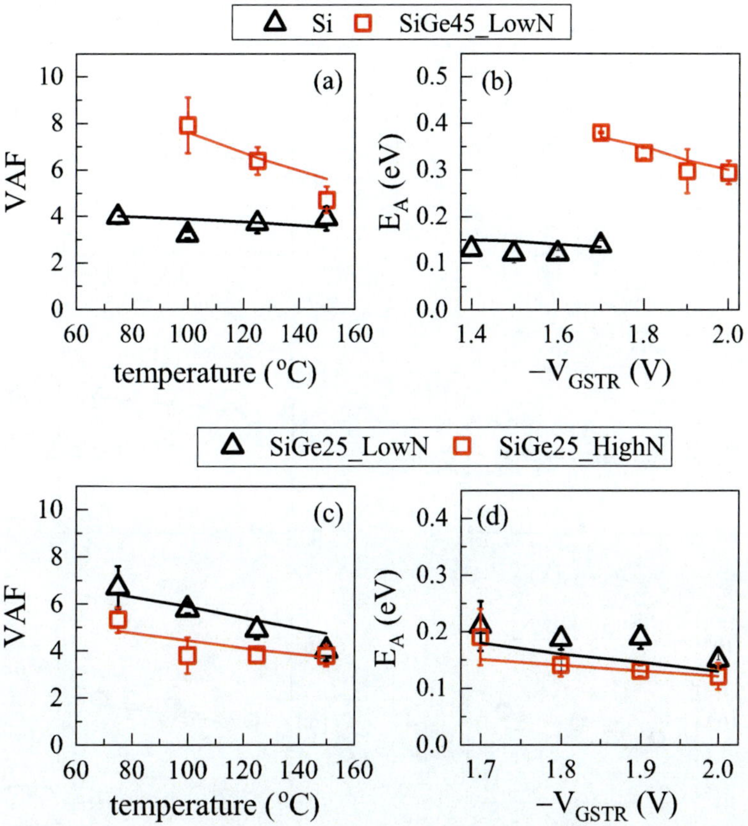

Fig. 11.12 Measured and model calculated (a, c) T dependence of VAF and (b, d) V_{GSTR} dependence of E_A, for different (a, b) Ge% and (b, d) N% p-FinFETs under DC stress. Symbols: experiment, lines: model calculation. Data from [21]

its V_{GSTR} dependence weaker at higher N%, see Fig. 11.12(d). Note that the V_{GSTR} dependence of E_A is consistent with the corresponding T dependence of VAF for different (Ge%, N%) devices.

11.6 Voltage and Temperature Dependence—AC Stress

Figure 11.13 shows the V_{GSTR} (=V_{GHIGH}) dependence of measured and modeled ΔV_T and the underlying ΔV_{IT} and ΔV_{OT} subcomponents at fixed T and t_{STR} = 1 Ks for Mode-B AC stress in Si and SiGe (different Ge% and N%) devices. The PDC and f are kept fixed, and all data are normalized to the modeled ΔV_T at −1.7 V/150 °C/10 Ks for the reference Si device at DC stress.

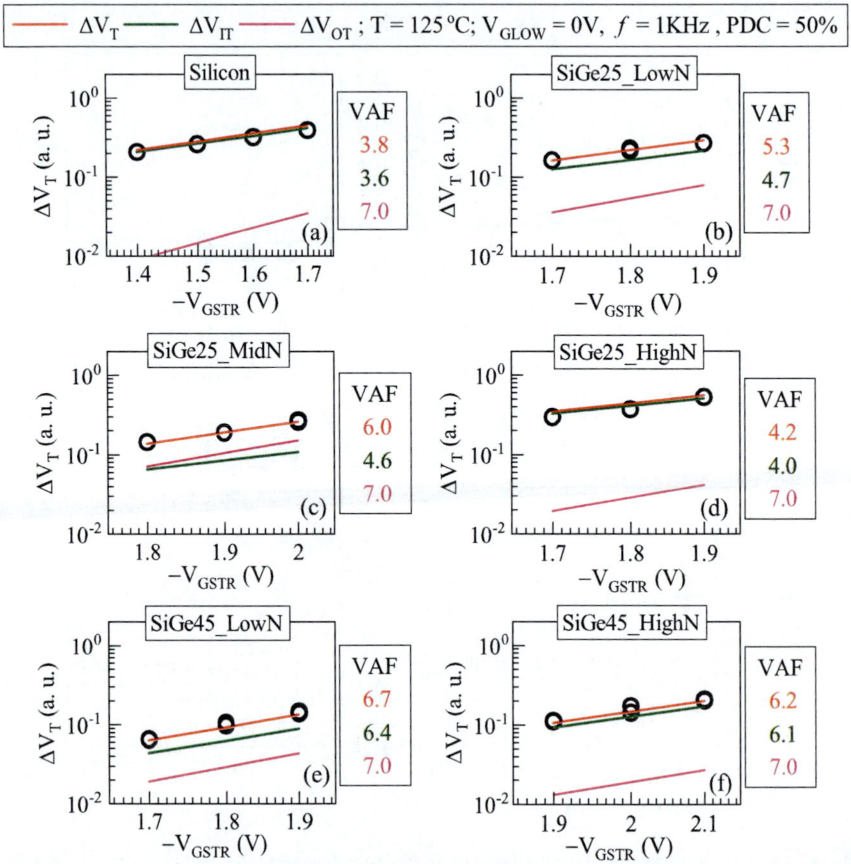

Fig. 11.13 Measured and modeled fixed time ($t_{STR} = 1$ Ks) ΔV_T as a function of V_{GSTR} for Si and SiGe (different Ge% and N%) p-FinFETs under Mode-B AC stress. The underlying ΔV_{IT} and ΔV_{OT} subcomponents are shown (ΔV_{HT} is negligible). All measured and modeled data including subcomponents are normalized to the Si device modeled ΔV_T during DC stress at −1.7 V, 150 °C, 10 Ks. The VAF values are mentioned for the overall model and underlying subcomponents. Symbols: experiment, lines: model calculation. Data from [22]

Note that only the K_{F30} pre-factor for ΔV_{OT} is different between AC and DC stress for a particular device (see Chap. 6, Sect. 6.3 and Chap. 14), all other parameters are the same as listed in Table 11.1 to model AC stress. The dependence of ΔV_T on V_{GSTR} for AC stress is very similar to that observed for DC stress (reduction at higher Ge% but increase at higher N%). The ΔV_{HT} contribution is negligible for Mode-B AC stress at PDC of 50% for all devices. The ΔV_{OT} contribution is negligible in the measurement window (it reduces at higher f, see Chap. 14) and therefore, ΔV_T depends only on ΔV_{IT} for the Si device. As Ge% is increased (Fig. 11.13(a), (b), (e)), the VAF of ΔV_T increases due to increase in the polarization factor α of the ΔV_{IT} subcomponent and relatively higher ΔV_{OT} contribution. Note, both ΔV_{IT} and ΔV_{OT}

reduce at higher Ge%, but the reduction in ΔV_{IT} is much more than that of ΔV_{OT}, which is evident for the SiGe45_LowN device. As N% is increased (Fig. 11.13(b)–(f)), the VAF of ΔV_T reduces as the VAF of ΔV_{IT} reduces due to higher ε_{IL} and the stress reduction effect related saturation, while relatively lower ΔV_{OT} contribution (note, ΔV_{IT} increases but ΔV_{OT} reduces at higher N%) also contributes. The VAF of the overall modeled ΔV_T and the underlying subcomponents are listed for all devices.

Figure 11.14 shows the T dependence of measured and modeled ΔV_T and the underlying ΔV_{IT} and ΔV_{OT} subcomponents at fixed V_{GSTR} and $t_{STR} = 1$ Ks for AC Mode-B stress in Si and SiGe (different Ge% and N%) devices. The PDC and f are kept fixed, and all data are normalized to the modeled ΔV_T at −1.7 V/150 °C/10 Ks for the reference Si device at DC stress. Note, the dependence of ΔV_T on T for AC stress is very similar to that observed for DC stress.

The impact of Ge% and N% on the T activation of ΔV_{IT} and ΔV_{OT} subcomponents is explained before in Sect. 11.3. Note, E_{AKF1} for ΔV_{IT} is process dependent, see Table 11.1, while E_{AOT} for ΔV_{OT} is not, see Chap. 6, Sect. 6.2. However, as also mentioned before, the obtained E_A for ΔV_{OT} is different due to different ranges of V_{GSTR} (E_{OX}) used for different devices and is due to the difference in stress reduction effect. The E_A of overall ΔV_T and E_{AKF1} increase with increase in Ge% and reduce with increase in N%. The relatively higher ΔV_{OT} at higher Ge% also helps increasing the E_A of overall ΔV_T. For devices with high N%, the E_A of overall ΔV_T is primarily governed by ΔV_{IT} (due to negligible ΔV_{OT}), and reduces at higher N%. The T activation E_A of the overall modeled ΔV_T and the underlying subcomponents are listed for all devices.

Figure 11.15 shows the measured and modeled V_{GSTR} (=V_{GHIGH}) dependence of ΔV_T at multiple T but fixed $t_{STR} = 1$ Ks under Mode-B AC stress for Si and SiGe (different Ge% and N%) devices. The PDC and f are kept fixed, and all data are normalized to the modeled ΔV_T at −1.7 V/150 °C/10 Ks for the reference Si device at DC stress. The VAF values for different processes (Ge% and N%) are similar to the respective values under DC stress, see Fig. 11.10. For a fixed T, the VAF increases with increase in Ge% and reduces with increase in N%. The VAF reduces with increase in T due to bond polarization factors associated with the ΔV_{IT} and ΔV_{OT} subcomponents.

Figure 11.16 shows the measured and modeled T dependence of ΔV_T at multiple V_{GSTR} but fixed $t_{STR} = 1$ Ks under Mode-B AC stress for Si and SiGe (different Ge% and N%) devices. The PDC and f are kept fixed, and all data are normalized to the modeled ΔV_T at −1.7 V/150 °C/10 Ks for the reference Si device at DC stress. The impact of Ge% and N% on E_A for Mode-B AC stress are similar to that for DC stress, see Fig. 11.11. Note that E_A increases with increase in Ge% and reduces with increase in N%. The magnitude of E_A depends on applied V_{GSTR} and it reduces with increase in V_{GSTR} due to the reasons mentioned in the earlier section.

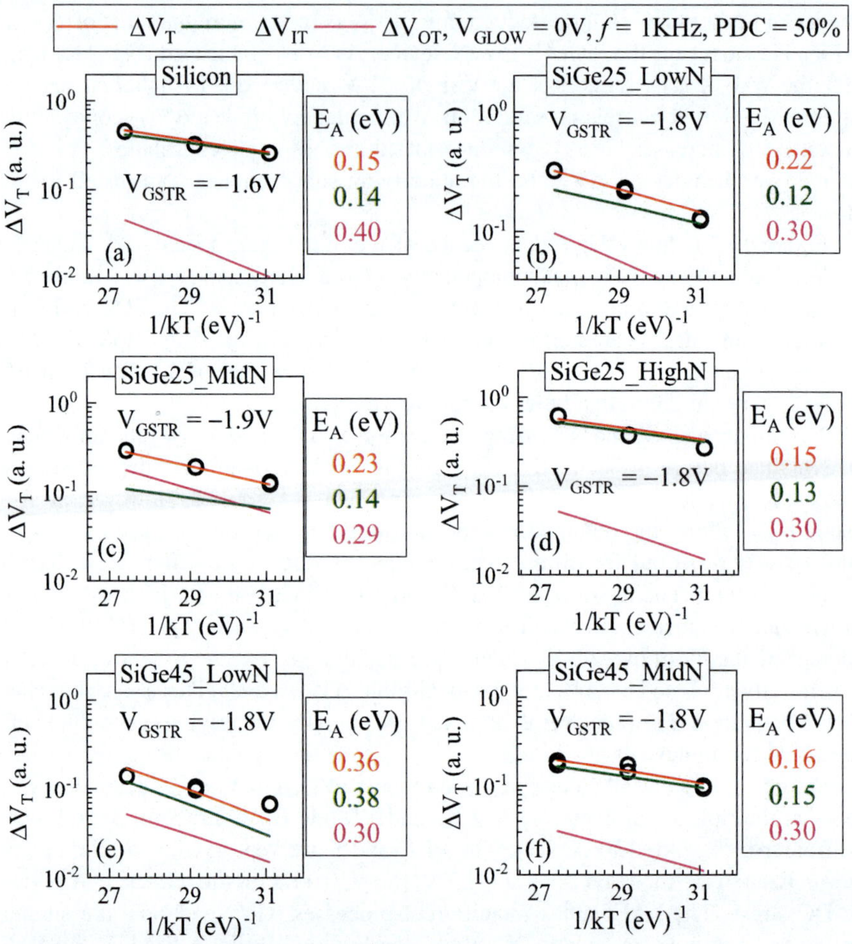

Fig. 11.14 Measured and modeled fixed time (t_{STR} = 1 Ks) ΔV_T as a function of T for Si and SiGe (different Ge% and N%) p-FinFETs under Mode-B AC stress. The underlying ΔV_{IT} and ΔV_{OT} subcomponents are shown (ΔV_{HT} is negligible). All measured and modeled data including subcomponents are normalized to the Si device modeled ΔV_T during DC stress at −1.7 V, 150 °C, 10 Ks. The E_A values are mentioned for the overall model and underlying subcomponents. Symbols: experiment, lines: model calculation. Data from [22]

11.7 Estimation of EOL Degradation

Figure 11.17 shows the extrapolated ΔV_T at end of life (EOL) of 10 years under use condition, calculated using the calibrated model for (a) DC and (b) Mode-B AC in Si and SiGe (different Ge% and N%) devices. The underlying ΔV_{IT}, ΔV_{HT}, and ΔV_{OT} subcomponents are also shown.

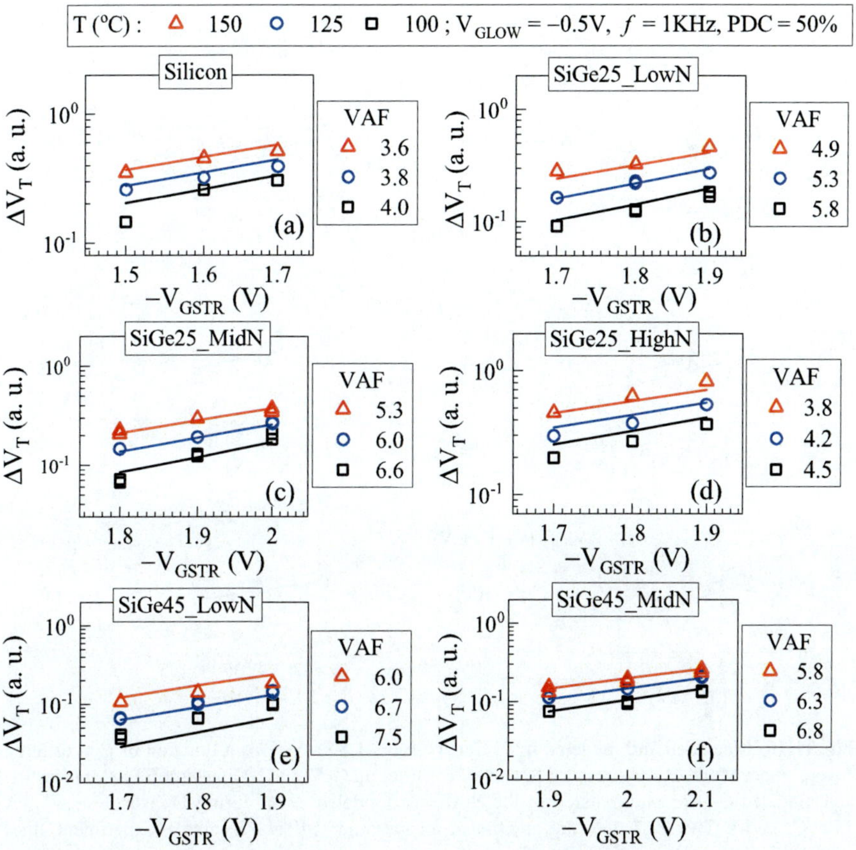

Fig. 11.15 Measured and modeled fixed time ($t_{STR} = 1$ Ks) ΔV_T as a function of V_{GSTR} at different T under Mode-B AC stress in Si and SiGe (different Ge% and N%) p-FinFETs. All measured and modeled data are normalized to the Si device modeled ΔV_T during DC stress at −1.7 V, 150 °C, 10 Ks. The VAF values are mentioned for the modeled data. Symbols: experiment, lines: model calculation. Data from [22]

The contribution from ΔV_{HT} is negligible in all devices except Si reference for DC stress, while it is negligible in all devices for Mode-B AC stress. Note that the EOL ΔV_T is dominated by ΔV_{IT} in all devices except in the SiGe45_LowN device. The contributions from ΔV_{IT} and ΔV_{OT} are similar for this device due to very high Ge%. All subcomponents and therefore overall ΔV_T reduce as Ge% is increased. With increase in N%, ΔV_{IT} increases while ΔV_{OT} reduces. The increase in ΔV_{IT} is larger than the reduction in ΔV_{OT}, and therefore, overall ΔV_T increases with N%.

Figure 11.17 also shows the T dependence of EOL ΔV_T under operating conditions together with the underlying subcomponents, calculated using the calibrated model for (c) Si and (d) SiGe25_MidN devices. Note that the ΔV_{IT} contribution is

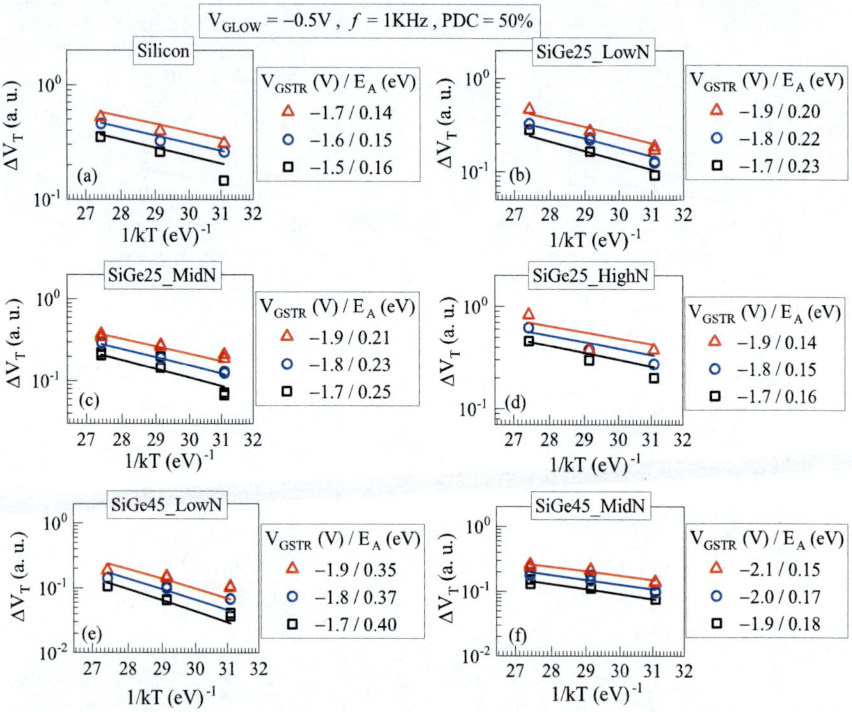

Fig. 11.16 Measured and modeled fixed time ($t_{STR} = 1$ Ks) ΔV_T as a function of T at different V_{GSTR} under Mode-B AC stress for Si and SiGe (different Ge% and N%) p-FinFETs. All measured and modeled data are normalized to the Si device modeled ΔV_T during DC stress at −1.7 V, 150 °C, 10 Ks. The E_A values are mentioned for the modeled data. Symbols: experiment, lines: model calculation. Data from [22]

larger than the other subcomponents at lower T, however, the relative ΔV_{OT} contribution increases at higher T, especially for the SiGe device. The EOL ΔV_T for the Si device increases by 2X with increase in T from 100 °C to 150 °C, while it increases by 4X for the SiGe device. This is due to larger reduction in the VAF of ΔV_{IT} at higher T for the SiGe device (due to larger polarization term α for bond dissociation) compared to Si. Moreover, a larger fractional ΔV_{OT} contribution also contributes to the SiGe device. Therefore, note that the improvement in EOL ΔV_T for SiGe device as compared to Si depends on T. However, even though the T impact is stronger for SiGe device, the EOL ΔV_T at 150 °C is still ~3X lower for SiGe compared to the Si device.

Figure 11.18 shows the comparison of EOL ΔV_T calculated using the calibrated model and conventional empirical regression (fit) based methods (the empirical methods are discussed in Chap. 1, Sect. 1.4). The comparison is done for (a, b) Si and SiGe devices having different Ge% (low N%), and for (c, d) SiGe25 devices having different N%, under (a, c) DC and (b, d) Mode-B AC condition. In the conventional empirical method, described in Chap. 1, Sect. 1.4, the shorter time measured

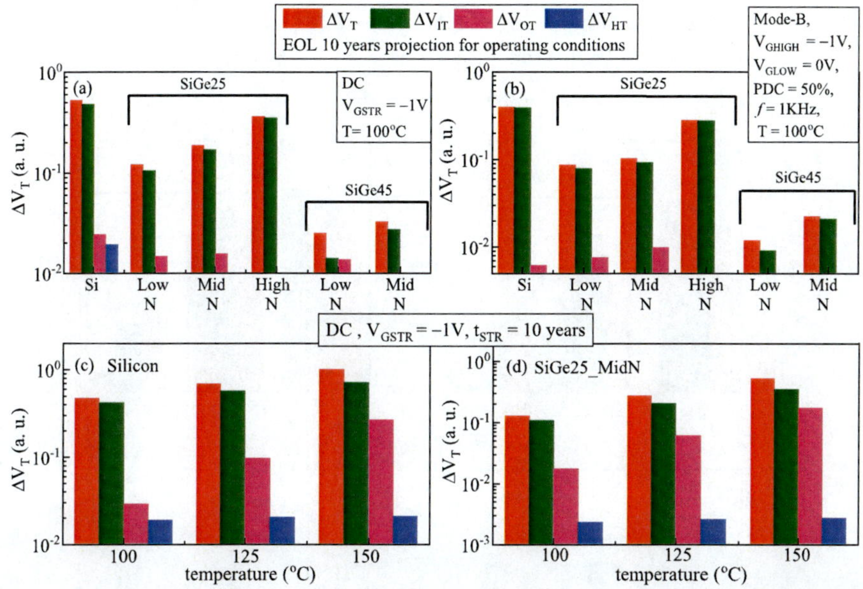

Fig. 11.17 Extrapolated ΔV_T together with the underlying subcomponents at EOL under use condition using the calibrated model. In (a, b) data are shown for (a) DC and (b) Mode-B AC condition for Si and SiGe (different Ge% and N%) p-FinFETs. In (c, d), data are shown for DC stress in (c) Si and (d) SiGe25_MidN p-FinFETs at different T

data are fitted using power-law dependence in time and extrapolated to EOL. This is repeated for multiple V_{GSTR} but fixed T. Subsequently, the EOL ΔV_T at different V_{GSTR} is fitted using either power-law or exponential V_{GSTR} dependence and extrapolated to operating (use) bias (V_{DD}), and the EOL ΔV_T at operating condition is obtained. In Chap. 6, Sect. 6.4, it is shown that the accuracy of the empirical fit based method depends on the range of V_{GSTR} (better if closer to V_{DD}) and t_{STR} (better if closer to EOL) used in this exercise. However, the use of very low V_{GTSR} for SiGe is challenging as the degradation level is very small, and the measurement bandwidth often limits the maximum usable stress duration.

In this analysis, data measured in the time interval of 1 s to 1 Ks are used for all devices. Note that the empirical method always overestimates the EOL ΔV_T for both DC and AC except for the SiGe45_LowN device. The EOL projected using the exponential V_{GSTR} dependence is more erroneous compared to the power-law V_{GSTR} dependence. Note that the different subcomponents (ΔV_{IT}, ΔV_{HT}, and ΔV_{OT}) of overall ΔV_T have different time kinetics, VAF, and E_A, and hence their relative contributions would be different for different stress conditions. The conventional fit based method is likely to fail unless the measured data are available close to the operating conditions. The difference between empirical and model calculated EOL ΔV_T is higher for devices with high Ge% and low N% having higher relative ΔV_{OT} contribution since these devices would show higher time slope n and hence result

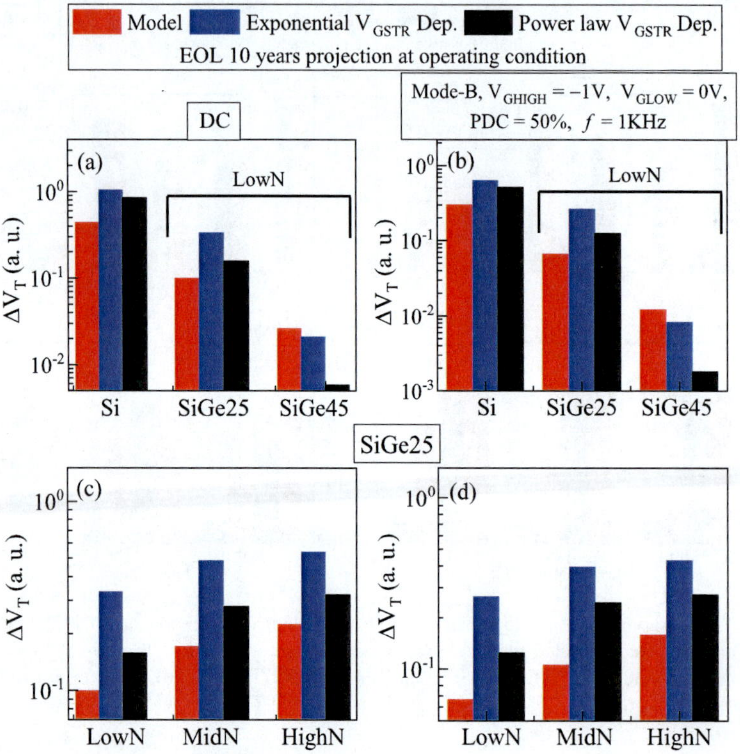

Fig. 11.18 Comparison of extrapolated ΔV_T at EOL under use condition obtained from the calibrated model and empirical methods for (a, c) DC and (b, d) Mode-B AC condition, in Si and SiGe p-FinFETs for variation in (a, b) Ge% and (c, d) N%

in higher empirically extrapolated ΔV_T during accelerated stress at high V_{GSTR}. As mentioned above, this issue can be addressed by stressing closer to the operating bias and over long time. However, the model-based extrapolation correctly isolates different subcomponents and extrapolates them to operating bias to calculate the EOL ΔV_T. Therefore, the model-based approach does not require the measurement to happen closer to operating voltage and for a very long stress time.

11.8 Summary

Ultra-fast NBTI measurements are done in Si and SiGe p-FinFETs. The impact of channel Ge% and gate stack N% on measured ΔV_T time kinetics during and after DC and AC stress is modeled by using the comprehensive framework explained in Chap. 4 through Chap. 6. It is verified that the measured ΔV_T time kinetics during DC stress is due to uncorrelated contributions from the ΔV_{IT}, $\Delta V_{HT,}$ and ΔV_{OT} subcomponents, the ΔV_{HT} contribution, however, is negligible for Mode-B AC stress.

It is shown that higher Ge% reduces all the subcomponents but increases the relative ΔV_{OT} contribution. Higher relative ΔV_{OT} contribution makes the longer-time slope higher during stress and makes recovery slower after stress. Relatively higher ΔV_{OT} contribution and higher bond polarization factor for ΔV_{IT} make the VAF of overall ΔV_{T} higher for higher Ge% in the channel. Relatively higher ΔV_{OT} and higher E_{A} of ΔV_{IT} increase the E_{A} of overall ΔV_{T} in these devices.

However, higher N% increases the ΔV_{IT} and ΔV_{HT} subcomponents but reduces ΔV_{OT}. Higher ΔV_{HT} and lower ΔV_{OT} contributions reduce the long-time slope during stress and make recovery faster. Lower ΔV_{OT} and higher dielectric constant of the IL reduce the VAF of overall ΔV_{T} at higher N%. Moreover, reduction in the E_{A} of the ΔV_{IT} subcomponent and relatively higher ΔV_{HT} reduce the E_{A} of overall ΔV_{T} in these devices.

The impact of Ge% and N% on the T dependence of VAF and V_{GSTR} dependence of E_{A} is explained using the polarization effect on ΔV_{IT} and ΔV_{OT} subcomponents.

The calibrated model is used to determine the extrapolated ΔV_{T} at EOL and use conditions for various processes (Ge% and N%). In most of the devices, the EOL ΔV_{T} is dominated by the ΔV_{IT} subcomponent, since ΔV_{HT} is negligible in SiGe devices and ΔV_{OT} is negligible at use conditions due to its higher VAF. The model extrapolated EOL ΔV_{T} at use condition is compared to the regression (fit) based empirical methods for different processes. The empirical methods provide inaccurate estimation of ΔV_{T} at EOL for both DC and AC stress. This necessitates proper modeling for reliable estimation of NBTI lifetime.

Acknowledgements All experimental data presented in this chapter are re-plotted from previously published reports. The authors acknowledge IBM for providing measurement facilities and FinFETs. Karansingh Thakor is acknowledged for help with manuscript preparation.

References

1. N. Kimizuka, K. Yamaguchi, K. Imai, T. Iizuka, C.T. Liu, R.C. Keller, T. Horiuchi, in *Symposium on VLSI Technology Digest of Technical Papers*, 92 (2000)
2. V. Huard, M. Denais, F. Perrier, N. Revil, C. Parthasarathy, A. Bravaix, E. Vincent, Microelectron. Reliab. **45**, 83 (2005)
3. S. Mahapatra, K. Ahmed, D. Varghese, A.E. Islam, G. Gupta, L. Madhav, D. Saha, M.A. Alam, in *IEEE International Reliability Physics Symposium Proceedings*, 1 (2007)
4. Y. Mitani, H. Satake, A. Toriumi, IEEE Trans. Device Mater. Reliab. **8**, 6 (2008)
5. S. Pae, M. Agostinelli, M. Brazier, R. Chau, G. Dewey, T. Ghani, M. Hattendorf, J. Hicks, J. Kavalieros, K. Kuhn, M. Kuhn, J. Maiz, M. Metz, K. Mistry, C. Prasad, S. Ramey, A. Roskowski, J. Sandford, C. Thomas, J. Thomas, C. Wiegand, J. Wiedemer, in *IEEE International Reliability Physics Symposium Proceedings*, 352 (2008)
6. S. Krishnan, U. Kwon, N. Moumen, M.W. Stoker, E.C.T. Harley, S. Bedell, D. Nair, B. Greene, W. Henson, M. Chowdhury, D.P. Prakash, E. Wu, D. Ioannou, E. Cartier, M.-H. Na, S. Inumiya,

K. Mcstay, L. Edge, R. Iijima, J. Cai, M. Frank, M. Hargrove, D. Guo, A. Kerber, H. Jagannathan, T. Ando, J. Shepard, S. Siddiqui, M. Dai, H. Bu, J. Schaeffer, D. Jaeger, K. Barla, T. Wallner, S. Uchimura, Y. Lee, G. Karve, S. Zafar, D. Schepis, Y. Wang, R. Donaton, S. Saroop, P. Montanini, Y. Liang, J. Stathis, R. Carter, R. Pal, V. Paruchuri, H. Yamasaki, J.-H. Lee, M. Ostermayr, J.-P. Han, Y. Hu, M. Gribelyuk, D.-G. Park, X. Chen, S. Samavedam, S. Narasimha, P. Agnello, M. Khare, R. Divakaruni, V. Narayanan, M. Chudzik, in *IEEE International Electron Devices Meeting Technical Digest*, 28.1.1 (2011)
7. K. Joshi, S. Hung, S. Mukhopadhyay, V. Chaudhary, N. Nanaware, B. Rajamohanan, T. Sato, M. Bevan, A. Wei, A. Noori, B. McDougal, C. Ni, G. Saheli, C. Lazik, P. Liu, D. Chu, L. Date, S. Datta, A. Brand, J. Swenberg, S. Mahapatra, in *IEEE International Reliability Physics Symposium Proceedings*, 4C.2.1 (2013).
8. J. Franco, B. Kaczer, P.J. Roussel, J. Mitard, M. Cho, L. Witters, T. Grasser, G. Groeseneken, IEEE Trans. Electron Devices **60**, 396 (2013)
9. P. Srinivasan, J. Fronheiser, K. Akarvardar, A. Kerber, L.F. Edge, R.G. Southwick, E. Cartier, H. Kothari, in *IEEE International Reliability Physics Symposium Proceedings*, 6A.3.1 (2014)
10. N. Parihar, N. Goel, S. Mukhopadhyay, S. Mahapatra, IEEE Trans. Electron Devices **65**, 392 (2018)
11. N. Parihar, S. Mahapatra, in *IEEE International Reliability Physics Symposium Proceedings*, TX.5.1 (2018)
12. V. Huard, C. Ndiaye, M. Arabi, N. Parihar, X. Federspiel, S. Mhira, S. Mahapatra, A. Bravaix, in *IEEE International Reliability Physics Symposium Proceedings*, TX.4.1 (2018)
13. N. Parihar, R. Tiwari, C. Ndiaye, M. Arabi, S. Mhira, H. Wong, S. Motzny, V. Moroz, V. Huard, S. Mahapatra, in *International Conference on Simulation of Semiconductor Processes and Devices*, 167 (2018)
14. S. Ramey, A. Ashutosh, C. Auth, J. Clifford, M. Hattendorf, J. Hicks, R. James, A. Rahman, V. Sharma, A. St. Amour, C. Wiegand, in *IEEE International Reliability Physics Symposium Proceedings*, 4C.5.1 (2013)
15. K.T. Lee, W. Kang, E.-A. Chung, G. Kim, H. Shim, H. Lee, H. Kim, M. Choe, N.-I. Lee, A. Patel, J. Park, J. Park, in *IEEE International Reliability Physics Symposium Proceedings*, 2D.1.1 (2013)
16. J. Franco, B. Kaczer, A. Chasin, H. Mertens, L.-A. Ragnarsson, R. Ritzenthaler, S. Mukhopadhyay, H. Arimura, P.J. Roussel, E. Bury, N. Horiguchi, D. Linten, G. Groeseneken, A. Thean, in *IEEE International Reliability Physics Symposium Proceedings*, 4B.2.1 (2016)
17. G. Jiao, M. Toledano-Luque, K.-J. Nam, N. Toshiro, S.-H. Lee, J.-S. Kim, T. Kauerauf, E. Chung, D. Bae, G. Bae, D.-W. Kim, K. Hwang, in *IEEE International Electron Devices Meeting Technical Digest*, 31.2.1 (2016)
18. N. Parihar, R.G. Southwick, U. Sharma, M. Wang, J.H. Stathis, S. Mahapatra, in *IEEE International Reliability Physics Symposium Proceedings*, 2D.4.1 (2017)
19. N. Parihar, R. Southwick, M. Wang, J.H. Stathis, S. Mahapatra, in *IEEE International Electron Devices Meeting Technical Digest*, 7.3.1 (2017)
20. N. Parihar, U. Sharma, R.G. Southwick, M. Wang, J.H. Stathis, S. Mahapatra, IEEE Trans. Electron Devices **65**, 23 (2018)
21. N. Parihar, R.G. Southwick, M. Wang, J.H. Stathis, S. Mahapatra, IEEE Trans. Electron Devices **65**, 1699 (2018)
22. N. Parihar, R.G. Southwick, M. Wang, J.H. Stathis, S. Mahapatra, IEEE Trans. Electron Devices **65**, 1707 (2018)
23. N. Parihar, R. Tiwari, S. Mahapatra, in *International Conference on Simulation of Semiconductor Processes and Devices*, 176 (2018)
24. R. Tiwari, N. Parihar, K. Thakor, H.Y. Wong, S. Motzny, M. Choi, V. Moroz, S. Mahapatra, IEEE Trans. Electron Devices **66**, 2086 (2019)
25. R. Tiwari, N. Parihar, K. Thakor, H.Y. Wong, S. Motzny, M. Choi, V. Moroz, S. Mahapatra, IEEE Trans. Electron Devices **66**, 2093 (2019)
26. N. Parihar, U. Sharma, R.G. Southwick, M. Wang, J.H. Stathis, S. Mahapatra, in *IEEE International Reliability Physics Symposium Proceedings* (2019)

27. M. Wang, J. Zhang, H. Zhou, R.G. Southwick, R. Hsin, K. Chao, X. Miao, V.S. Basker, T. Yamashita, D. Guo, G. Karve, H. Bu, in *IEEE International Reliability Physics Symposium Proceedings* (2019)
28. N. Choudhury, U. Sharma, H. Zhou, R.G. Southwick, M. Wang, S. Mahapatra, in *IEEE International Reliability Physics Symposium Proceedings* (2020)
29. H. Zhou, M. Wang, J. Zhang, K. Watanabe, C. Durfee, S. Mochizuki, R. Bao, R. Southwick, M. Bhuiyan, B. Veeraraghavan, in *IEEE International Reliability Physics Symposium Proceedings* (2020)
30. N. Choudhury, T. Samadder, R. Tiwari, H. Zhou, R.G. Southwick, M. Wang, S. Mahapatra, in *IEEE International Reliability Physics Symposium Proceedings* (2021)
31. S. Mahapatra, N. Goel, S. Desai, S. Gupta, B. Jose, S. Mukhopadhyay, K. Joshi, A. Jain, A.E. Islam, M.A. Alam, IEEE Trans. Electron Devices **60**, 901 (2013)
32. A.E. Islam, N. Goel, S. Mahapatra, M.A. Alam, in *Fundamentals of Bias Temperature Instability in MOS Transistors* (Springer India, 2015), pp. 181–207
33. S. Mukhopadhyay, K. Joshi, V. Chaudhary, N. Goel, S. De, R.K. Pandey, K.V.R.M. Murali, S. Mahapatra, in *IEEE International Reliability Physics Symposium Proceedings*, GD 3.1 (2014)
34. N. Choudhury, N. Parihar, N. Goel, A. Thirunavukkarasu, S. Mahapatra, IEEE J. Electron Devices Society **8**, 1281 (2020)
35. T. Samadder, N. Choudhury, S. Kumar, D. Kochar, N. Parihar, S. Mahapatra, IEEE Trans. Electron Devices **68**, 485 (2021)
36. L.K. Han, S. Crowder, M. Hargrove, E. Wu, S.H. Lo, F. Guarin, E. Crabbe, L. Su, in *IEEE International Electron Devices Meeting Technical Digest*, 643 (1997)
37. S.S. Tan, T.P. Chen, C.H. Ang, L. Chan, IEEE Electron Device Lett. **25**, 504 (2004)
38. S. Kumar, R. Anandkrishnan, N. Parihar, S. Mahapatra, IEEE Trans. Electron Devices **67**, 4741 (2020)
39. L. Zhou, Q. Zhang, H. Yang, Z. Ji, Z. Zhang, Q. Liu, H. Xu, B. Tang, E. Simoen, X. Ma, X. Wang, Y. Li, H. Yin, J. Luo, C. Zhao, W. Wang, IEEE Electron Device Lett. **41**, 965 (2020)
40. R. Ranjan, Y. Liu, T. Nigam, A. Kerber, B. Parameshwaran, in *IEEE International Reliability Physics Symposium Proceedings*, DG.10.1 (2017)
41. M. Rafik, A.P. Nguyen, X. Garros, M. Arabi, X. Federspiel, C. Diouf, in *IEEE International Reliability Physics Symposium Proceedings*, 4A.3.1 (2018)

Chapter 12
BAT Framework Modeling of RMG HKMG GAA-SNS FETs

Nilotpal Choudhury, Tarun Samadder, Richard Southwick, Huimei Zhou, Miaomiao Wang, and Souvik Mahapatra

12.1 Introduction

As discussed in the earlier chapters, Negative Bias Temperature Instability (NBTI) became a serious reliability issue with the migration from Silicon Dioxide (SiO_2) to Silicon Oxynitride (SiON) gate insulator planar MOSFETs [1–4]. It continues to remain so for dual-layer (SiO_2 or SiON Interlayer (IL) and HfO_2 High-K layer) High-K Metal Gate (HKMG)-based bulk [5–10] and Fully Depleted Silicon On Insulator (FDSOI) [11, 12] planar MOSFETs, bulk and SOI FinFETs [12–25], and Gate All Around Stacked Nanosheet (GAA-SNS) FETs [26–29], with either Silicon (Si) or Silicon Germanium (SiGe) channel. The key features of NBTI are summarized below (reproduced from Chap. 3, Sect. 3.1).

As shown in Chap. 1, Sect. 1.3, NBTI results in a gradual buildup of positive gate insulator charges when the gate bias (V_G) is held at a negative value during stress ($V_G = V_{GSTR}$) and shift the device parameters, e.g., threshold voltage shift (ΔV_T), over time. The ΔV_T magnitude gets accelerated at more negative V_{GSTR} and higher temperature (T), and are, respectively, governed by the voltage acceleration factor (VAF) and Arrhenius T activation energy (E_A). However, the gate insulator charges and the associated ΔV_T reduce when the V_G is reduced or removed after stress ($V_G = V_{GREC}$ or 0 V). Hence, AC stress results in lower ΔV_T than DC stress. Moreover, NBTI recovery necessitates the use of ultra-fast methods for estimation of device degradation without artifacts, see Chap. 1, Sect. 1.2. Therefore, the ultra-fast Measure Stress Measure (MSM) method is used in this book.

N. Choudhury · T. Samadder · S. Mahapatra (✉)
Department of Electrical Engineering, Indian Institute of Technology Bombay, Powai, Mumbai 400076, India
e-mail: souvik@ee.iitb.ac.in

R. Southwick · H. Zhou · M. Wang
IBM Research Division, Albany Nanotech, Albany, NY 12203, USA

S. Mahapatra (ed.), *Recent Advances in PMOS Negative Bias Temperature Instability*,
https://doi.org/10.1007/978-981-16-6120-4_12

The stress conditions (V_{GSTR} and T) also impact the time kinetics during stress and recovery after stress (i.e., the power law slope (n) for longer stress time (t_{STR}), and Fraction Remaining (FR) during recovery after stress at a particular time (t_{REC}); FR is defined as the ratio of ΔV_T during recovery at a time $t = t_{REC}$ to that after the end of stress at $t = t_{STR}$), as well as VAF, E_A, and T dependence of VAF (i.e., the reduction of VAF at higher T). As mentioned in Chap. 2, process changes, i.e., the introduction of Nitrogen (N) in the gate insulator or migration from the Si to SiGe channel impact NBTI. Higher Nitrogen content (N%) increases ΔV_T, but reduces n, FR (i.e., the recovery becomes faster), VAF, E_A, and T dependence of VAF (i.e., lower reduction of VAF at higher T). However, higher Germanium content (Ge%) reduces ΔV_T, but increases n, FR (i.e., the recovery becomes slower), VAF, E_A, and T dependence of VAF (i.e., higher reduction of VAF at higher T).

The above features are explained and modeled for different device architectures in the previous chapters of this book: Si bulk MOSFETs, Chap. 7, Si-capped SiGe bulk MOSFETs, Chap. 8, Si and SiGe FDSOI MOSFETs, Chap. 9, SOI FinFETs, Chap. 10, as well as Si and SiGe bulk FinFETs, Chap. 11, using the BTI Analysis Tool (BAT) framework described in Chaps. 4–6. In Chap. 2, it is also shown that changes in the mechanical stress in the channel, due to changes in layout or device dimensions, also impact NBTI. The layout impact is modeled by the BAT framework for FDSOI MOSFETs in Chap. 9.

The GAA-SNS FETs are being actively considered at present to continue with the CMOS technology scaling for sub-3 nm technology nodes, due to their superior electrostatics and short channel effect control over FinFETs [26]. In this chapter, the impact of sheet dimension (length and width) scaling on NBTI is studied in GAA-SNS FETs [28, 29]. Further analysis of the mechanism responsible for the device dimension scaling impact on NBTI in FinFETs and GAA-SNS FETs is discussed in Chap. 13. The BAT framework is also used to perform a detailed analysis of AC NBTI in Chap. 14. The framework is briefly explained hereinafter (reproduced from Chap. 6, Sect. 6.3).

Figure 12.1 illustrates the BAT framework [9] used to model the time kinetics of ΔV_T during and after stress. It uses uncorrelated contributions from generated interface (ΔV_{IT}) and bulk (ΔV_{OT}) gate insulator traps, and hole trapping in process-related preexisting bulk gate insulator traps (ΔV_{HT}). The Reaction Diffusion (RD) model is used to calculate the time kinetics of interface traps (density ΔN_{IT}) [9, 30, 31]. In Chap. 4, the RD model is described and independently validated by measured data from Direct Current IV (DCIV) method [32]. The Transient Trap Occupancy Model (TTOM) is used to calculate the occupancy of generated interface traps and their contribution (ΔV_{IT}) [9], which is described and validated in Chap. 5, Sect. 5.3. The ΔV_{HT} and ΔV_{OT} kinetics are modeled by the Activated Barrier Bouble Well Thermionic (ABDWT) model [33] and Reaction Diffusion Drift (RDD) model [34], respectively, and these are described and validated in Chap. 5, Sect. 5.4 and Chap. 6, Sect. 6.2.

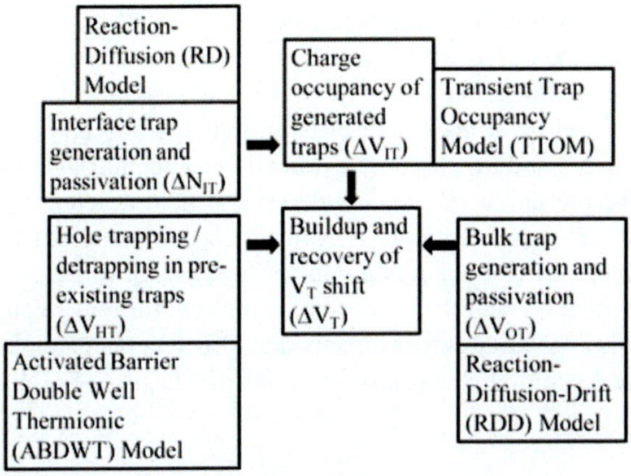

Fig. 12.1 Schematic of the BTI Analysis Tool (BAT) framework used in this book to model measured ΔV_T kinetics during and after DC and AC NBTI stress, reproduced from Chap. 4

12.2 Device Details and Model Parameters

Measured data from GAA-SNS FETs, fabricated using a proprietary IBM process, are modeled in this chapter. The devices have three sheets of Si channel with varying gate length (L_{GATE}), defined as Sheet Length (SL) in this chapter, and Sheet Width (SW), and a Replacement Metal Gate (RMG) HKMG gate insulator stack, featuring standard chemical oxide-based IL and HfO_2 based high-K that result in an equivalent oxide thickness of approximately 1.1 nm.

DC stressing is done with varying V_{GSTR} and T, recovery after stress is at V_{GREC} = 0 V, and all data are measured using the One Point Drop Down (OPDD) MSM method with 10 μs measurement delay, see Chap. 1, Sect. 1.2.

As mentioned before, the measured ΔV_T time kinetics is modeled using uncorrelated contributions from ΔV_{IT}, ΔV_{HT}, and ΔV_{OT} subcomponents using the framework illustrated in Fig. 12.1. The model parameters are listed in Table 12.1 and are described below (reproduced from Chap. 6, Sect. 6.3). The process- dependent RD model parameters are related to the pre-factor (K_{F10}), T-independent field acceleration (Γ_0), bond polarization (α), and T activation energy (E_{AKF1}) of the inversion layer hole-assisted bond dissociation process, see Chap. 4, Fig. 4.5. As mentioned in Chap. 4, Sect. 4.4, the channel/IL defect precursor densities are different for the (100) top surface and (110) sidewalls, and a suitable average value is used based on the SL and SW of a particular device. Note, unlike FinFETs where the (110) sidewall dominates, the ratio of (100) top to (110) side areas can be similar in GAA-SNS FETs, more so in devices having smaller SW. Only K_{F10} and Γ_0 are varied with SL and SW. The process-dependent TTOM parameters are related to fast fraction of traps that undergo electron capture (f_{FAST}) and the electron capture time constant (τ_{EC}), see Chap. 5, Sect. 5.3. The parameters remain unchanged with changes in SL and SW. The process-dependent ABDWT model parameters are related to the density of preexisting defects (N_{0HT}), varied with dimensions, as well as the energy barrier (E_{BM}) and factors associated with oxide electric field (E_{OX}) dependence of the barrier (γ_B) and trap energy level ($\gamma_{E2}=m\gamma_B$), all of these remain constant across dimensions, see Chap. 5, Sect. 5.4 and Fig. 5.10. The process-dependent RDD model parameters are related to the pre-factor (K_{F30}), T-independent field acceleration (Γ_{0OT}) and T activation energy (E_{AOT}) of the anode hole injection (AHI) mechanism assisted bond dissociation (note, the bond polarization factor (α_{OT}) is not varied across devices but listed for completeness), and the forward reaction rate for ions (K_{F50}), see Chap. 6, Sect. 6.2. Only K_{F30} is varied across dimensions. Other model parameters are process agnostic and are listed in the respective sections of Chaps. 4–6 (Tables 4.1, 5.1, 5.2 and 6.1).

12.3 Description of Model Subcomponents

Figure 12.2(a, b) shows the time evolution of measured and modeled ΔV_T (a) during and (b) after stress together with the underlying ΔV_{IT}, ΔV_{HT}, and ΔV_{OT} subcomponents. As explained in the earlier chapters, ΔV_{HT} evolves rapidly at the onset of stress but saturates at longer t_{STR}, while ΔV_{IT} and ΔV_{OT} have a more gradual buildup and show power law time dependence at longer t_{STR} (~1 s and higher) with n ~ 1/6 and ~ 1/3, respectively. The saturated ΔV_{HT} magnitude is small in these devices due to the production quality gate insulator stacks. ΔV_{HT} recovers rapidly after the stoppage of stress but ΔV_{OT} is semi-permanent. The recovery of ΔV_{IT} proceeds over several decades in time, and is governed by the ΔV_{IT_FAST} (due to fast electron capture) and ΔV_{IT_SLOW} (due to interface trap passivation) processes.

Figure 12.2(c, d) shows the measured and modeled ΔV_T at fixed time (t_{STR} = 1Ks) during stress as a function of (c) V_{GSTR} and (d) T, the underlying model subcomponents are also shown. The VAF and E_A values for the ΔV_{OT} are highest and for the ΔV_{HT} are lowest, while intermediate values are seen for the ΔV_{IT} subcomponent. Therefore, the relative ΔV_{OT} contribution increases at higher V_{GSTR} and/or T, although ΔV_{IT} dominates overall ΔV_T for most stress conditions, including that at low V_G use condition shown in Sect. 12.6.

Note, higher relative ΔV_{OT} contribution would increase the n of overall ΔV_T at higher V_{GSTR} and/or T, due to higher value of the long time n associated with the ΔV_{OT} subcomponent, see Chap. 3, Sect. 3.4. However, the stress reduction effect associated with increase in ΔV_T (see Chap. 4, Sect. 4.3) would be more prominent at higher V_{GSTR} and T, and would result in a reduction in n. Therefore, these processes sort of cancel out each other, and the n does not show any changes with V_{GSTR} and T. Moreover, the recovery after stress at higher V_{GSTR} and/or T would be slower, as ΔV_{OT} is semi-permanent. These aspects are observed in the following section, where the modeling of measured data is done at multiple V_{GSTR} and T.

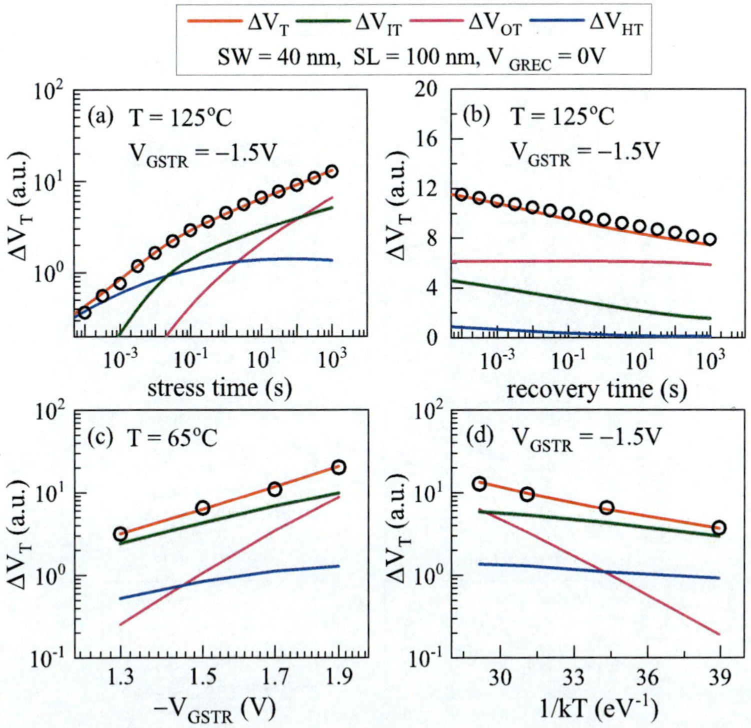

Fig. 12.2 Measured and modeled ΔV_T: (a, b) time evolution (a) during and (b) after stress, and (c, d) at fixed time (t_{STR} = 1Ks) during stress as a function of (c) V_{GSTR} and (d) T. The model subcomponents (ΔV_{IT}, ΔV_{HT}, and ΔV_{OT}) are also shown. Symbols: experiment, lines: model calculation. Data from [29]

12.4 Stress and Recovery Time Kinetics

Measured and modeled ΔV_T time kinetics is shown in Figs. 12.3, 12.4, 12.5 and 12.6 (left panels: stress, right panels: recovery), for changes in SL in Figs. 12.3, 12.4 and SW in Figs. 12.5 and 12.6, and for different V_{GSTR} in Figs. 12.3 and 12.5 and T in Figs. 12.4 and 12.6.

For every device, the stress-recovery time kinetics is measured at four different T and four different V_{GSTR} at each T, and this is done in four different devices having varying SL (but fixed SW) and also four different devices having varying SW (but fixed SL). Hence, 32 datasets (16 stress and 16 recovery) are measured for each device, resulting in a total of 256 stress-recovery datasets across all devices. However, for the ease of plotting, the time kinetics of stress and recovery from different devices is plotted at different V_{GSTR} but fixed T, and also at different T but fixed V_{GSTR}.

As mentioned before, only four parameters are adjusted to model the 32 stress-recovery datasets for each device, see Table 12.1. All the four parameters reduce as SL is scaled. However, the parameters K_{F10}, N_{0HT}, and K_{F30} increase, while Γ_0

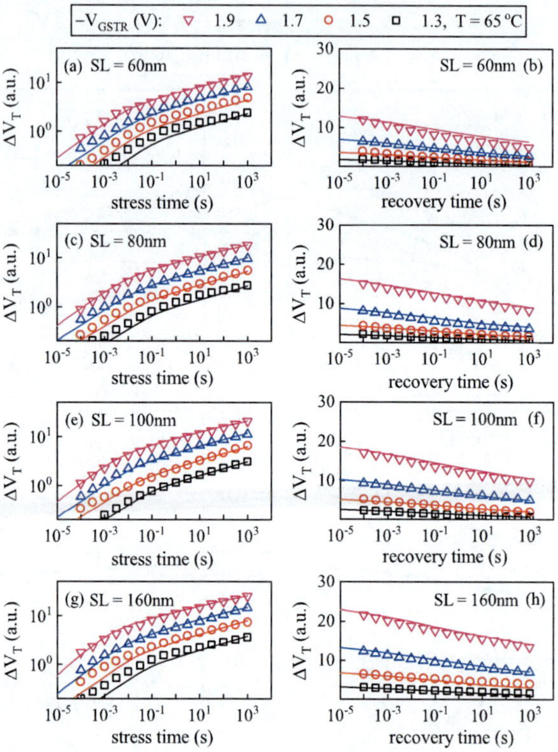

Fig. 12.3 Measured and modeled time kinetics of ΔV_T during stress (left panels) and after stress (right panels) at different V_{GSTR} but fixed T. Results are shown from devices having lowest SL (top panels) through highest SL (bottom panels). Symbols: experiment, lines: model calculation. Data from [29]

reduces as SW is scaled. Besides any possible changes in the gate insulator quality, the parameter changes are due to changes in sheet dimensions that in turn changes the compressive mechanical strain in the channel. The impact of mechanical strain on NBTI degradation is discussed in Chap. 13.

Note, for a given V_{GSTR} and T, ΔV_T reduces with reduction in SL but increases with reduction in SW. As mentioned before, for a given device, there is no significant changes in n with varying V_{GSTR} and T. However, FR slightly increases (and recovery slightly slows down) with the increase in V_{GSTR} and T, due to increase in the ΔV_{OT} subcomponent.

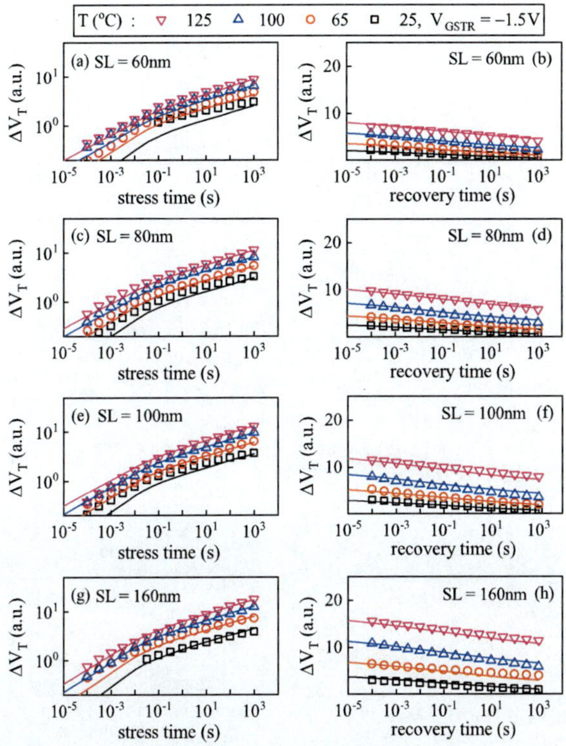

Fig. 12.4 Measured and modeled time kinetics of ΔV_T during stress (left panels) and after stress (right panels) at different T but fixed V_{GSTR}. Results are shown from devices having lowest SL (top panels) through highest SL (bottom panels). Symbols: experiment, lines: model calculation. Data from [29]

12.5 Voltage and Temperature Dependence

The model accuracy is further verified by noting the measured and modeled ΔV_T at fixed time as a function of V_{GSTR} at different T (left panels: during stress at $t_{STR} = 1$Ks, right panels: during recovery after stress at $t_{REC} = 1$Ks), for changes in SL, Fig. 12.7 and changes in SW, Fig. 12.8. All 256 datasets (128 stress and 128 recovery, for $V_{GSTR} \times T$ (4×4) conditions in four SL and four SW devices) are shown in these figures and are modeled with only four adjustable parameters as stated before.

Note that the T sensitivity of VAF (i.e., the reduction in VAF at higher T) is small in these devices and is similar to the Si channel devices modeled in Chaps. 7 and 11. This is because of the dominant contribution of the ΔV_{IT} subcomponent to overall ΔV_T, and lower polarization factor α associated with the generated interface traps for Si channel devices, see Table 12.1.

Figure 12.9 shows the measured and modeled fixed time ($t_{STR} = 1$Ks) ΔV_T during stress as a function of (a, b) V_{GSTR} but at fixed T and (c, d) T but at fixed V_{GSTR}, to

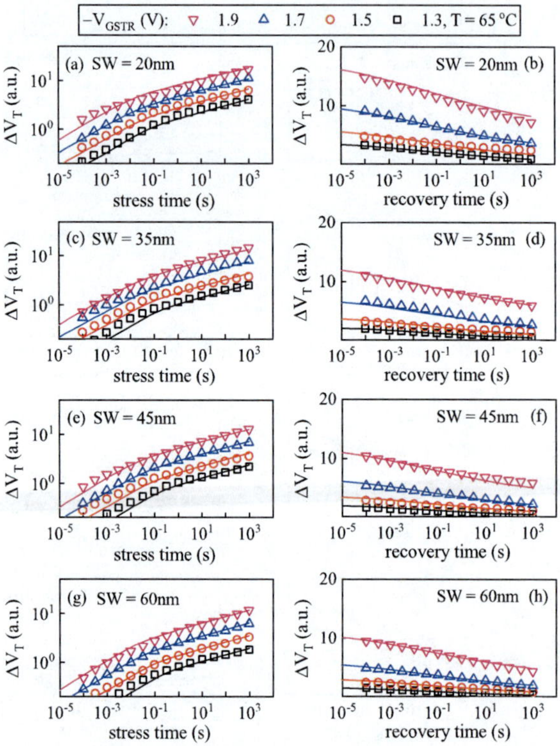

Fig. 12.5 Measured and modeled time kinetics of ΔV_T during stress (left panels) and after stress (right panels) at different V_{GSTR} but fixed T. Results are shown from devices having lowest SW (top panels) through highest SW (bottom panels). Symbols: experiment, lines: model calculation. Data from [29]

highlight the impact of (a, c) SL and (b, d) SW scaling. Note, the ΔV_T magnitude increases for increase in SL but reduction in SW. The VAF increases with increase in SL (but reduces at higher SL, the reason for this is explained later) and reduces with reduction in SW. The E_A increases slightly at higher SL, and does not vary for changes in SW. Similar comparisons also hold at other values of T (for VAF) and V_{GSTR} (for E_A), not explicitly shown.

Figure 12.10 shows the measured and modeled fixed time ($t_{STR} = 1$Ks) ΔV_T during stress at fixed V_{GSTR} and T, together with the underlying ΔV_{IT}, ΔV_{HT}, and ΔV_{OT} subcomponents, as a function of (a) SL and (b) SW. As mentioned before, there is some difference in the processes used to fabricate the devices used for SL and SW scaling experiments.

Note that ΔV_{IT} and ΔV_{OT} similarly contribute at longer SL and reduce with SL scaling, and hence, ΔV_T reduces, although the reduction in ΔV_{OT} is larger than that of ΔV_{IT}. The ΔV_{HT} contribution is very small in these devices. Therefore, the ΔV_{IT} subcomponent dominates overall ΔV_T at smaller SL. Also note that the VAF and

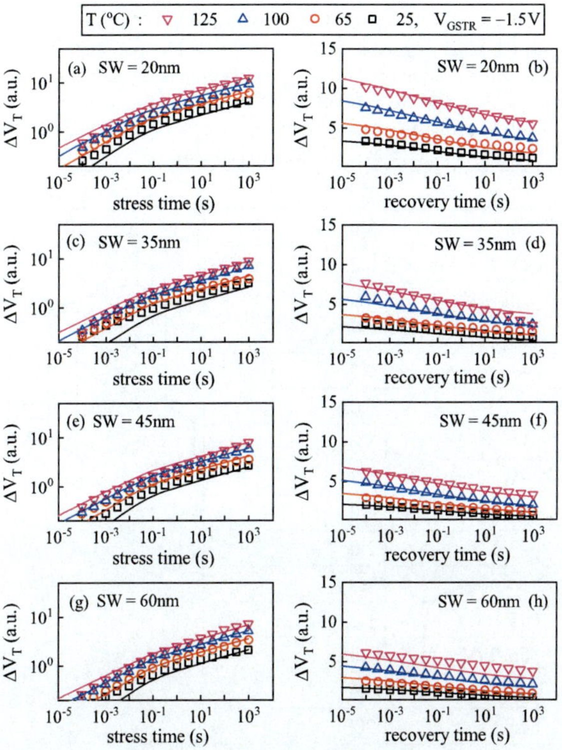

Fig. 12.6 Measured and modeled time kinetics of ΔV_T during stress (left panels) and after stress (right panels) at different T but fixed V_{GSTR}. Results are shown from devices having lowest SW (top panels) through highest SW (bottom panels). Symbols: experiment, lines: model calculation. Data from [29]

E_A values of the ΔV_{IT} subcomponent are lower than those for ΔV_{OT}, see Fig. 12.2. Therefore, the relatively larger reduction of ΔV_{OT} compared to ΔV_{IT} at smaller SL reduces the VAF of overall ΔV_T, see Fig. 12.9(a). However, as mentioned before, the longest SL device shows lower VAF. This in part is due to the stress reduction effect, i.e., the reduction in NBTI stress at higher V_{GSTR} and/or T due to higher ΔV_T. Moreover, lower VAF (~Γ_0) of the ΔV_{IT} subcomponent at larger SL is also responsible (see Table 12.1). Moreover, the larger relative reduction in ΔV_{OT} than ΔV_{IT} also causes a slight reduction in E_A at lower SL, see Fig. 12.9(c).

However, the contributions from ΔV_{HT} and ΔV_{OT} are similar for SW scaling, although that from ΔV_{IT} is much larger and dominates overall ΔV_T for all SW. All subcomponents increase at smaller SW (but the relative increase in ΔV_{IT} is slightly higher than that of the others) and ΔV_T increases. Therefore, the relative contribution from ΔV_{IT} becomes larger than ΔV_{OT} at smaller SW, which can explain the reduction in VAF of overall ΔV_T, see Fig. 12.9 (b). The stress reduction effect due to higher

Table 12.1. Process-dependent RD, TTOM, ABDWT, and RDD parameters (as per classification done in Chap. 4 through Chap. 6) used in this chapter. **The parameters K_{F10}, Γ_0, N_{OHT}, and K_{F30} are normalized to their values for the largest SL or SW device to maintain confidentiality (SW is 40 nm for SL scaling and SL is 60 nm for SW scaling, and the devices for SL scaling are fabricated using a slightly different process than those for SW scaling). The device schematic is shown (reproduced from Chap. 1). Device details in [26, 28]

Parameter	Unit	Value
K_{F10}	cm/Vs	Vary
E_{AKF1}	eV	0.25
Γ_0	cm/MV	Vary
α	qÅ	1.23
f_{FAST}	–	0.67
τ_{EC}	s	0.03
N_{OHT}	$1/cm^2$	Vary
E_{BM}	eV	1.3
γ_B	C.cm	6.0×10^{-9}
m	–	2.0
K_{F30}	1/s	Vary
E_{AOT}	eV	1.2
Γ_{0OT}	cm/MV	1.35
α_{OT}	qÅ	3.6
K_{F50}	cm^3/s	200

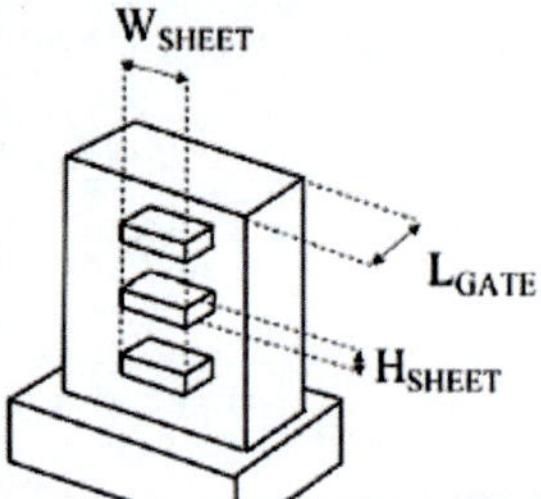

Dimension	K_{F10}	Γ_0	N_{OHT}	K_{F30}
**SL = 160	1.0	1.0	1.0	1.0
SL = 100	0.88	0.89	0.97	1.1×10^{-1}
SL = 80	0.64	0.84	0.94	5.0×10^{-2}
SL = 60	0.60	0.8	0.64	8.0×10^{-3}
**SW = 60	1.0	1.0	1.0	1.0
SW = 45	3.6	0.82	1.20	1.6
SW = 35	14.5	0.5	1.48	3.4
SW = 20	56.5	0.4	2.22	30.0

ΔV_T and lower VAF (~Γ_0) of the ΔV_{IT} component (see Table 12.1) also contribute. The ΔV_{OT} and ΔV_{HT} subcomponents remain small compared to ΔV_{IT} across all SW, and hence no noticeable change is observed in E_A, see Fig. 12.9 (d).

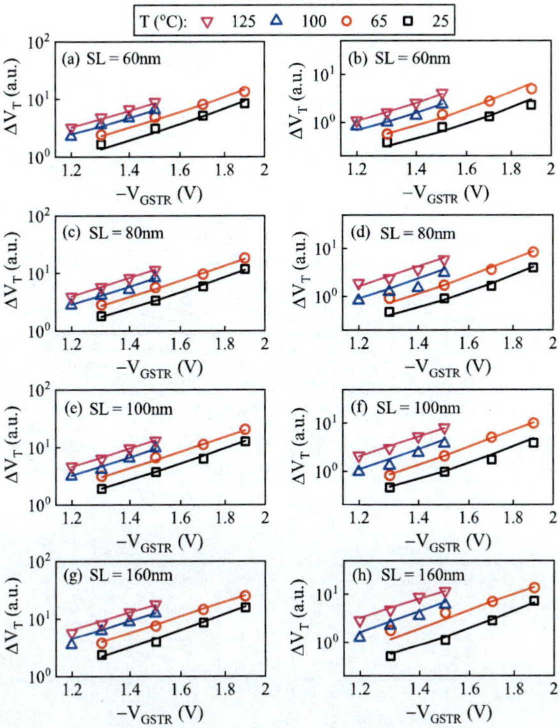

Fig. 12.7 Measured and modeled ΔV_T at fixed time during stress ($t_{STR} = 1$Ks, left panels) and after stress ($t_{REC} = 1$Ks, right panels) as a function of V_{GSTR} at different T. Results are shown from devices having lowest SL (top panels) through highest SL (bottom panels). Symbols: experiment, lines: model calculation. Data from [29]

12.6 Estimation of EOL Degradation

Figure 12.11 shows the projected ΔV_T at End Of Life (EOL) under use condition as obtained from the calibrated model, together with the underlying ΔV_{IT}, ΔV_{HT}, and ΔV_{OT} subcomponents, for different (a) SL and (b) SW. Note that ΔV_{IT} dominates overall ΔV_T for all cases (although the ΔV_{OT} contribution is not negligible for the longer SL devices). This is because ΔV_{OT} has higher VAF, and therefore, it reduces at low V_G use condition. Moreover, ΔV_{HT} is not significant in these devices, and in any case, it saturates at longer ($t_{STR} > 1$s) time and is never a significant contributor to the EOL ΔV_T. Therefore, from a practical standpoint of technology qualification, the contribution from the ΔV_{IT} subcomponent has the most significant impact. The impact of SL and SW scaling on ΔV_{IT} is analyzed in detail in Chap. 13.

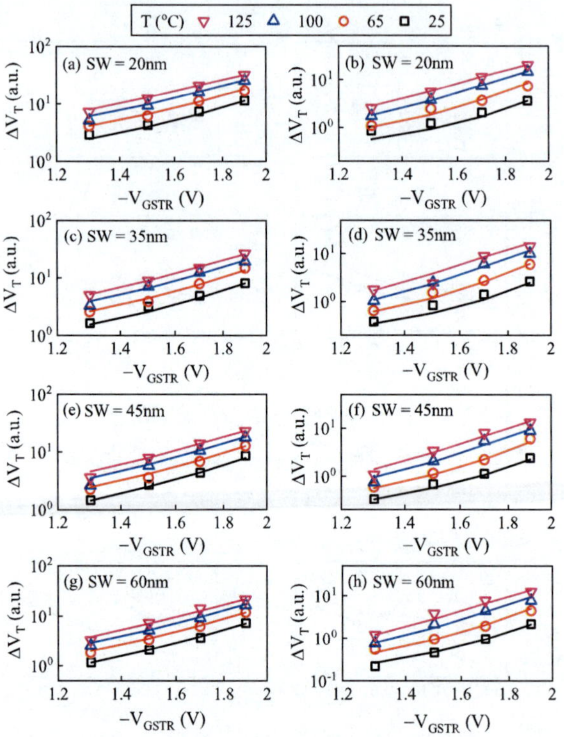

Fig. 12.8 Measured and modeled ΔV_T at fixed time during stress ($t_{STR} = 1$Ks, left panels) and after stress ($t_{REC} = 1$Ks, right panels) as a function of V_{GSTR} at different T. Results are shown from devices having lowest SW (top panels) through highest SW (bottom panels). Symbols: experiment, lines: model calculation. Data from [29]

12.7 Summary

The ultra-fast measured time kinetics of ΔV_T during and after DC NBTI stress is modeled using the BAT framework in RMG HKMG GAA-SNS FETs having different SL and SW. Stress and recovery data obtained at multiple $V_{GSTR} \times T$ values (32 datasets for each device, a total of 256 datasets over 8 devices) are modeled by using only four process-dependent model parameters. The overall ΔV_T and the underlying ΔV_{IT}, ΔV_{HT}, and ΔV_{OT} subcomponents reduce at lower SL and increase at lower SW. The ΔV_{IT} contribution dominates overall ΔV_T for all SW and shorter SL devices, especially for stress at moderate V_{GSTR} and T. The ΔV_{OT} contribution becomes appreciable in longer SL devices especially at high V_{GSTR} and T. However, the ΔV_{HT} contribution is negligible in all devices. The EOL ΔV_T at use condition is dominated by ΔV_{IT} for all devices. The reduction in VAF at lower SL and SW

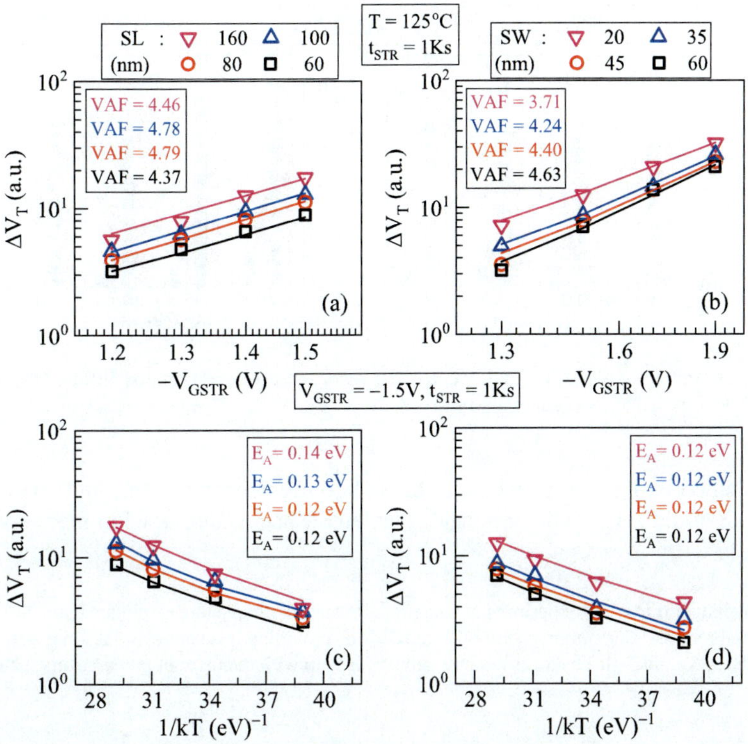

Fig. 12.9 Measured and modeled fixed time ($t_{STR} = 1Ks$) ΔV_T as a function of (a, b) V_{GSTR} at fixed T and (c, d) T at fixed V_{GSTR} during stress in devices having different (a, c) SL and (b, d) SW. Symbols: experiment, lines: model calculation. The VAF and E_A values are shown for the modeled lines. Data from [29]

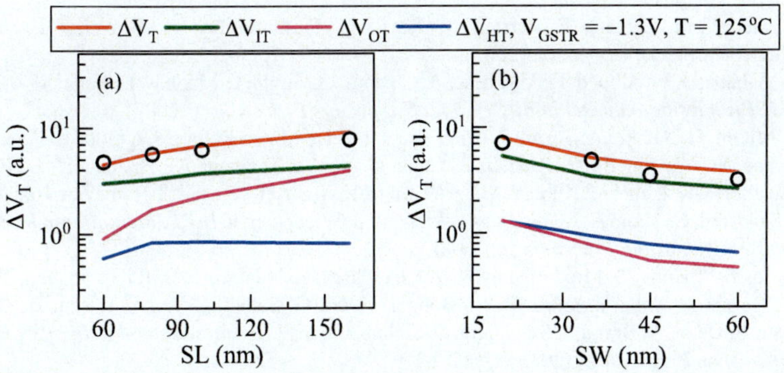

Fig. 12.10 Measured and modeled fixed time ($t_{STR} = 1Ks$) ΔV_T as a function of (a) SL and (b) SW during stress at fixed V_{GSTR} and T. The model subcomponents (ΔV_{IT}, ΔV_{HT} and ΔV_{OT}) are also shown. Symbols: experiment, lines: model calculation. Data from [29]

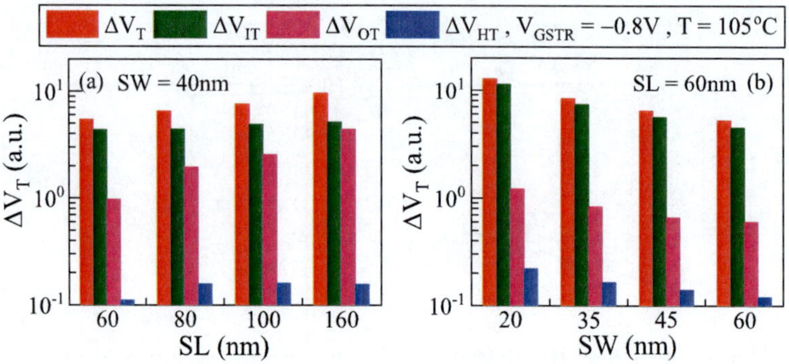

Fig. 12.11 Projected V_T at the EOL value of 10 years and use condition, for different (a) SL and (b) SW devices. The model subcomponents (ΔV_{IT}, ΔV_{HT}, and ΔV_{OT}) are also shown

is explained by relative changes in the underlying subcomponents and changes in the VAF of the ΔV_{IT} subcomponent. The later aspect is discussed in more detail in Chap. 13.

Acknowledgements All experimental data presented in this chapter are re-plotted from previously published reports. The authors acknowledge IBM for providing measurement facilities and GAA-SNS FETs. Karansingh Thakor is acknowledged for help with manuscript preparation.

References

1. N. Kimizuka, K. Yamaguchi, K. Imai, T. Iizuka, C.T. Liu, R.C. Keller, T. Horiuchi, in *Symposium on VLSI Technology Digest of Technical Papers* (2000), p. 92
2. V. Huard, M. Denais, F. Perrier, N. Revil, C. Parthasarathy, A. Bravaix, E. Vincent, Microelectron. Reliab. **45**, 83 (2005)
3. S. Mahapatra, K. Ahmed, D. Varghese, A.E. Islam, G. Gupta, L. Madhav, D. Saha, M.A. Alam, in *IEEE International Reliability Physics Symposium Proceedings* (2007), p. 1
4. Y. Mitani, H. Satake, A. Toriumi, IEEE Trans. Device Mater. Reliab. **8**, 6 (2008)
5. S. Pae, M. Agostinelli, M. Brazier, R. Chau, G. Dewey, T. Ghani, M. Hattendorf, J. Hicks, J. Kavalieros, K. Kuhn, M. Kuhn, J. Maiz, M. Metz, K. Mistry, C. Prasad, S. Ramey, A. Roskowski, J. Sandford, C. Thomas, J. Thomas, C. Wiegand, J. Wiedemer, in *IEEE International Reliability Physics Symposium Proceedings* (2008), p. 352
6. K. Joshi, S. Hung, S. Mukhopadhyay, V. Chaudhary, N. Nanaware, B. Rajamohanan, T. Sato, M. Bevan, A. Wei, A. Noori, B. McDougal, C. Ni, G. Saheli, C. Lazik, P. Liu, D. Chu, L. Date, S. Datta, A. Brand, J. Swenberg, S. Mahapatra, in *IEEE International Reliability Physics Symposium Proceedings* (2013), p. 4C.2.1
7. J. Franco, B. Kaczer, P.J. Roussel, J. Mitard, M. Cho, L. Witters, T. Grasser, G. Groeseneken, IEEE Trans. Electron Devices **60**, 396 (2013)
8. P. Srinivasan, J. Fronheiser, K. Akarvardar, A. Kerber, L.F. Edge, R.G. Southwick, E. Cartier, H. Kothari, in *IEEE International Reliability Physics Symposium Proceedings* (2014), p. 6A.3.1
9. N. Parihar, N. Goel, S. Mukhopadhyay, S. Mahapatra, IEEE Trans. Electron Devices **65**, 392 (2018)

10. N. Parihar, S. Mahapatra, in *IEEE International Reliability Physics Symposium Proceedings* (2018), p. TX.5.1
11. V. Huard, C. Ndiaye, M. Arabi, N. Parihar, X. Federspiel, S. Mhira, S. Mahapatra, A. Bravaix, in *IEEE International Reliability Physics Symposium Proceedings* (2018), p. TX.4.1
12. N. Parihar, R. Tiwari, C. Ndiaye, M. Arabi, S. Mhira, H. Wong, S. Motzny, V. Moroz, V. Huard, S. Mahapatra, in *International Conference on Simulation of Semiconductor Processes and Devices* (2018), p. 167
13. S. Ramey, A. Ashutosh, C. Auth, J. Clifford, M. Hattendorf, J. Hicks, R. James, A. Rahman, V. Sharma, A. St. Amour, C. Wiegand, in *IEEE International Reliability Physics Symposium Proceedings* (2013), p. 4C.5.1
14. K.T. Lee, W. Kang, E.-A. Chung, G. Kim, H. Shim, H. Lee, H. Kim, M. Choe, N.-I. Lee, A. Patel, J. Park, J. Park, in *IEEE International Reliability Physics Symposium Proceedings* (2013), p. 2D.1.1
15. J. Franco, B. Kaczer, A. Chasin, H. Mertens, L.-A. Ragnarsson, R. Ritzenthaler, S. Mukhopadhyay, H. Arimura, P.J. Roussel, E. Bury, N. Horiguchi, D. Linten, G. Groeseneken, A. Thean, in *IEEE International Reliability Physics Symposium Proceedings* (2016), p. 4B.2.1
16. G. Jiao, M. Toledano-Luque, K.-J. Nam, N. Toshiro, S.-H. Lee, J.-S. Kim, T. Kauerauf, E. Chung, D. Bae, G. Bae, D.-W. Kim, K.H. Wang, in *IEEE International Electron Devices Meeting Technical Digest* (2016), p. 31.2.1
17. N. Parihar, R.G. Southwick, U. Sharma, M. Wang, J.H. Stathis, S. Mahapatra, in *IEEE International Reliability Physics Symposium Proceedings* (2017), p. 2D.4.1
18. N. Parihar, R. Southwick, M. Wang, J.H. Stathis, S. Mahapatra, in *IEEE International Electron Devices Meeting Technical Digest* (2017), p. 7.3.1
19. N. Parihar, U. Sharma, R.G. Southwick, M. Wang, J.H. Stathis, S. Mahapatra, IEEE Trans. Electron. Devices **65**, 23 (2018)
20. N. Parihar, R.G. Southwick, M. Wang, J.H. Stathis, S. Mahapatra, IEEE Trans. Electron. Devices **65**, 1699 (2018)
21. N. Parihar, R.G. Southwick, M. Wang, J.H. Stathis, S. Mahapatra, IEEE Trans. Electron. Devices **65**, 1707 (2018)
22. N. Parihar, R. Tiwari, S. Mahapatra, in *International Conference on Simulation of Semiconductor Processes and Devices* (2018), p. 176
23. R. Tiwari, N. Parihar, K. Thakor, H.Y. Wong, S. Motzny, M. Choi, V. Moroz, S. Mahapatra, IEEE Trans. Electron. Devices **66**, 2086 (2019)
24. R. Tiwari, N. Parihar, K. Thakor, H.Y. Wong, S. Motzny, M. Choi, V. Moroz, S. Mahapatra, IEEE Trans. Electron. Devices **66**, 2093 (2019)
25. N. Parihar, U. Sharma, R.G. Southwick, M. Wang, J.H. Stathis, S. Mahapatra, in *IEEE International Reliability Physics Symposium Proceedings* (2019)
26. M. Wang, J. Zhang, H. Zhou, R.G. Southwick, R. Hsin, K. Chao, X. Miao, V.S. Basker, T. Yamashita, D. Guo, G. Karve, H. Bu, in *IEEE International Reliability Physics Symposium Proceedings* (2019)
27. N. Choudhury, U. Sharma, H. Zhou, R.G. Southwick, M. Wang, S. Mahapatra, in *IEEE International Reliability Physics Symposium Proceedings* (2020)
28. H. Zhou, M. Wang, J. Zhang, K. Watanabe, C. Durfee, S. Mochizuki, R. Bao, R. Southwick, M. Bhuiyan, B. Veeraraghavan, *IEEE International Reliability Physics Symposium Proceedings* (2020)
29. N. Choudhury, T. Samadder, R. Tiwari, H. Zhou, R.G. Southwick, M. Wang, S. Mahapatra, in *IEEE International Reliability Physics Symposium Proceedings* (2021)
30. S. Mahapatra, N. Goel, S. Desai, S. Gupta, B. Jose, S. Mukhopadhyay, K. Joshi, A. Jain, A.E. Islam, M.A. Alam, IEEE Trans. Electron. Devices **60**, 901 (2013)
31. A.E. Islam, N. Goel, S. Mahapatra, M.A. Alam, in *Fundamentals of Bias Temperature Instability in MOS Transistors* (Springer India, 2015), pp. 181–207

32. S. Mukhopadhyay, K. Joshi, V. Chaudhary, N. Goel, S. De, R.K. Pandey, K.V.R.M. Murali, S. Mahapatra, in *IEEE International Reliability Physics Symposium Proceedings* (2014), p. GD 3.1
33. N. Choudhury, N. Parihar, N. Goel, A. Thirunavukkarasu, S. Mahapatra, IEEE J. Electron. Devices Soc. **8**, 1281 (2020)
34. T. Samadder, N. Choudhury, S. Kumar, D. Kochar, N. Parihar, S. Mahapatra, IEEE Trans. Electron. Devices **68**, 485 (2021)

Chapter 13
BAT Framework Modeling of Dimension Scaling in FinFETs and GAA-SNS FETs

Souvik Mahapatra, Narendra Parihar, Nilotpal Choudhury, Ravi Tiwari, and Tarun Samadder

13.1 Introduction

As discussed in the earlier chapters, Negative Bias Temperature Instability (NBTI) became a serious reliability issue with the migration from Silicon Dioxide (SiO_2) to Silicon Oxynitride (SiON) planar MOSFETs [1–4], and it continues to remain so for dual layer (SiO_2 or SiON Interlayer (IL) and HfO_2 High-K layer) High-K Metal Gate (HKMG) based bulk [5–10] and Fully Depleted Silicon On Insulator (FDSOI) [11, 12] planar MOSFETs, bulk and SOI FinFETs [12–25], and Gate All Around Stacked Nanosheet (GAA-SNS) FETs [26–29], having either Silicon (Si) or Silicon Germanium (SiGe) channel. The following is a summary of the key NBTI features (reproduced from Chap. 3, Sect. 3.1).

As shown in Chap. 1, Sect. 1.3, NBTI results in the buildup of positive charges in the gate insulator of the device, when the gate bias (V_G) is held at a negative value during stress ($V_G = V_{GSTR}$). The gradual accrual of charges shifts the device parameters, *e.g.*, threshold voltage shift (ΔV_T), over time. The ΔV_T magnitude gets accelerated at more negative V_{GSTR} and higher temperature (T), respectively governed by the Voltage Acceleration Factor (VAF) and Arrhenius T activation energy (E_A). However, the accrued gate insulator charges and the resulting ΔV_T reduce if the V_G is reduced or removed after stress ($V_G = V_{GREC}$ or 0 V). Hence, AC stress results in lower ΔV_T as compared to DC stress. The AC to DC ratio depends on the pulse duty cycle (PDC), pulse low bias (V_{GLOW}), AC stress mode (A or B), and may or may not depend on pulse frequency (f). NBTI recovery also necessitates the use of ultra-fast measurements, see Chap. 1, Sect. 1.2. The ultra-fast Measure Stress Measure (MSM) method is used in this book.

S. Mahapatra (✉) · N. Parihar · N. Choudhury · R. Tiwari · T. Samadder
Department of Electrical Engineering, Indian Institute of Technology Bombay, Powai, Mumbai 400076, India
e-mail: souvik@ee.iitb.ac.in

S. Mahapatra (ed.), *Recent Advances in PMOS Negative Bias Temperature Instability*,
https://doi.org/10.1007/978-981-16-6120-4_13

The DC or AC conditions (V_{GSTR} (or V_{GHIGH} during AC), V_{GREC} (or V_{GLOW} during AC), T, PDC, f, and AC stress mode) impact the ΔV_T magnitude and its time kinetics during stress and recovery after stress (*i.e.*, the power-law slope (n) during stress for longer stress time (t_{STR}), and the Fraction Remaining (FR) during recovery after stress at a particular time (t_{REC}); FR is defined as the ratio of ΔV_T during recovery at a time $t = t_{REC}$ to that after the end of stress at $t = t_{STR}$), VAF, E_A and T dependence of VAF (*i.e.*, the reduction of VAF at higher T). As shown in Chap. 2, different process changes, *i.e.*, the introduction of Nitrogen (N) in the gate insulator or migration from the Si to SiGe channel impact NBTI. Higher Nitrogen content (N%) increases ΔV_T but reduces n, FR (*i.e.*, the recovery becomes faster), VAF, E_A, and T dependence of VAF (*i.e.*, lower reduction of VAF at higher T). On the other hand, higher Germanium content (Ge%) reduces ΔV_T but increases n, FR (*i.e.*, the recovery becomes slower), VAF, E_A, and T dependence of VAF (*i.e.*, higher reduction of VAF at higher T).

The above features are explained and modeled for different device architectures in the previous chapters of this book: Si bulk MOSFETs, Chap. 7, Si-capped SiGe bulk MOSFETs, Chap. 8, Si and SiGe FDSOI MOSFETs, Chap. 9, SOI FinFETs, Chap. 10, Si and SiGe bulk FinFETs, Chap. 11 and GAA-SNS FETs, Chap. 12, using the BTI Analysis Tool (BAT) framework described in Chap. 4 through Chap. 6. The framework is briefly explained below (reproduced from Chap. 6, Sect. 6.3).

Figure 13.1 illustrates the BAT framework [9] used to model the time kinetics of ΔV_T during and after stress. It uses uncorrelated contributions from generated interface (ΔV_{IT}) and bulk (ΔV_{OT}) gate insulator traps, and hole trapping in process related pre-existing bulk gate insulator traps (ΔV_{HT}). The Reaction Diffusion (RD) model is used to calculate the time kinetics of interface traps (density ΔN_{IT}) [9, 30, 31]. In Chap. 4, the RD model is described and independently validated by measured data from Direct Current IV (DCIV) method [32]. The Transient Trap Occupancy Model (TTOM) is used to calculate the occupancy of generated interface traps and their contribution (ΔV_{IT}) [9], which is described and validated in Chap. 5, Sect. 5.3.

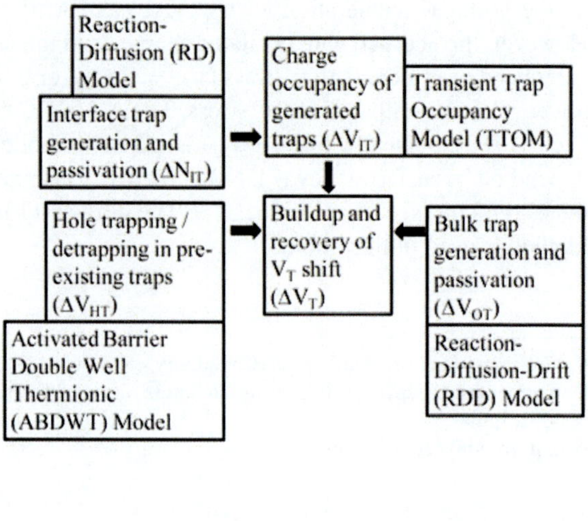

Fig. 13.1 Schematic of the BTI Analysis Tool (BAT) framework used in this book to model measured ΔV_T kinetics during and after DC and AC NBTI stress, reproduced from Chap. 4

The ΔV_{HT} and ΔV_{OT} kinetics are modeled by the Activated Barrier Double Well Thermionic (ABDWT) model [33] and Reaction Diffusion Drift (RDD) model [34] respectively, and these are described and validated in Chap. 5, Sect. 5.4 and Chap. 6, Sect. 6.2.

In Chaps. 7, 10, 11, and 12, a detailed analysis of the stress and recovery time kinetics has been done for DC stress at different V_{GSTR} and T. The same is done for Mode-B AC stress in Chaps. 7, 10, and 11 at different V_{GHIGH} and T but at fixed PDC, *f*, and $V_{GLOW} = 0$ V. The impact of PDC, V_{GLOW}, *f*, and AC stress mode is analyzed in Chap. 14.

In Chaps. 7, 9, and 11, the impact of N% on ΔV_T is shown to be due to the increase in ΔV_{IT} and ΔV_{HT} (with relatively higher increase in ΔV_{HT}) and reduction in the ΔV_{OT} subcomponents, although ΔV_{IT} dominates overall ΔV_T for different V_{GSTR} and T. In Chaps. 8, 9 and 11, the impact of Ge% on ΔV_T is shown to be due to the reduction in ΔV_{IT}, ΔV_{HT} and ΔV_{OT} subcomponents (with relatively higher reduction in ΔV_{HT} and lower reduction in ΔV_{OT}), although once again, ΔV_{IT} dominates overall ΔV_T across V_{GSTR} and T, unless Ge%, and/or V_{GSTR} and/or T are very high, when both ΔV_{IT} and ΔV_{OT} equally contribute.

Moreover, as shown in Chap. 2, Sect. 2.4, the transistor layout (*i.e.*, the spacing between the Shallow Trench Isolation (STI) to the device active) and channel dimensions (length and width) also impact the ΔV_T magnitude and VAF. The layout effect is analyzed for FDSOI MOSFETs using the BAT framework in Chap. 9 and is attributed to the changes in mechanical strain in the channel, which primarily impacts the ΔV_{IT} subcomponent. The BAT framework is used to model the NBTI stress and recovery kinetics in GAA-SNS FETs with different sheet dimensions in Chap. 12. The modeling of fin dimension changes in FinFETs is done in this chapter. The mechanism responsible for dimension changes in FinFETs and GAA-SNS FETs is explained.

13.2 Device Details and Model Parameters

Figure 13.2 illustrates the schematic of FinFETs and GAA-SNS FETs, the fin and sheet length and width are shown. Replacement Metal Gate (RMG) HKMG based Si channel SOI FinFETs with different fin length (FL), SiGe channel (Ge = 25%) bulk FinFETs with different FL and fin width (FW), and Si channel GAA-SNS FETs with different sheet length and width (SL and SW) are analyzed in this chapter. All devices have the standard Chemical Oxide-based IL and HfO_2 High-K that results in an equivalent oxide thickness of approximately 1.1 nm. Measurements in all devices are done using the One Point Drop Down (OPDD) MSM method with 10 μs measurement delay, refer to Chap. 1, Sect. 1.2 for further details.

A detailed analysis of the SOI and SiGe FinFETs has been done in Chaps. 10 and 11 respectively but at fixed FL and FW. The FL and FW dependencies studied in this chapter are in devices that are made using a slightly different process, and the adjustable model parameters are listed in Table 13.1. Note, the FL and FW

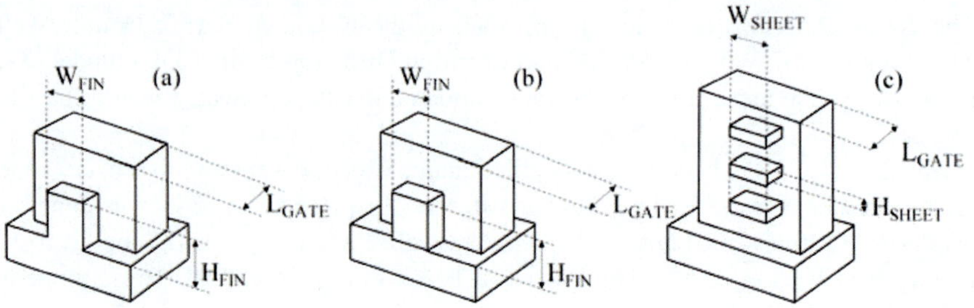

Fig. 13.2 Schematic of (a) bulk and (b) SOI FinFETs and (c) GAA-SNS FET. The gate length (L_{GATE}) is denoted as fin length (FL) or sheet length (SL) in this chapter

Table 13.1 Process dependent RD, TTOM, ABDWT, and RDD parameters (as per classification done in Chap. 4 through Chap. 6) used in this chapter. As per Chap. 4, Sect. 4.4, the channel/IL precursor density for the RD model is suitably adjusted according to Ge% in the channel for FinFETs and the ratio of (100)–(110) channel area for GAA-SNS FETs. Only stress experiments are analyzed for the SiGe devices, so TTOM parameters are not shown. ΔV_T is fully due to ΔV_{IT} for SiGe25-MidN, so only the RD model parameters are shown.

Parameter	Unit	SOI	SiGe-MidN	SiGe-HighN	GAA
Scaling	–	FL	FL	FW	SL, SW
K_{F10}	cm/Vs	Vary	Vary	Vary	Vary
E_{AKF1}	eV	0.2	0.6	0.5	0.25
Γ_0	cm/MV	Vary	Vary	Vary	Vary
α	qÅ	1.8	2.3	1.8	1.23
f_{FAST}	–	0.45	–	–	0.67
τ_{EC}	s	0.03	–	–	0.03
N_{0HT}	1/cm^2	Vary	–	Vary	Vary
E_{BM}	eV	0.9	–	1.3	1.3
γ_B	C.cm	6.7 x 10^{-9}	–	3.2 x 10^{-9}	6.0 x 10^{-9}
m	–	2.4	–	2.4	2.0
K_{F30}	1/s	Vary	–	Vary	Vary
E_{AOT}	eV	0.9	–	1.14	1.2
Γ_{0OT}	cm/MV	0.8	–	0.13	1.35
α_{OT}	qÅ	3.6	–	3.6	3.6
K_{F50}	cm^3/s	23	–	23	200

dependencies in SiGe FinFETs are also from different processes, respectively having medium and high N% in the gate insulator. Note that the GAA-SNS FETs analyzed in this chapter are exactly the same as in Chap. 12, and the parameters are re-listed again (from Table 12.1).

As mentioned before, the measured ΔV_T time kinetics is modeled using uncorrelated contributions from ΔV_{IT}, $\Delta V_{HT,}$ and ΔV_{OT} subcomponents using the framework illustrated in Fig. 13.1. The adjustable model parameters are described below (reproduced from Chap. 6, Sect. 6.3). The process dependent RD model parameters are related to the pre-factor (K_{F10}), T independent field acceleration (Γ_0), bond polarization (α), and T activation energy (E_{AKF1}) of the inversion layer hole assisted bond dissociation process, see Chap. 4, Fig. 4.5. As mentioned in Chap. 4, Sect. 4.4, the channel/IL defect precursor densities are different for Si and SiGe channels and a suitable value is used based on the Ge% in the channel. The same is also different for (100) versus (110) channel or surface orientation, and an average value is used for GAA-SNS FETs based on SL and SW. Only K_{F10} and Γ_0 are varied with dimensions.

The process dependent TTOM parameters are related to the fast fraction of traps that undergo electron capture (f_{FAST}) and the electron capture time constant (τ_{EC}), see Chap. 5, Sect. 5.3, and they remain unchanged with dimensions. The process dependent ABDWT model parameters are related to the density of pre-existing defects (N_{0HT}), as well as the energy barrier (E_{BM}) and factors associated with oxide electric field (E_{OX}) dependence of the barrier (γ_B) and trap energy level ($\gamma_{E2} = m\gamma_B$), see Chap. 5, Sect. 5.4 and Fig. 5.10. Only N_{0HT} is varied with dimensions. The process dependent RDD model parameters are related to the pre-factor (K_{F30}), T independent field acceleration (Γ_{0OT}), and T activation energy (E_{AOT}) of the Anode Hole Injection (AHI) process assisted bond dissociation (note, the bond polarization factor (α_{OT}) is not varied across devices but listed for completeness), and the forward reaction rate for ions (K_{F50}), see Chap. 6, Sect. 6.2. Only K_{F30} is varied with dimensions. Other model parameters are process agnostic and are listed in the respective sections of Chap. 4 through Chap. 6 (Tables 4.1, 5.1, 5.2, and 6.1).

As mentioned before, only 4 parameters vary with dimensions, and are listed in Table 13.2 for FL, FW, and SL, SW changes respectively in FinFETs and GAA-SNS FETs. K_{F10}, $N_{0HT,}$ and K_{F30} reduce with reduction in FL and FW in FinFETs, they also reduce with reduction in SL but increase with reduction in SW in GAA-SNS FETs. On the other hand, Γ_0 increases with reduction in FL and FW but reduces with reduction in SL and SW respectively in FinFETs and GAA-SNS FETs.

13.3 Fin Length Scaling in SOI FinFETs

Figure 13.3 shows the measured and modeled time evolution of ΔV_T during (left panels) and after (right panels) stress at different V_{GSTR} and T but $V_{GREC} = 0$ V, in SOI FinFETs having different FL. The underlying ΔV_{IT}, $\Delta V_{HT,}$ and ΔV_{OT} subcomponents are shown in each panel corresponding to the dataset having lowest degradation magnitude. The time evolution of ΔV_T at the initiation of stress is different for these SOI FinFETs than those analyzed in Chap. 10 and is primarily due to the difference in the early time kinetics of the ΔV_{HT} subcomponent.

Table 13.2 Process dependent RD, ABDWT, and RDD parameters that vary with scaling of device dimensions, see Table 13.1 for additional details

	K_{F10}	Γ_0	N_{0HT}	K_{F30}
SOI FL (nm)				
[a]FL = 160	1.0	1.0	1.0	1.0
FL = 80	1.0 x 10^{-1}	2.67	0.88	0.25
FL = 20	5.0 x 10^{-3}	3.33	0.63	0.10
SiGe FL (nm)				
[a]FL = 200	1.0	1.0	–	–
FL = 60	9.1 x 10^{-2}	1.2	–	–
FL = 20	6.1 x 10^{-4}	7.7	–	–
SiGe FW (nm)				
[a]FW = 1000	1.0	1.0	1.0	1.0
FW = 20	1.7 x 10^{-1}	4.33	0.76	0.45
FW = 10	9.6 x 10^{-2}	6.0	0.47	0.30
GAA SL (nm)				
[a]SL = 160	1.0	1.0	1.0	1.0
SL = 100	0.88	0.89	0.97	1.1 x 10^{-1}
SL = 80	0.64	0.84	0.94	5.0 x 10^{-2}
SL = 60	0.60	0.8	0.64	8.0 x 10^{-3}
GAA SW (nm)				
[a]SW = 60	1.0	1.0	1.0	1.0
SW = 45	3.6	0.82	1.20	1.6
SW = 35	14.5	0.5	1.48	3.4
SW = 20	56.5	0.4	2.22	30.0

[a] For each category of device, the parameters are normalized to their values for the highest dimension device. The SL and SW scaling are done on slightly differently processed GAA-SNS FETs (see Chap. 12)

Nevertheless, power-law time kinetics is observed at longer t_{STR} (>1s) like that in Chap. 10, and the slope n is governed by the relative contributions of the underlying subcomponents. ΔV_{HT} saturates at longer t_{STR} (n ~ 0 in a log–log plot), while ΔV_{IT} and ΔV_{OT} show power-law time kinetics with n ~ 1/6 and ~1/3 respectively. The saturated ΔV_{HT} contribution is not significant in these devices due to the use of production quality gate insulators. The overall ΔV_T and the underlying subcomponents reduce as FL is reduced, although, ΔV_{IT} reduces more than ΔV_{OT}. Therefore, although ΔV_{IT} dominates overall ΔV_T, the larger relative ΔV_{OT} contribution slightly increases the long time n of overall ΔV_T for shorter FL devices.

The ΔV_T recovery after stress proceeds over several decades in time and shows a logistic time dependence. ΔV_{HT} recovers fast and ΔV_{OT} is semi-permanent. ΔV_{IT} recovery is due to the contributions from the fast electron capture (ΔV_{IT_FAST}) and

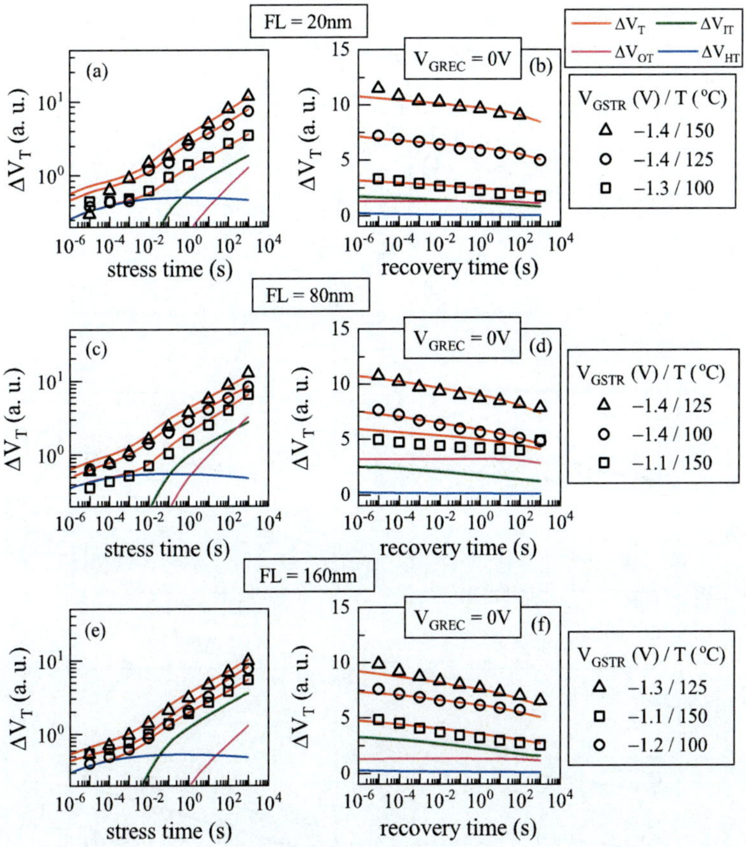

Fig. 13.3 Time evolution of measured and modeled ΔV_T during (left panels) and after (right panels) DC stress in SOI p-FinFETs having different FL. The underlying model subcomponents (ΔV_{IT}, ΔV_{HT}, and ΔV_{OT}) are shown for the lowest degradation dataset. Symbols: experiment, lines: model calculation. Data from [22]

the slow trap passivation processes (ΔV_{IT_SLOW}), see Chap. 5, Sect. 5.3. Note that the larger relative ΔV_{OT} contribution slightly increases FR (*i.e.*, recovery becomes slower) at shorter FL devices.

Figure 13.4 shows the measured and modeled fixed time ($t_{STR} = 1$Ks) ΔV_T as a function of V_{GSTR} (a) with the underlying ΔV_{IT}, ΔV_{HT}, and ΔV_{OT} subcomponents at fixed T and (b) at different T. Note that ΔV_{IT} dominates the overall ΔV_T except at high V_{GSTR} when both ΔV_{IT} and ΔV_{OT} similarly contribute, whereas ΔV_{HT} contribution is negligible across all V_{GSTR}, for all FL devices. The VAF of ΔV_{OT} is much higher than that of the other subcomponents. The relatively higher ΔV_{OT} contribution slightly increases the VAF of overall ΔV_T for shorter FL devices. Moreover, the increase in Γ_0 (hence VAF) of the ΔV_{IT} subcomponent at shorter FL (see Table 13.2) also contributes.

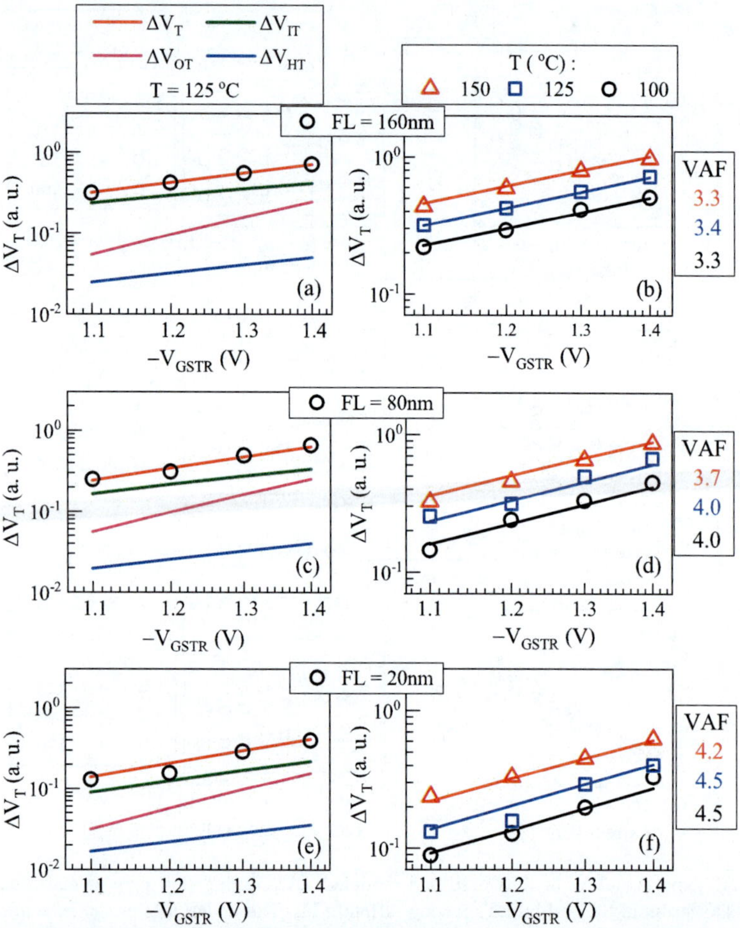

Fig. 13.4 Measured and modeled fixed time ($t_{STR} = 1$Ks) ΔV_T as a function of V_{GSTR} during DC stress in SOI p-FinFETs having different FL, with the underlying model subcomponents (ΔV_{IT}, ΔV_{HT}, and ΔV_{OT}) at fixed T (left panels) and at different T (right panels). Symbols: experiment, lines: model calculation. The VAF values are listed for simulated lines. Data from [22]

The VAF reduction at higher T is due to the bond polarization terms associated with the ΔV_{IT} and ΔV_{OT} subcomponents and also the stress reduction effect (higher ΔV_T at longer t_{STR} reduces the effective NBTI stress). The stress reduction effect is higher at longer FL as ΔV_T is higher, so the relative contribution due to polarization would be smaller, the opposite is true for shorter FL devices. The VAF reduction is slightly higher at shorter FL due to relatively higher contribution from the ΔV_{OT} subcomponent (and the associated polarization term).

Figure 13.5 shows the measured and modeled fixed time ($t_{STR} = 1$Ks) ΔV_T as a function of T (a) with the underlying ΔV_{IT}, ΔV_{HT}, and ΔV_{OT} subcomponents at fixed V_{GSTR} and (b) at different V_{GSTR}. Note that ΔV_{IT} dominates the overall

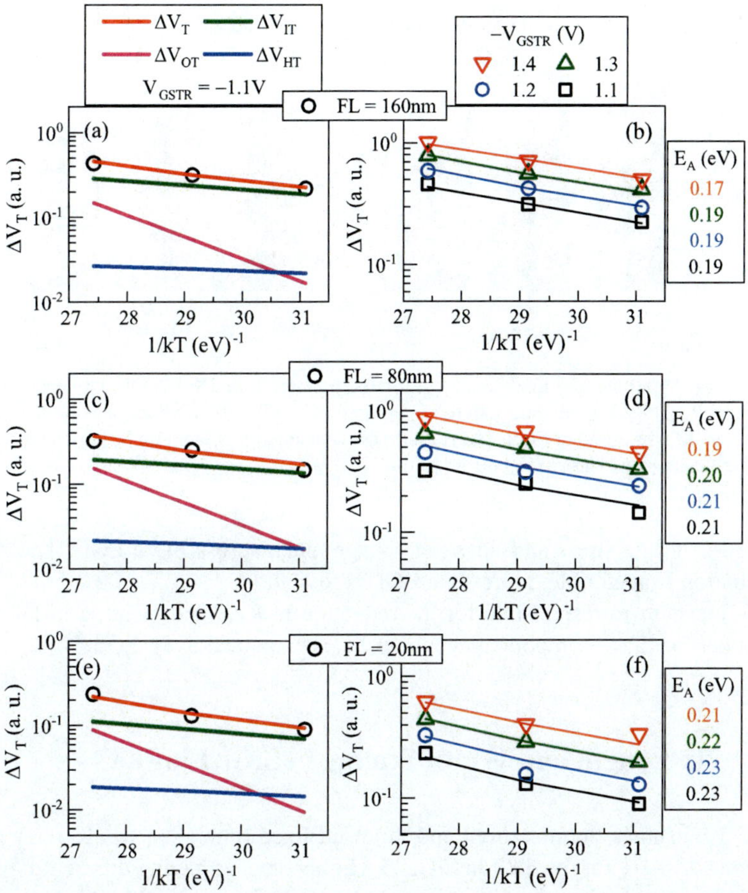

Fig. 13.5 Measured and modeled fixed time ($t_{STR} = 1$Ks) ΔV_T as a function of T during DC stress in SOI p-FinFETs having different FL, with the underlying model subcomponents (ΔV_{IT}, ΔV_{HT}, and ΔV_{OT}) at fixed V_{GSTR} (left panels) and at different V_{GSTR} (right panels). Symbols: experiment, lines: model calculation. The E_A values are listed for simulated lines. Data from [22]

ΔV_T except at high T when both ΔV_{IT} and ΔV_{OT} similarly contribute, however, ΔV_{HT} contribution is negligible across all T, for all FL devices. The E_A values are highest for ΔV_{OT} and lowest for ΔV_{HT}, *i.e.*, $E_A\ (\Delta V_{OT}) > E_A\ (\Delta V_{IT}) > E_A\ (\Delta V_{HT})$. Relatively higher ΔV_{OT} contribution slightly increases E_A in shorter FL devices.

Figure 13.6 (a) shows the measured and modeled ΔV_T during stress at a fixed V_{GSTR}, T, and time ($t_{STR} = 1$ Ks) together with the underlying ΔV_{IT}, ΔV_{HT}, and ΔV_{OT} subcomponents versus FL. The calibrated BAT framework is used to determine the extrapolated ΔV_T at the end of life (EOL) value of 10 years at a fixed (low) V_G and T. Figure 13.6 (b) shows the model extrapolated EOL ΔV_T and the underlying subcomponents. The reduction of ΔV_T at shorter FL is primarily due to that of the ΔV_{IT} subcomponent in both cases. The ΔV_{OT} subcomponent reduces much less at

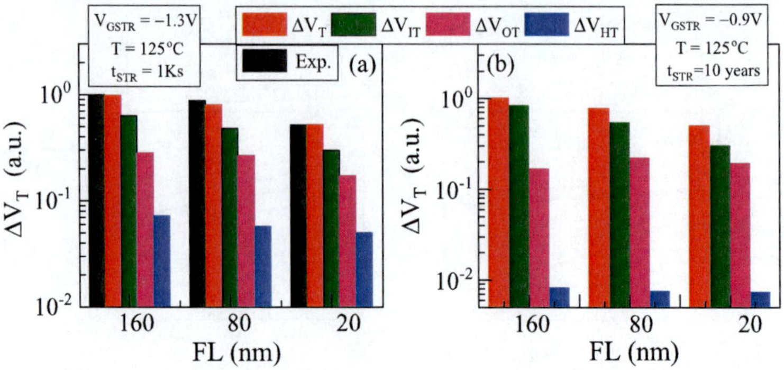

Fig. 13.6 (a) Measured and modeled ΔV_T at fixed time ($t_{STR} = 1$Ks) during DC stress at fixed V_{GSTR} and T, and (b) extrapolated EOL ΔV_T using the calibrated BAT framework at fixed use (low) V_G and T, for different FL in SOI p-FinFETs. The underlying model subcomponents (ΔV_{IT}, ΔV_{HT}, and ΔV_{OT}) are shown. Data from [22]

shorter FL during stress and does not change much across FL at EOL. The ΔV_{HT} contribution is negligible, more so at EOL across all FL.

The mechanism responsible for the reduction in K_{F10} and increase in Γ_0 of the RD model for ΔV_{IT} component at shorter FL is explained in Sect. 13.6.

13.4 Fin Length and Width Scaling in SiGe FinFETs

Figure 13.7 shows the measured and modeled fixed time ($t_{STR} = 1$Ks) ΔV_T as a function of V_{GSTR} (at fixed T) in SiGe25_HighN devices having different FW. The underlying model subcomponents are also shown. Unlike the SOI FinFETs analyzed above, these devices show higher relative contribution from ΔV_{HT} compared to ΔV_{OT}, although ΔV_{IT} dominates the overall ΔV_T across V_{GSTR} and all FW devices. However, like before, the VAF of ΔV_{OT} is much higher compared to the other subcomponents, while the VAF of ΔV_{IT} is slightly higher than that of ΔV_{HT}.

Figure 13.8 shows the measured and modeled fixed time ($t_{STR} = 1$Ks) ΔV_T as a function of V_{GSTR} at fixed T in (a) SiGe25_MidN devices having different FL and (b) SiGe25_HighN devices having different FW. As mentioned before (see Table 13.1), the overall ΔV_T for the SiGe25_MidN devices is fully due to the ΔV_{IT} subcomponent across all FL, and as shown above, ΔV_{IT} dominates the overall ΔV_T for the SiGe25_HighN devices across all FW. The magnitude of ΔV_T reduces and VAF increases for both devices as FL and FW are reduced. Since ΔV_{IT} is the dominating component and ΔV_{OT} is negligible, this is due to the reduction in K_{F10} and increase in Γ_0 as FL and FW are scaled, see Table 13.2. The FL scaling impact is the same for the SOI and SiGe bulk FinFETs.

Once again, the calibrated BAT framework is used to determine the extrapolated ΔV_T at the EOL value of 10 years at a fixed (low) V_G and T. Figure 13.9 shows the

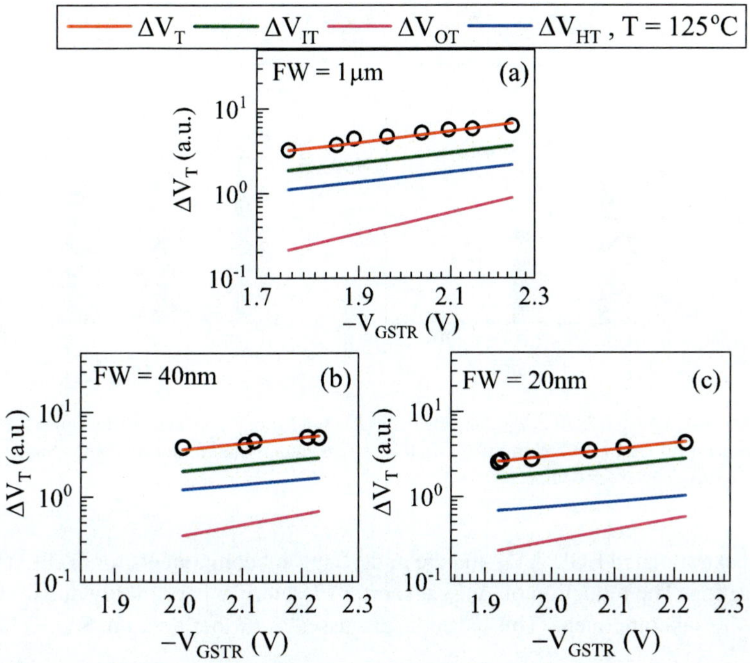

Fig. 13.7 Measured and modeled fixed time ($t_{STR} = 1$Ks) ΔV_T together with the underlying model subcomponents (ΔV_{IT}, $\Delta V_{HT,}$ and ΔV_{OT}) as a function of V_{GSTR} during DC stress at fixed T in SiGe p-FinFETs having different FW. Symbols: experiment, lines: model calculation. Data from [24]

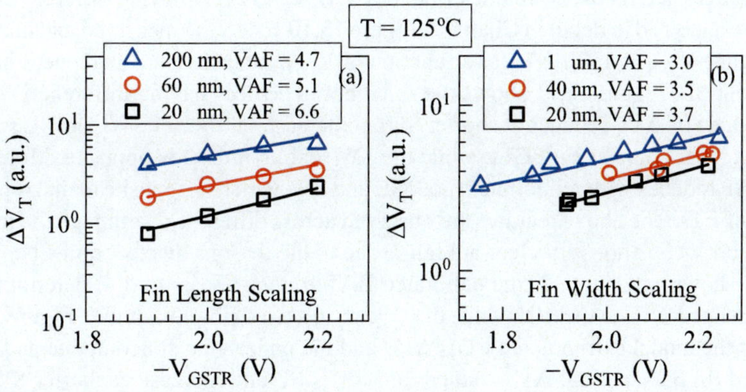

Fig. 13.8 Measured and modeled fixed time ($t_{STR} = 1$ Ks) ΔV_T as a function of V_{GSTR} during DC stress at fixed T in SiGe p-FinFETs having different (a) FL and (b) FW. Symbols: experiment, lines: model calculation. Data from [24]

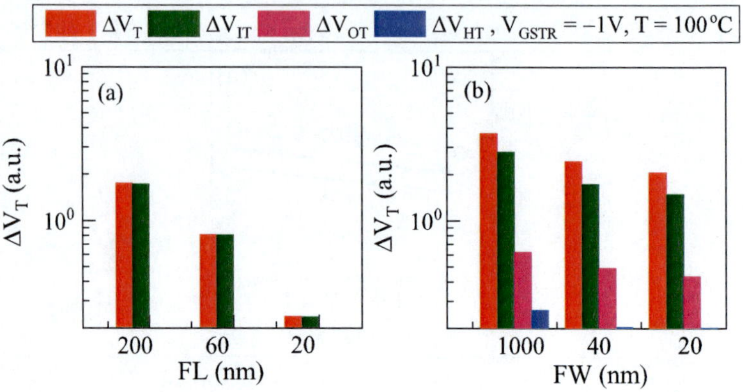

Fig. 13.9 Extrapolated EOL ΔV_T using the calibrated BAT framework at fixed use (low) V_G and T, for different (a) FL and (b) FW in SiGe p-FinFETs. The underlying model subcomponents (ΔV_{IT}, $\Delta V_{HT,}$ and ΔV_{OT}) are shown

model extrapolated EOL ΔV_T and the underlying subcomponents for (a) FL and (b) FW scaling. The reduction of ΔV_T at shorter FL and FW is primarily due to that of the ΔV_{IT} subcomponent. This aspect is discussed in further detail in Sect. 13.6.

13.5 Sheet Length and Width Scaling in GAA-SNS FETs

NBTI stress and recovery kinetics from the GAA-SNS FETs having different SL and SW are analyzed in detail in Chap. 12. Figure 13.10 shows the measured and modeled fixed time ($t_{STR} = 1\text{Ks}$) ΔV_T as a function of V_{GSTR} at different T in devices having different SL. Figure 13.11 shows the same but in devices having different SW. The magnitude of ΔV_T reduces at shorter SL but increases at shorter SW. The SL scaling impact is the same as FinFETs, while the SW scaling impact is opposite. Moreover, the VAF reduces with reduction in both SL and SW, which is opposite to that reported for FinFETs. The above features are observed across different T, although the impact of SL on VAF is not very clear at high T due to the reasons discussed in Chap. 12.

As shown in Chap. 12, the calibrated BAT framework is used to determine the extrapolated ΔV_T at the EOL value of 10 years at a fixed (low) V_G and T. Figure 13.12 shows the model extrapolated EOL ΔV_T and the underlying subcomponents for (a) SL and (b) SW scaling. ΔV_T reduces at shorter SL and increase at shorter SW, but is always dominated by the ΔV_{IT} subcomponent.

As discussed in Chap. 12, the reduction in ΔV_T at shorter SL is due to the reduction of both ΔV_{IT} and ΔV_{OT} subcomponents, although ΔV_{IT} always dominates the overall ΔV_T across all SL. The reduction of VAF at shorter SL is due to the reduction in Γ_0 of the ΔV_{IT} subcomponent, and relatively lower contribution from ΔV_{OT} (the

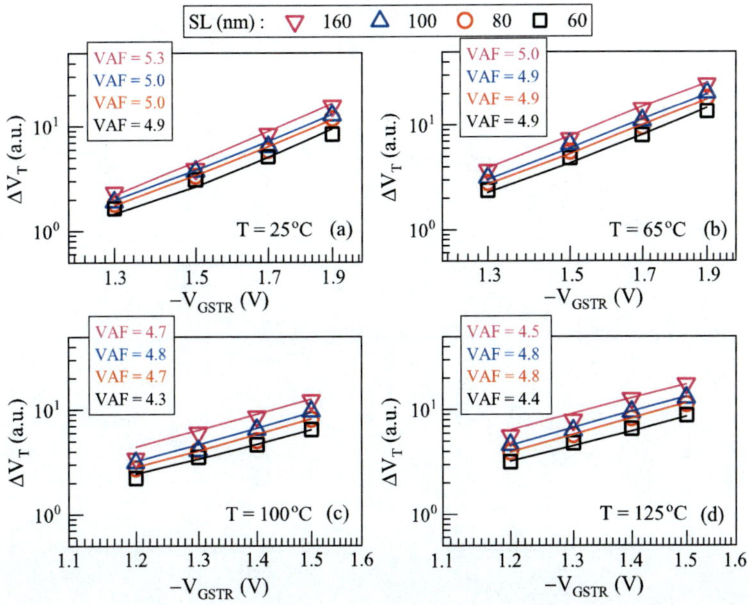

Fig. 13.10 Measured and modeled fixed time ($t_{STR} = 1$ Ks) ΔV_T as a function of V_{GSTR} at different T during DC stress in GAA-SNS p-FETs having different SL. Symbols: experiment, lines: model calculation. Data from [22]

reduction in ΔV_{OT} is more than that of ΔV_{IT} as SL is scaled). ΔV_{HT} is negligible in devices used for SL scaling experiments.

As also discussed in Chap. 12, the increase in ΔV_T at shorter SW is due to the increase of all subcomponents. ΔV_{IT} always dominates the overall ΔV_T, while the contributions from ΔV_{HT} and ΔV_{OT} are similar (but much smaller than ΔV_{IT}) for the devices used in SW scaling experiments. Note, the relative increase in ΔV_{IT} is much more than the other subcomponents as SW is reduced. The VAF reduction at shorter SW is primarily due to the reduction in Γ_0 of the ΔV_{IT} subcomponent.

The impact of SL and SW scaling on the K_{F10} and Γ_0 parameters of the RD model is discussed next.

13.6 Explanation of Fin and Sheet Dimension Scaling

As mentioned before, the RD model parameters K_{F10} and Γ_0 depend on the inversion layer hole assisted interfacial bond dissociation process, see Chap. 4, Sect. 4.3, Fig. 4.5, reproduced below as Fig. 13.13. These RD parameters in turn depend on the effective mass (m_T) and barrier (φ_B) of the hole tunneling process and can be determined using band structure calculations. However, since the material interface (channel/IL) is difficult to model, rather than their absolute values, changes in the

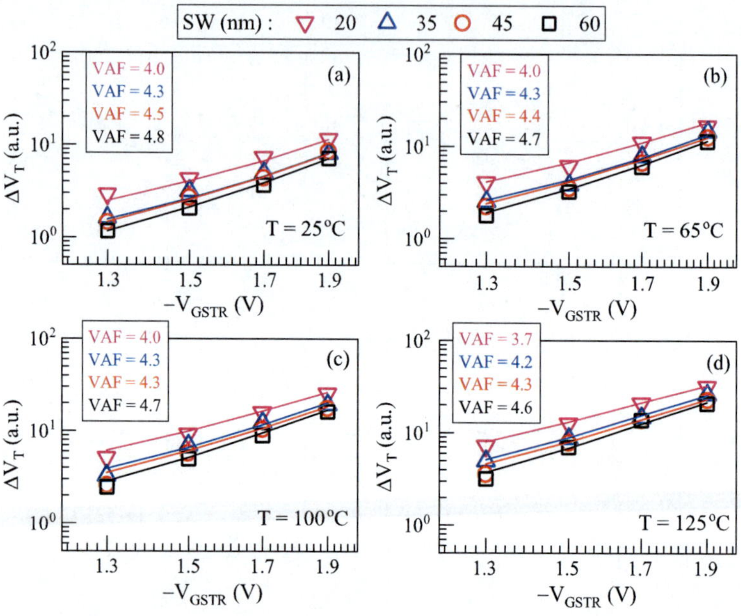

Fig. 13.11 Measured and modeled fixed time ($t_{STR} = 1$ Ks) ΔV_T as a function of V_{GSTR} at different T during DC stress in GAA-SNS p-FETs having different SW. Symbols: experiment, lines: model calculation. Data from [22]

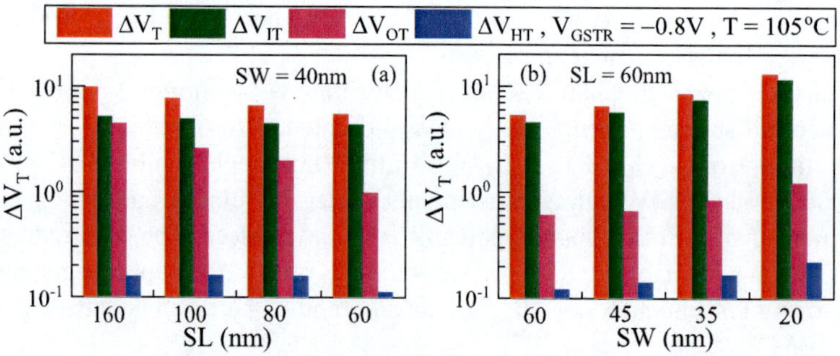

Fig. 13.12 Extrapolated EOL ΔV_T using the calibrated BAT framework at fixed use (low) V_G and T, for different (a) SL and (b) SW in GAA-SNS p-FETs. The underlying model subcomponents (ΔV_{IT}, $\Delta V_{HT,}$ and ΔV_{OT}) are shown

m_T and φ_B due to changes in transistor process (mechanical strain in the channel in this case) can be used to estimate the changes in the above parameters as channel (fin or sheet length or width) dimension is changed.

Technology CAD (TCAD) [35] process simulation is used to generate the FinFET and GAA-SNS FET structures and also calculate the mechanical stress in the channel due to changes in the fin or sheet dimensions. Since the actual structures are difficult

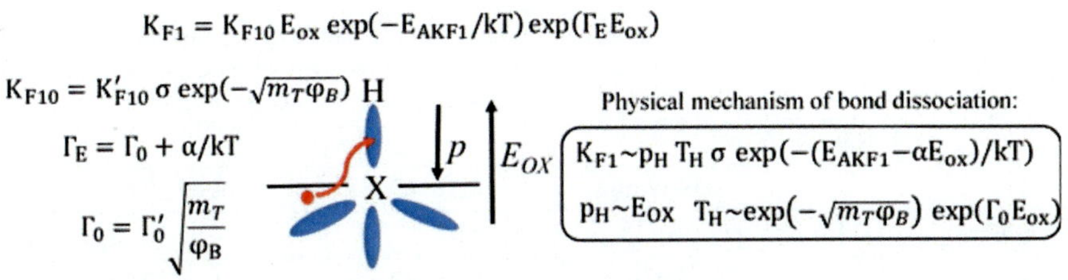

Fig. 13.13 Schematic of the inversion layer hole and oxide electric field induced dissociation of H passivated defects at the channel/gate insulator interface, see Chap. 4, Sect. 4.3

to simulate (need information on the exact pitch for the multiple fins and stacks of nanosheets, spacing between the nanosheets in a stack, structure of the epitaxial SiGe Source-Drain, etc.), the goal is to provide a qualitative estimation of the impact of mechanical stress on the band structure parameters and in turn on the RD model parameters.

Figure 13.14 illustrates the TCAD process simulated FinFET and GAA-SNS FET structures and the calculated mechanical strain, only one fin or sheet is simulated for simplicity. Table 13.3 lists the calculated mechanical strain in the channel perpendicular to the (110) sidewalls for FinFET and the (100) top and (110) side surface for GAA-SNS FETs, for changes in fin or sheet dimensions. Note that the stress is compressive in nature due to the epitaxial SiGe Source-Drain regions. The stress increases with reduction in FL and FW for FinFETs and with reduction in SL in GAA-SNS FETs for both surface orientations. However, with reduction in SW, the stress increases slightly for (100) but reduce for (110) surface orientation in GAA-SNS FETs.

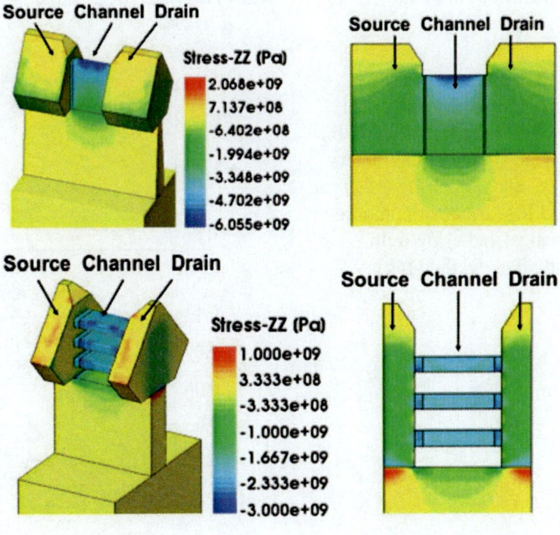

Fig. 13.14 Simulated FinFET (top) and GAA-SNS FET (bottom) structures together with the calculated compressive mechanical stress, the isometric view (left), and cross sectional view (right) are shown, from TCAD [35]

Table 13.3 TCAD simulated average mechanical stress in the channel perpendicular to the fin sidewalls for FinFETs and the sheet top and side surfaces for GAA-SNS FETs, for FL or SL and FW or SW scaling. For GAA-SNS FETs, the value is averaged over 3 sheets

Device	Surface		Stress (GPa)	
		FW (nm)	FL = 40 nm	FL = 20 nm
FinFET	(110)	10	2.0	3.39
		SW (nm)	SL = 40 nm	SL = 20 nm
GAA	(100)	15	1.56	2.22
GAA	(110)	15	1.68	2.45
		FL (nm)	FW = 20 nm	FW = 10 nm
FinFET	(110)	20	3.21	3.39
		SL (nm)	SW = 30 nm	SW = 15 nm
GAA	(100)	20	2.14	2.22
GAA	(110)	20	2.62	2.45

Figure 13.15 shows the changes in m_T and φ_B for the (100) and (110) surface orientations due to changes in strain related to the compressive mechanical stress, as obtained from band structure calculations using the tight-binding method [36]. Note that increase in mechanical stress increases m_T for the (110) surface but has no significant impact on φ_B, while it increases φ_B for the (100) surface but has no significant impact on m_T. As shown in Table 13.3, the stress increases with the FL and FW scaling in FinFETs, dominated by (110) surface orientation. Therefore, m_T increases, and as a consequence, K_{F10} reduces but Γ_0 increases, consistent with the parameter values shown in Table 13.2.

The stress increases with SL reduction in GAA-SNS FETs for both surface orientations. Increase in φ_B for the (100) surface and m_T for the (110) surface reduces

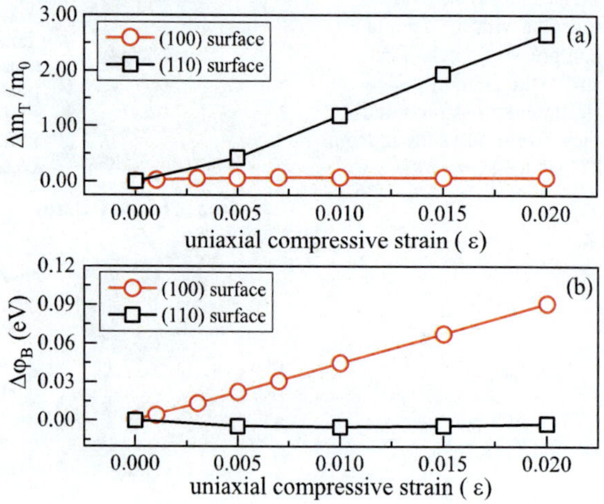

Fig 13.15 Impact of compressive mechanical stress-induced strain on the band structure parameters m_T and φ_B for the (100) and (110) surface orientations, calculated using tight binding method [36]

K_{F10} for both surfaces, reduces Γ_0 for the (100) surface but increase Γ_0 for the (110) surface orientations. The net result is a reduction in K_{F10}, however, a slight reduction in Γ_0 has been observed since the (100) surface dominates in this case, consistent with the parameter values shown in Table 13.2.

The stress increases only slightly for the (100) surface but reduces for the (110) surface for SW reduction in GAA-SNS FETs. Moreover, the relative dominance of the (110) surface becomes larger at smaller SW. Therefore, although φ_B slightly increases for the (100) surface, the m_T would reduce for the (110) surface, resulting in increased K_{F10} (the reduction in m_T dominates over the increase in φ_B) and reduced Γ_0 (increase in φ_B and reduction in m_T) consistent with Table 13.2.

13.7 Summary

The BAT framework is utilized to model the ultra-fast measured NBTI data from RMG HKMG SOI and SiGe FinFETs and GAA-SNS FETs having different channel dimensions (length and width). The ΔV_T magnitude reduces and VAF increases with reduction in FL and FW for FinFETs. On the other hand, the ΔV_T magnitude reduces with reduction in SL but increases with reduction in SW, while VAF reduces with reduction in both SL and SW for GAA-SNS FETs. These changes are primarily due to changes in the magnitude and VAF of the underlying ΔV_T subcomponent for both device architectures. TCAD and band structure simulations are done to qualitatively explained the observed trends. The compressive stress in the channel increases with reduction in FL, FW, and SL but reduces with reduction in SW. Changes in strain have different implications for the band structure parameters for (110) sidewall dominated FinFETs and (100) top surface dominated GAA-SNS FETs. These experiments verify the channel/IL bond dissociation mechanism invoked in the RD model.

Acknowledgements All experimental data presented in this chapter are re-plotted from previously published reports. The authors acknowledge IBM for providing measurement facilities, FinFETs and GAA-SNS FETs. The authors Narendra Parihar and Nilotpal Choudhury acknowledge Richard Southwick, Miaomiao Wang, Huimei Zhou and James Stathis for help with measurement facilities. Karansingh Thakor is acknowledged for help with manuscript preparation.

References

1. N. Kimizuka, K. Yamaguchi, K. Imai, T. Iizuka, C. T. Liu, R. C. Keller, T. Horiuchi, in *Symposium on VLSI Technology Digest of Technical Papers*, **92** (2000)
2. V. Huard, M. Denais, F. Perrier, N. Revil, C. Parthasarathy, A. Bravaix, E. Vincent, Microelectron. Reliab. **45**, 83 (2005)
3. S. Mahapatra, K. Ahmed, D. Varghese, A. E. Islam, G. Gupta, L. Madhav, D. Saha, M.A. Alam, in *IEEE International Reliability Physics Symposium Proceedings*, **1** (2007)
4. Y. Mitani, H. Satake, A. Toriumi, IEEE Trans. Device Mater. Reliab. **8**, 6 (2008)

5. S. Pae, M. Agostinelli, M. Brazier, R. Chau, G. Dewey, T. Ghani, M. Hattendorf, J. Hicks, J. Kavalieros, K. Kuhn, M. Kuhn, J. Maiz, M. Metz, K. Mistry, C. Prasad, S. Ramey, A. Roskowski, J. Sandford, C. Thomas, J. Thomas, C. Wiegand, J. Wiedemer, in *IEEE International Reliability Physics Symposium Proceedings*, **352** (2008)
6. K. Joshi, S. Hung, S. Mukhopadhyay, V. Chaudhary, N. Nanaware, B. Rajamohanan, T. Sato, M. Bevan, A. Wei, A. Noori, B. McDougal, C. Ni, G. Saheli, C. Lazik, P. Liu, D. Chu, L. Date, S. Datta, A. Brand, J. Swenberg, S. Mahapatra, *in IEEE International Reliability Physics Symposium Proceedings*, 4C.2.1 (2013).
7. J. Franco, B. Kaczer, P.J. Roussel, J. Mitard, M. Cho, L. Witters, T. Grasser, G. Groeseneken, IEEE Trans. Electron Devices **60**, 396 (2013)
8. P. Srinivasan, J. Fronheiser, K. Akarvardar, A. Kerber, L. F. Edge, R.G. Southwick, E. Cartier, H. Kothari, in *IEEE International Reliability Physics Symposium Proceedings*, 6A.3.1 (2014)
9. N. Parihar, N. Goel, S. Mukhopadhyay, S. Mahapatra, IEEE Trans. Electron Devices **65**, 392 (2018)
10. N. Parihar, S. Mahapatra, in *IEEE International Reliability Physics Symposium Proceedings*, TX.5.1 (2018)
11. V. Huard, C. Ndiaye, M. Arabi, N. Parihar, X. Federspiel, S. Mhira, S. Mahapatra, A. Bravaix, in *IEEE International Reliability Physics Symposium Proceedings*, TX.4.1 (2018)
12. N. Parihar, R. Tiwari, C. Ndiaye, M. Arabi, S. Mhira, H. Wong, S. Motzny, V. Moroz, V. Huard, S. Mahapatra, in *International Conference on Simulation of Semiconductor Processes and Devices*, **167** (2018)
13. S. Ramey, A. Ashutosh, C. Auth, J. Clifford, M. Hattendorf, J. Hicks, R. James, A. Rahman, V. Sharma, A. S. Amour, and C. Wiegand, in *IEEE International Reliability Physics Symposium Proceedings*, 4C.5.1 (2013).
14. K.T. Lee, W. Kang, E.-A. Chung, G. Kim, H. Shim, H. Lee, H. Kim, M. Choe, N.-I. Lee, A. Patel, J. Park, J. Park, in *IEEE International Reliability Physics Symposium Proceedings*, 2D.1.1 (2013)
15. J. Franco, B. Kaczer, A. Chasin, H. Mertens, L.-A. Ragnarsson, R. Ritzenthaler, S. Mukhopadhyay, H. Arimura, P.J. Roussel, E. Bury, N. Horiguchi, D. Linten, G. Groeseneken, A. Thean, in *IEEE International Reliability Physics Symposium Proceedings*, 4B.2.1 (2016)
16. G. Jiao, M. Toledano-Luque, K.-J. Nam, N. Toshiro, S.-H. Lee, J.-S. Kim, T. Kauerauf, E. Chung, D. Bae, G. Bae, D.-W. Kim, K. Hwang, in *IEEE International Electron Devices Meeting Technical Digest*, 31.2.1 (2016).
17. N. Parihar, R.G. Southwick, U. Sharma, M. Wang, J.H. Stathis, S. Mahapatra, in *IEEE International Reliability Physics Symposium Proceedings*, 2D.4.1 (2017)
18. N. Parihar, R. Southwick, M. Wang, J.H. Stathis, S. Mahapatra, in *IEEE International Electron Devices Meeting Technical Digest*, 7.3.1 (2017)
19. N. Parihar, U. Sharma, R.G. Southwick, M. Wang, J.H. Stathis, S. Mahapatra, IEEE Trans. Electron Devices **65**, 23 (2018)
20. N. Parihar, R.G. Southwick, M. Wang, J.H. Stathis, S. Mahapatra, IEEE Trans. Electron Devices **65**, 1699 (2018)
21. N. Parihar, R.G. Southwick, M. Wang, J.H. Stathis, S. Mahapatra, IEEE Trans. Electron Devices **65**, 1707 (2018)
22. N. Parihar, R. Tiwari, S. Mahapatra, in *International Conference on Simulation of Semiconductor Processes and Devices*, **176** (2018)
23. R. Tiwari, N. Parihar, K. Thakor, H.Y. Wong, S. Motzny, M. Choi, V. Moroz, S. Mahapatra, IEEE Trans. Electron Devices **66**, 2086 (2019)
24. R. Tiwari, N. Parihar, K. Thakor, H.Y. Wong, S. Motzny, M. Choi, V. Moroz, S. Mahapatra, IEEE Trans. Electron Devices **66**, 2093 (2019)
25. N. Parihar, U. Sharma, R.G. Southwick, M. Wang, J.H. Stathis, S. Mahapatra, in *IEEE International Reliability Physics Symposium Proceedings*, (2019)
26. M. Wang, J. Zhang, H. Zhou, R. G. Southwick, R. Hsin, K. Chao, X. Miao, V. S. Basker, T. Yamashita, D. Guo, G. Karve, H. Bu, in *IEEE International Reliability Physics Symposium Proceedings*, (2019)

27. N. Choudhury, U. Sharma, H. Zhou, R.G. Southwick, M. Wang, S. Mahapatra, in *IEEE International Reliability Physics Symposium Proceedings*, (2020)
28. H. Zhou, M. Wang, J. Zhang, K. Watanabe, C. Durfee, S. Mochizuki, R. Bao, R. Southwick, M. Bhuiyan, B. Veeraraghavan, *IEEE International Reliability Physics Symposium Proceedings*, (2020)
29. N. Choudhury, T. Samadder, R. Tiwari, H. Zhou, R. G. Southwick, M. Wang, S. Mahapatra, in *IEEE International Reliability Physics Symposium Proceedings*, (2021)
30. S. Mahapatra, N. Goel, S. Desai, S. Gupta, B. Jose, S. Mukhopadhyay, K. Joshi, A. Jain, A.E. Islam, M.A. Alam, IEEE Trans. Electron Devices **60**, 901 (2013)
31. A.E. Islam, N. Goel, S. Mahapatra, M.A. Alam, in *Fundamentals of Bias Temperature Instability in MOS Transistors* (Springer India, 2015), pp. 181–207
32. S. Mukhopadhyay, K. Joshi, V. Chaudhary, N. Goel, S. De, R.K. Pandey, K.V.R.M. Murali, S. Mahapatra, in *IEEE International Reliability Physics Symposium Proceedings*, GD 3.1 (2014)
33. N. Choudhury, N. Parihar, N. Goel, A. Thirunavukkarasu, S. Mahapatra, IEEE J of Electron Devices Soc. **8**, 1281 (2020)
34. T. Samadder, N. Choudhury, S. Kumar, D. Kochar, N. Parihar, S. Mahapatra, IEEE Trans. Electron Devices **68**, 485 (2021)
35. Sentaurus Process user manual, Synopsys
36. https://nanohub.org/resources/bandstrlab

Chapter 14
BAT Framework Modeling of AC NBTI: Stress Mode, Duty Cycle and Frequency

Souvik Mahapatra, Narendra Parihar, Nilesh Goel, Nilotpal Choudhury, Tarun Samadder, and Uma Sharma

14.1 Introduction

As shown in the previous chapters, Negative Bias Temperature Instability (NBTI) became an important reliability concern during the migration from Silicon Dioxide (SiO_2) to Silicon Oxynitride (SiON) gate insulator MOSFETs [1–4]. It continues to remain an issue for the dual layer (SiO_2 or SiON Interlayer (IL) and HfO_2 High-K layer) High-K Metal Gate (HKMG) gate insulator based bulk [5–10] and Fully Depleted Silicon On Insulator (FDSOI) [11, 12] planar MOSFETs, bulk and SOI FinFETs [12–25], as well as Gate All Around Stacked Nanosheet (GAA-SNS) FETs [26–29], having either Silicon (Si) or Silicon Germanium (SiGe) channel. The key features of NBTI are listed below (reproduced from Chap. 3, Sect. 3.1):

As mentioned in Chap. 1, Sect. 1.3, NBTI results in a gradual buildup of positive gate insulator charges and shifts transistor parameters, e.g., threshold voltage shift (ΔV_T), over time. It gets accelerated at more negative gate bias (V_G) during stress ($V_G = V_{GSTR}$) and at higher temperature (T), governed by the Voltage Acceleration Factor (VAF) and Arrhenius T activation energy (E_A) respectively. However, the positive charges accrued during stress reduce if the magnitude of V_G is reduced or removed ($V_G = V_{GREC}$ or 0 V), which reduces ΔV_T. Therefore, NBTI for AC stress results in lower ΔV_T than DC stress. The ratio of AC to DC NBTI ΔV_T depends on the pulse duty cycle (PDC), pulse low bias (V_{GLOW}), AC stress mode (A or B), and may or may not depend on the pulse frequency (f). On the other hand, the recovery of NBTI necessitates the use of ultra-fast methods to measure the device parametric shift without any artifacts, see Chap. 1, Sect. 1.2. Hence, the ultra-fast Measure Stress Measure (MSM) method is used throughout this book.

S. Mahapatra (✉) · N. Parihar · N. Goel · N. Choudhury · T. Samadder · U. Sharma
Department of Electrical Engineering, Indian Institute of Technology Bombay, Powai, Mumbai 400076, India
e-mail: souvik@ee.iitb.ac.in

S. Mahapatra (ed.), *Recent Advances in PMOS Negative Bias Temperature Instability*,
https://doi.org/10.1007/978-981-16-6120-4_14

Besides impacting the magnitude of ΔV_{T}, the DC or AC stress conditions, i.e., V_{GSTR} or V_{GHIGH} during AC, V_{GREC} or V_{GLOW} during AC, T, as well as PDC, f and stress modes for AC impact the time kinetics of ΔV_{T} during stress and recovery after stress (i.e., the power law slope (n) for longer stress time (t_{STR}) during stress, and fraction remaining (FR) during recovery at a particular time (t_{REC}) after stress; FR is defined as the ratio of ΔV_T during recovery at a time $t = t_{\mathrm{REC}}$ to that after the end of stress at $t = t_{\mathrm{STR}}$). These stress conditions also impact the VAF, $E_{A,}$ and T dependence of VAF (i.e., the reduction of VAF at higher T).

As also shown in the earlier chapters, the ΔV_T magnitude increases with higher Nitrogen content (N%) in the gate insulator, for both SiON [1–4] and HKMG [6, 9, 12, 17, 18, 20, 21] devices. On the other hand, the SiGe channel devices show reduced ΔV_T as compared to their Si channel counterparts, shown in bulk [7, 8, 10] and FDSOI [11, 12] planar MOSFETs and FinFETs [17, 18, 20, 21, 23, 25]. Besides ΔV_T magnitude, changes in N% in the gate insulator and Germanium content (Ge%) in the channel also impact the different parameters listed above. Increase in N% reduces n at longer t_{STR}, FR for a particular ratio of t_{REC} to t_{STR} (*i.e.*, the recovery becomes faster), VAF, $E_{\mathrm{A},}$ and the T dependence of VAF (i.e., the VAF reduction at higher T becomes smaller). On the other hand, increase in Ge% increases n at longer t_{STR}, FR for a particular ratio of t_{REC} to t_{STR} (i.e., the recovery becomes slower), VAF, $E_{\mathrm{A},}$ and the T dependence of VAF (i.e., VAF reduction at higher T becomes higher). Moreover, changes in the compressive mechanical stress in the channel because of changes in the layout or device dimensions impact NBTI, shown for FDSOI planar MOSFETs [11, 12], as well as for FinFETs and GAA-SNS FETs [22, 24, 29].

The BTI Analysis Tool (BAT) framework described in Chap. 4 through Chap. 6 has been used to model the above listed features in Chap. 7 through Chap. 13 of this book. Data measured during and after DC and Mode-B AC stress at different V_{GSTR} (V_{GHIGH} for AC stress) and T have been modeled, but only using fixed PDC, f, and digital ($V_{\mathrm{GLOW}} = 0$ V) pulse for AC stress. In this chapter, the impact of stress mode (A or B, see Chap. 1, Sect. 1.2), PDC, f, and V_{GLOW} of the AC pulse on NBTI kinetics are modeled across different device architectures. The BAT framework is briefly described below (reproduced from Chap. 6, Sect. 6.3).

Figure 14.1 illustrates the BAT framework [9] used to model the time kinetics of ΔV_{T} during and after stress. It uses uncorrelated contributions from generated interface (ΔV_{IT}) and bulk (ΔV_{OT}) gate insulator traps, and hole trapping in process related pre-existing bulk gate insulator traps (ΔV_{HT}). The Reaction Diffusion (RD) model is used to calculate the time kinetics of interface traps (density ΔN_{IT}) [9, 30, 31]. In Chap. 4, the RD model is described and independently validated by measured data from Direct Current IV (DCIV) method [32]. The Transient Trap Occupancy Model (TTOM) is used to calculate the occupancy of generated interface traps and their contribution (ΔV_{IT}) [9], which is described and validated in Chap. 5, Sect. 5.3. The ΔV_{HT} and ΔV_{OT} kinetics are modeled by the Activated Barrier Double Well Thermionic (ABDWT) model [33] and Reaction Diffusion Drift (RDD) model [34] respectively, and these are described and validated in Chap. 5, Sect. 5.4 and Chap. 6, Sect. 6.2.

Fig. 14.1 Schematic of the BTI Analysis Tool (BAT) framework used in this book to model measured ΔV_T kinetics during and after DC and AC NBTI stress, reproduced from Chap. 4

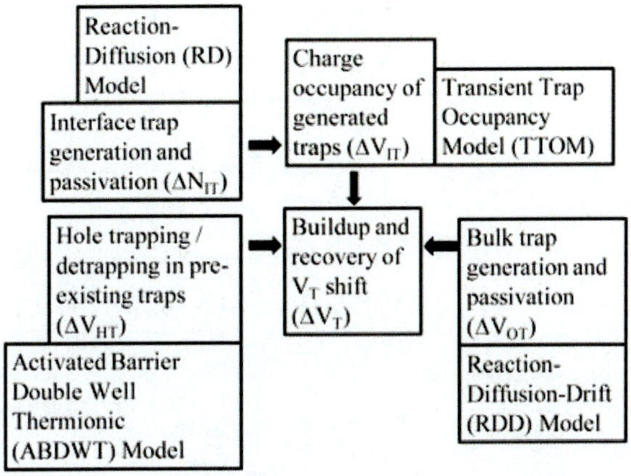

14.2 Device Details and Model Parameters

Measured data from Gate First (GF) HKMG planar MOSFET (see device D1 of Chap. 7), Replacement Metal Gate (RMG) HKMG SOI FinFET (see Chap. 10), and RMG HKMG bulk FinFETs having Si and SiGe channels and different N% in the gate stack for the SiGe channel devices (see Chap. 11) are used. Measurements are done by the full sweep MSM method for GF MOSFET and One Point Drop Down (OPDD) MSM method for RMG FinFETs. Stressing is done for DC, Mode-A, and Mode-B AC conditions. Refer to Chap. 1, Sect. 1.2 for stress and measurement details. The total stress time for AC stress includes the pulse on and off phases, and hence the actual stress depends on the PDC of the gate pulse.

The BAT model parameters are listed in the respective chapters, and the same values are also used in this chapter, except for one. Note, the pre-factor of the RDD model (the K_{F30} parameter, see Chap. 6, Table 6.2) is re-adjusted between DC and AC stress. Although the exact reason is not yet known, the ΔV_{OT} subcomponent is found to reduce AC stress. Therefore, K_{F30} depends on f but not on any other conditions of the AC gate pulse. This aspect has been mentioned before in Chaps. 6, 7, 10, and 11 to model the Mode-B AC data at 50% PDC.

14.3 Impact of AC Stress Mode, PDC, V_{GLOW}, and Frequency

Figure 14.2 shows the time evolution of measured and modeled ΔV_T (a, c) during and (b, d) after AC stress in GF MOSFET under (a, b) Mode-A and (c, d) Mode-B conditions, the underlying ΔV_{IT}, $\Delta V_{HT,}$ and ΔV_{OT} subcomponents are also shown. Mode-A (measurements are done after the pulse on phase or the last half cycle) and Mode-B (measurements are done after the pulse off phase or the last full cycle) AC

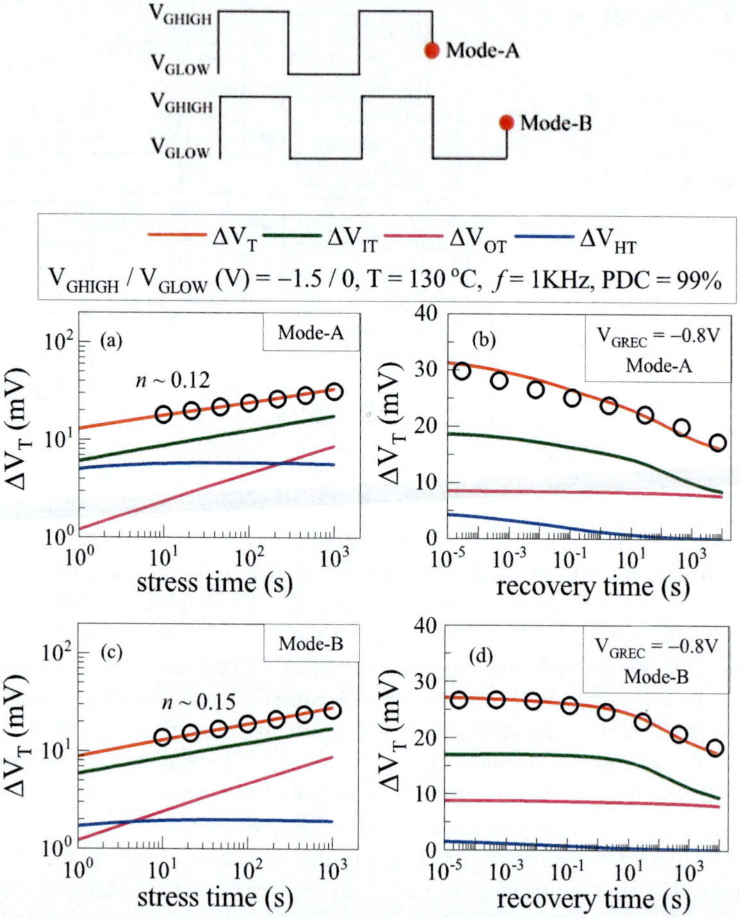

Fig. 14.2 Time evolution of measured and modeled ΔV_T kinetics in GF p-MOSFET (a, c) during and (b, d) after (a, b) Mode-A and (c, d) Mode-B AC stress. The AC stress modes are illustrated using the schematic on top of the figure. The underlying subcomponents are shown. Symbols: experiment, lines: model calculation. Data from [9]

stress conditions are illustrated using the schematic, see Chap. 1, Sect. 1.2 for details of the actual gate waveforms.

Figure 14.3 shows the time evolution of measured and modeled ΔV_T (a, c) during and (b, d) after AC stress in RMG SOI FinFET under (a, b) Mode-A and (c, d) Mode-B conditions, the underlying ΔV_{IT}, $\Delta V_{HT,}$ and ΔV_{OT} subcomponents are also shown. Note, although ΔV_{IT} dominates overall ΔV_T in both devices and for both DC and AC stress conditions, the relative contribution from ΔV_{HT} is larger in GF MOSFET while that from ΔV_{OT} is larger in RMG SOI FinFET.

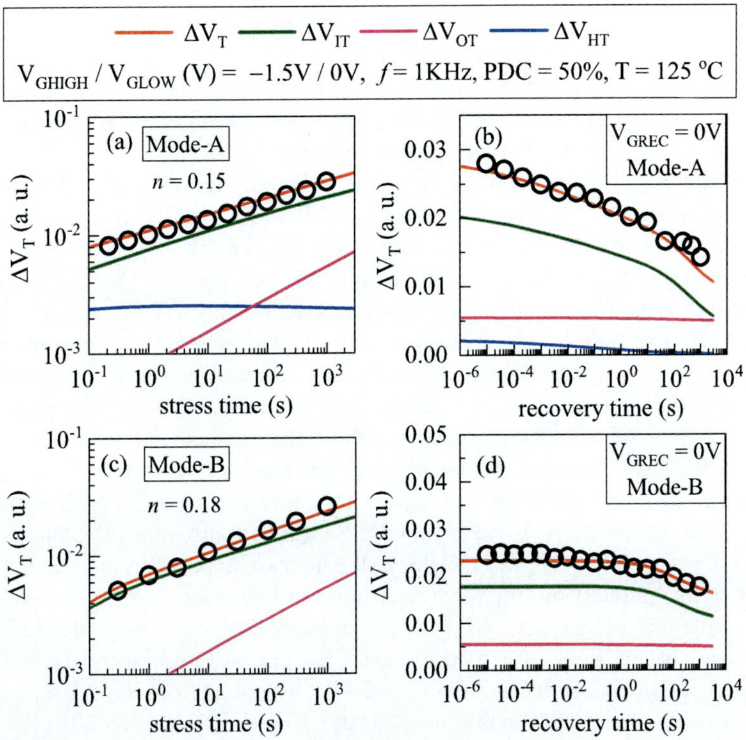

Fig. 14.3 Time evolution of measured and modeled ΔV_T kinetics in SOI p-FinFET (a, c) during and (b, d) after (a, b) Mode-A and (c, d) Mode-B AC stress. The underlying subcomponents are shown. Symbols: experiment, lines: model calculation. Data from [19]

Devices are stressed at fixed V_{GSTR} (V_{GHIGH} in this case) and T, while V_{GREC} is held at a fixed value mentioned in the figures. Note, Mode-A AC stress has non-negligible ΔV_{HT} contribution, as shown in Figs. 14.2(a) and 14.3(a), since the trapped holes in the last pulse on cycle do not detrap before the onset of the measurement phase. However, Mode-B AC stress has negligible ΔV_{HT} contribution, as shown in Figs. 14.2(c) and 14.3(c), since the trapped holes recover during the pulse off phase before the onset of measurement. The impact of ΔV_{HT} is clearly reflected in the time evolution of ΔV_T; since the Mode-A AC stress has lower time slope n due to non-negligible ΔV_{HT} (the explanation for the impact of ΔV_{HT} on n is provided in Chap. 3, Sect. 3.3). However, the difference between the Mode-A and Mode-B AC stress reduces at higher f, as shown later in this section (also Chap. 1, Sect. 1.3). Note that the negligible ΔV_{HT} contribution during Mode-B AC stress is consistent with earlier discussions in Chap. 1, Fig. 1.14, Chap. 5, Sect. 5.3, Chap. 7, Sect. 7.8, Chap. 10, Sect. 10.5 and Chap. 11, Sect. 11.4.

The ΔV_{IT} contribution is slightly smaller for Mode-B compared to Mode-A AC stress, due to reduction in the TTOM enabled fast fraction (ΔV_{IT_FAST}) during the pulse off phase before measurement (see Chap. 5, Sect. 5.3 for the role of electron capture in generated interface traps during AC stress). However, the ΔV_{OT} contribution is similar for both stress modes since it shows negligible recovery (~semi-permanent). The impact of ΔV_{OT} is clearly reflected in higher slope n for the SOI FinFET compared to GF MOSFET (the explanation of ΔV_{OT} on n is provided in Chap. 3, Sect. 3.4), which is observed for both stress modes.

The recovery after Mode-A AC stress, as shown in Figs. 14.2(b) and 14.3(b), is impacted by all the subcomponents (ΔV_{HT}, ΔV_{IT_FAST}, and ΔV_{IT_SLOW} for ΔV_{IT} and ΔV_{OT}, see Chap. 5, Sect. 5.3 for a description of the ΔV_{IT_FAST} and ΔV_{IT_SLOW} components of ΔV_{IT} recovery). On the other hand, recovery after Mode-B AC stress, as shown in Figs. 14.2(d) and 14.3(d), is impacted by only ΔV_{IT_SLOW} and ΔV_{OT}, as the ΔV_{HT} and ΔV_{IT_FAST} contributions at the end of pulse on phase get recovered during the pulse off phase before the onset of measurement. Therefore, the start of recovery is delayed and the fractional recovery is smaller (i.e., FR is higher) after Mode-B compared to Mode-A AC stress. Note that the delayed onset of recovery after Mode-B AC stress has also been reported in Chaps. 6, 10 and 11.

Figure 14.4 shows the time evolution of measured and modeled ΔV_T during (a) Mode-A and (b) Mode-B AC stress in GF MOSFET for different PDC and V_{GLOW}. Note, ΔV_T increases at larger PDC due to the increase in pulse on time. It increases at higher (magnitude) of V_{GLOW} as well, due to lower recovery in the pulse off phase. Note that the impact of V_{GLOW} on recovery during the AC pulse off phase is same as that of V_{GREC} for recovery after DC stress. Lower recovery at higher magnitude of V_{GLOW} is due to the increased contributions from the ΔV_{IT_FAST} and ΔV_{HT} subcomponents, which is explained later. Note that for non-zero V_{GLOW}, even the Mode-B AC stress shows some contributions from these two subcomponents. As mentioned before, Mode-A AC stress always shows lower slope n due to higher ΔV_{HT} contribution than Mode-B AC stress.

Figure 14.5 shows the measured and modeled ΔV_T at fixed time (t_{STR} = 1Ks) as a function of PDC at different V_{GLOW} but under fixed V_{GSTR} (V_{GHIGH}), T and f,

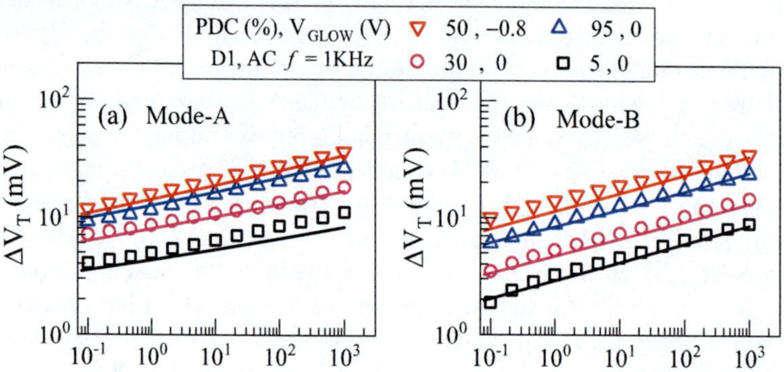

Fig. 14.4 Time evolution of measured and modeled ΔV_T kinetics in GF p-MOSFET during (a) Mode-A and (b) Mode-B AC stress at different PDC and V_{GLOW}. Symbols: experiment, lines: model calculation

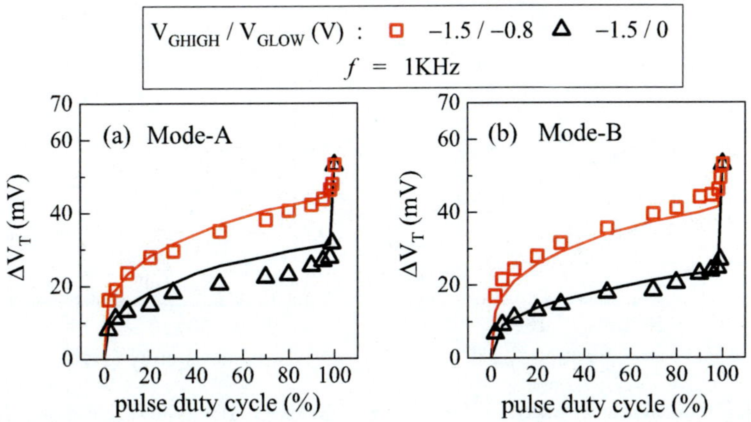

Fig. 14.5 Measured and modeled fixedtime ($t_{STR} = 1$Ks) ΔV_T in GF p-MOSFET during (a) Mode-A and (b) Mode-B AC stress at different V_{GLOW}. The DC value is also shown. Symbols: experiment, lines: model calculation. Data from [9]

for (a) Mode-A and (b) Mode-B AC stress in GF MOSFET. The DC value is shown as reference. ΔV_T increases at larger PDC due to increased stress (pulse on) phase. For a fixed PDC and V_{GLOW}, Mode-A shows higher ΔV_T compared to Mode-B AC stress. Moreover, a large kink or jump is seen near DC. The kink is larger for lower magnitude of V_{GLOW} for both stress modes, and for Mode-B when compared to Mode-A AC stress.

Note that the ΔV_{OT} subcomponent is not significant in this device for AC stress at high f and under moderate V_{GSTR} and T. Moreover, the ΔV_{IT} subcomponent has much larger impact than ΔV_{HT} for Mode-A and Mode-B AC stress at higher PDC, while ΔV_{IT} completely dominates ΔV_T for Mode-B AC stress at low to moderately high PDC (note, due to negligible ΔV_{HT}, Mode-B AC stress does not suffer from measurement delay artifacts, see Chap. 1, Figs. 1.6 and 1.15).

Therefore, higher ΔV_T for Mode-A as compared to Mode-B AC stress at fixed PDC and V_{GLOW} is due to additional contributions from ΔV_{IT_FAST} fraction of the ΔV_{IT} subcomponent and also ΔV_{HT}. At higher V_{GLOW} magnitude, the fraction f_{FAST} reduces, and therefore ΔV_{IT} increases, resulting in higher ΔV_T that is observed for both Mode-A and Mode-B AC stress. Although ΔV_{HT} recovery slows at higher magnitude of V_{GLOW}, its impact on the overall ΔV_T is less significant. Also note, the impact of V_{GLOW} is larger for Mode-B compared to Mode-A AC stress, due to its impact on reduction in the ΔV_{IT_FAST} and ΔV_{HT} contributions in the last half-cycle before measurement.

Finally, the jump or kink near DC is higher for Mode-B compared to Mode-A AC stress. This is because of recovery due to the hole de-trapping and fast electron capture processes during the pulse off phase in Mode-B AC stress. These processes are absent for DC and are somewhat less for Mode-A AC stress.

Figure 14.6 shows the measured and modeled ΔV_T at fixed time ($t_{STR} = 1$Ks) as a function of f for different (a) PDC (fixed V_{GLOW}) and (b) V_{GLOW} (fixed PDC) for

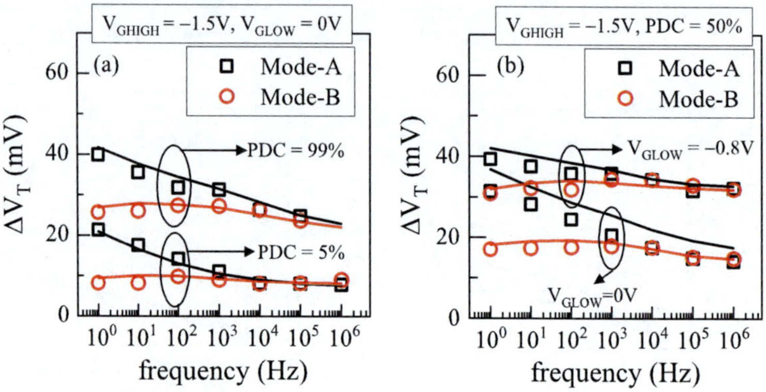

Fig. 14.6 Measured and modeled fixed time (t_{STR} = 1Ks) ΔV_T in GF p-MOSFET during Mode-A and Mode-B AC stress at different (a) PDC (for V_{GLOW} = 0V) and (b) V_{GLOW} (for PDC = 50%). Symbols: experiment, lines: model calculation. Data from [9]

Mode-A and Mode-B AC stress in GF MOSFET, the V_{GSTR} (V_{GHIGH}) and T values are kept fixed. ΔV_T shows f independence for Mode-B AC stress under different PDC and V_{GLOW}. In this case, ΔV_T is primarily governed by the ΔV_{IT_SLOW} fraction of the ΔV_{IT} subcomponent (see Fig. 14.2, Chap. 1, Fig. 1.15 and Chap. 5, Sect. 5.3), as ΔV_{OT} is negligible in this device for AC stress at high f (see Sect. 14.5 for the impact of ΔV_{OT} on Mode-B AC stress). This is consistent with the DCIV measurements that show f independence of measured ΔN_{IT}, see Chap. 4, Fig. 4.10. Note that Mode-A AC stress shows f dependence at low to moderate f due to additional contribution from the ΔV_{IT_FAST} fraction of the ΔV_{IT} subcomponent and ΔV_{HT}. However, their contributions reduce at high f, and hence ΔV_T for Mode-A merges with Mode-B AC stress.

14.4 Mode-B AC Stress, Impact of PDC and Frequency

Note that the Mode-B AC stress is relevant from the perspective of actual circuit operation. For a pure digital pulse (V_{GLOW} = 0 V), ΔV_T would be governed by ΔV_{IT} (TTOM enabled RD model) alone in most experiments. Even if ΔV_{OT} (RDD model) is present in some cases, its contribution is not significant at higher f, and also if the V_{GSTR} and/or T values are not very large. Measured data for such cases are modeled in this section.

Figure 14.7 shows the measured and modeled fixed time (t_{STR} = 1Ks) ΔV_T and the underlying ΔV_{IT} and ΔV_{OT} subcomponents as a function of frequency in RMG HKMG (a, b) Si and (c, d) SiGe25 (Ge = 25%) High-N FinFETs stressed using (a, c) high and (b, d) moderate values of V_{GSTR}. It is clearly evident that the ΔV_{IT} subcomponent dominates the overall ΔV_T and f independence is observed.

Figure 14.8 shows the measured and modeled fixed time (t_{STR} = 1Ks) ΔV_T as

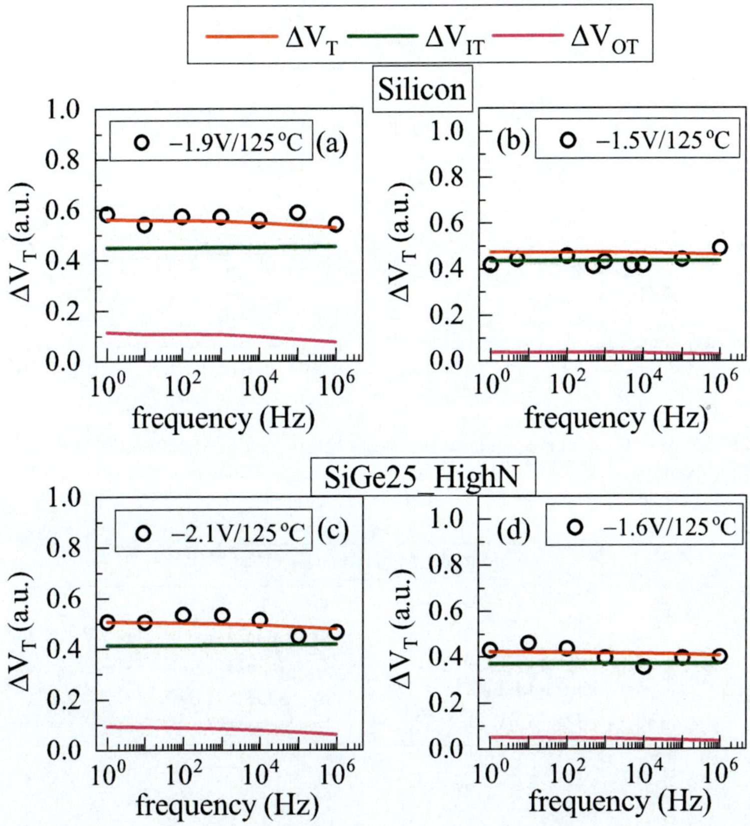

Fig. 14.7 Measured and modeled fixed time (t_{STR} = 1Ks) ΔV_T and the underlying ΔV_{IT} and ΔV_{OT} subcomponents as a function of f at fixed PDC of 50% under Mode-B AC stress in (a, b) Si and (c, d) SiGe25_HighN p-FinFETs. The V_{GSTR} and T conditions are mentioned in the panels. All AC data are normalized to the reference DC stress. Symbols: experiment, lines: model calculation. Data from [25]

a function of (a) PDC at fixed f and (b) f at fixed PDC for Mode-B AC stress in SOI FinFET at different T but fixed V_{GSTR} (=V_{GHIGH}). The measured and modeled AC data at different PDC and f are normalized to the corresponding DC stress value [the DC data are specifically shown in panel (a)]. The PDC dependence shows the usual "S" shaped characteristics with a large kink near DC, and f independence is obtained, which is similar to the data from GF MOSFETs shown in the previous section. The ΔV_T is dominated by the ΔV_{IT} subcomponent in this case.

Figure 14.9(a) shows the measured and modeled PDC dependence of ΔV_T (at fixed t_{STR} = 1Ks, f = 1 KHz) and the underlying ΔV_{IT}, $\Delta V_{HT,}$ and ΔV_{OT} subcomponents for a SiGe25 device having medium N%. The overall ΔV_T is dominated by ΔV_{IT} for different PDC since moderate V_{GSTR} and T values are used that limits the ΔV_{OT} contribution (more on this in the next section). The PDC dependence of ΔV_T

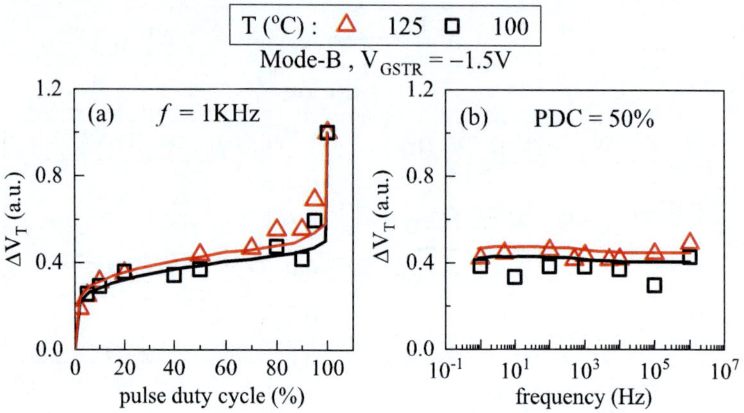

Fig. 14.8 Measured and modeled fixed time ($t_{STR} = 1$Ks) ΔV_T in SOI p-FinFET as a function of (a) PDC at fixed f and (b) f at fixed PDC. All AC data are normalized to the reference DC stress. Symbols: experiment, lines: model calculation. Data from [19]

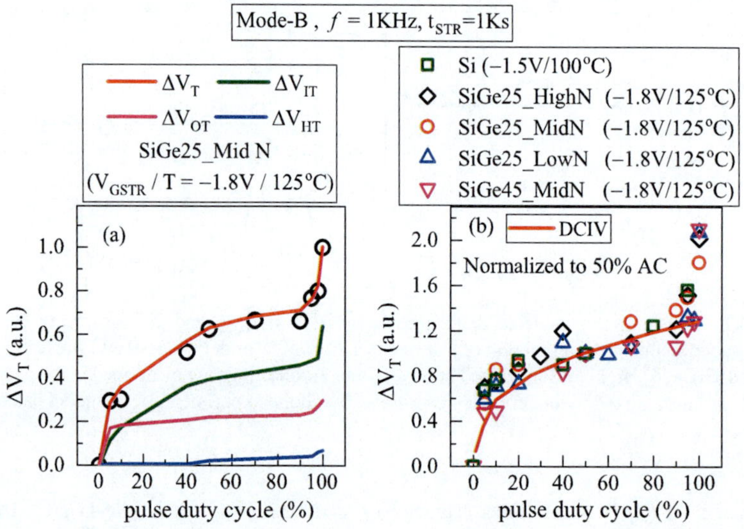

Fig. 14.9 (a) Measured and modeled fixed time ($t_{STR} = 1$Ks) ΔV_T along with the underlying model subcomponents as a function of PDC at fixed $f = 1$KHz during Mode-B AC stress in SiGe25 MidN p-FinFET. All AC data are normalized to the reference DC stress. Symbols: experiment, lines: model calculation. (b) Measured ΔV_T as a function of PDC at fixed $f = 1$KHz during Mode-B AC stress in Si and SiGe (different Ge%, N%) p-FinFETs. All AC data are normalized to the value at 50% AC stress. The DCIV measured data (also normalized to 50% AC) from Si p-FinFET are shown as the reference (line). Data from [21]

shows a kink or jump near DC, which depends on the time constant associated with the occupancy of interface traps, as explained in Chap. 5, Sect. 5.3. Note that the ΔV_{HT} contribution is small in this device for both DC and AC stress.

Figure 14.9(b) shows the PDC dependence of measured ΔV_T (at fixed $t_{STR} = 1$Ks, $f = 1$KHz) for Si and SiGe (different Ge% and N%) devices under moderate V_{GSTR} and T stress conditions. The PDC dependence of each device, when normalized to their PDC = 50% value, shows a universal "S" shape [35]. The universal PDC dependence is due to the ΔV_{IT} domination of ΔV_T for low to moderately high PDC, which is consistent with the duty cycle dependence of ΔN_{IT} measured using the DCIV method, see Chap. 4, Fig. 4.10. Interestingly, the DCIV measured ΔN_{IT} (if normalized to 50% PDC) overlaps with ΔV_T up to high PDC values as shown. The kink or jump near DC is not observed for ΔN_{IT} but is observed for ΔV_T and is different for different devices. Note that the difference in this kink in ΔV_T is due to the difference in the f_{FAST} component for electron capture for ΔV_{IT} (handled using TTOM) and contribution from ΔV_{HT} (only for Si) for different devices near DC.

14.5 Mode-B AC Stress, Conditions for Frequency Dependence

Literature shows a conflicting account of f dependence [36–38] and f independence [4, 19, 21, 39–41] of measured ΔV_T during AC NBTI stress. The impact of f is investigated for different devices in the previous sections. It has been demonstrated that when ΔV_T for AC stress is measured in Mode-B condition (i.e., it is measured after the end of full cycle or the pulse off phase), it always shows f independence for various PDC and pulse low bias conditions. However, f dependence is observed for Mode-A AC stress (i.e., when measured at the end of half cycle or pulse on phase), especially at lower f range. However, even Mode-A stress becomes f independent at higher f range. The mechanism responsible for the difference in the f impact between the Mode-A and Mode-B AC stress is explained.

However, it is recently reported that the Mode-B AC stress can also show f dependence, especially if high V_{GSTR} and T are used for stress [25, 42]. This aspect is analyzed in this section. All data are from SiGe channel based RMG bulk FinFETs with either low or moderate N% in the gate stack.

Figure 14.10 shows the measured and modeled ΔV_T at fixed $t_{STR} = 1$Ks under Mode-B AC stress in SiGe FinFET with moderate N, as a function of (a) V_{GSTR} (at fixed T) and (b) T (at fixed V_{GSTR}), at two different f (at fixed PDC = 50%). The underlying ΔV_{IT} and ΔV_{OT} subcomponents are also shown (ΔV_{HT} is negligible for Mode-B stress). The V_{GSTR} and T dependencies can be modeled by identical ΔV_{IT} but different ΔV_{OT} at different f (note, ΔV_{OT} reduces at higher f, which is shown in [25, 42], consistent with AC Time Dependent Dielectric Breakdown (TDDB) reports [43, 44]). As shown in Chap. 11, Fig. 11.13, the fractional ΔV_{OT} contribution can

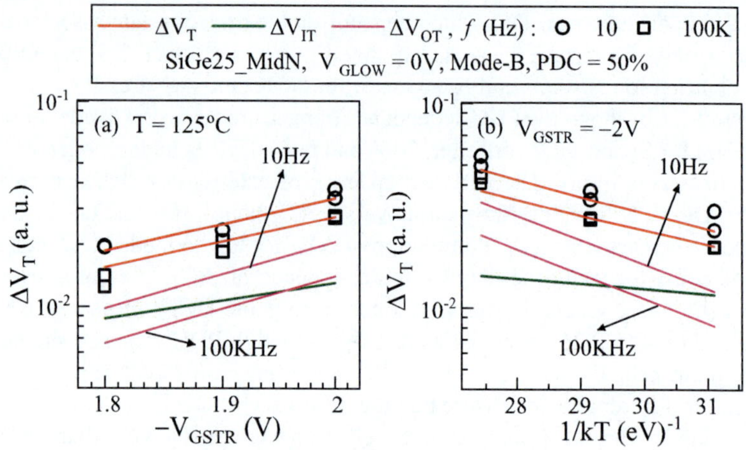

Fig. 14.10 Measured and modeled fixed time (t_{STR} = 1Ks) ΔV_T as a function of (a) V_{GSTR} (=V_{GHIGH}) at fixed T and (b) T at fixed V_{GSTR} (=V_{GHIGH}) under Mode-B AC stress at two different f (PDC = 50%) in SiGe p-FinFET. Symbols: experiment, lines: model calculation. Data from [25]

become significant during Mode-B AC stress at high V_{GSTR} and T, especially in SiGe devices having low to moderate N% in the gate stack.

Figure 14.11 shows the V_{GSTR} dependence of measured and modeled ΔV_T at different T but fixed t_{STR} = 1Ks in SiGe FinFETs having (a, b) Ge = 25% and (c, d) Ge = 45% and moderate N%, for (a, c) low f and (b, d) high f Mode-B AC stress. The PDC is kept constant at 50%. Note that the ΔV_{OT} contribution is appreciable in these devices under high V_{GSTR} and T, see Chap. 11, Fig. 11.13. The model accuracy is verified by noting that only the RDD model K_{F30} pre-factor related to the ΔV_{OT} subcomponent is reduced at higher f, all other model parameters are the same as shown in Chap. 11, Table 11.1 for these devices.

Note that the f dependence of ΔV_{OT} (ΔV_{IT} is f independent, see Chap. 4, Sect. 4.4 for direct measurement using the DCIV method) would impact the f dependence of overall ΔV_T if the ΔV_{OT} contribution becomes appreciable during Mode-B AC stress. This aspect is verified next.

Figure 14.12 shows the (a) measured and modeled ΔV_T and (b) the underlying ΔV_{IT} and ΔV_{OT} subcomponents as a function of f for different V_{GSTR} and T (fixed t_{STR} = 1Ks, PDC = 50%) for Mode-B AC stress in SiGe FinFET having moderate N. Measured ΔV_T shows f independence at low V_{GSTR} and T, while f dependence is observed at higher V_{GSTR} and T. This can be explained by noting the f independence (for ΔV_{IT}) and f dependence (for ΔV_{OT}) of the underlying subcomponents and their relative contributions at different V_{GSTR} and T. Due to higher relative ΔV_{OT} contribution, the f dependence of overall ΔV_T is observed at higher V_{GSTR} and T. However, for realistic operating conditions (low V_G), ΔV_T would be dominated by ΔV_{IT} for all processes (Ge% and N%), see Chap. 11, Fig. 11.17, and hence it would show f independence (as shown in the previous section).

Figure 14.13 shows the measured and modeled ΔV_T and the underlying ΔV_{IT}

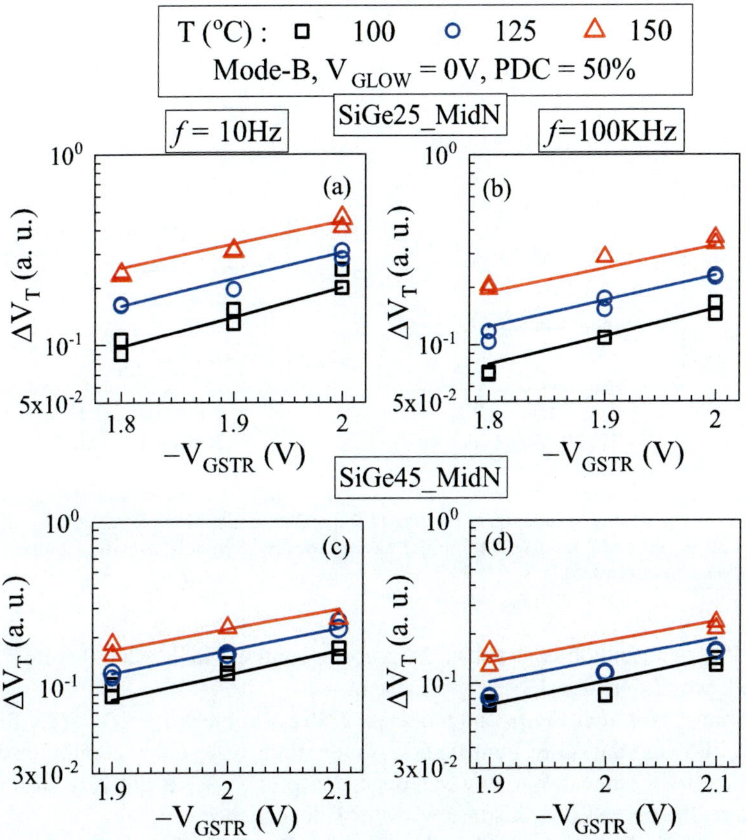

Fig. 14.11 Measured and modeled fixed time ($t_{\text{STR}} = 1\text{Ks}$) ΔV_T as a function of V_{GSTR} (= V_{GHIGH}) at different T for Mode-B AC stress at (a, c) low f and (b, d) high f (PDC = 50%) in SiGe p-FinFET having (a, b) Ge = 25% and (c, d) Ge = 45% and moderate N%. Symbols: experiment, lines: model calculation. Data from [25]

and ΔV_{OT} subcomponents as a function of f (at fixed $t_{\text{STR}} = 1\text{Ks}$, PDC = 50%) for Mode-B AC stress in (a, b) SiGe25_LowN and (c, d) SiGe45_LowN FinFETs. Each device is stressed at high V_{GSTR} but moderately high T (left panels) and also moderately high V_{GSTR} and T (right panels). Figure 14.13 also shows the same in (e, f) SiGe25_LowN FinFET for high V_{GSTR} but (e) very high T and (f) moderately high T stress.

As shown in Sect. 14.4, the ΔV_{OT} contribution is negligible in Fig. 14.7 for (a, b) Si and (c, d) SiGe25_HighN devices for the values of V_{GSTR} and T used in those experiments. Therefore, the ΔV_{IT} dominates overall ΔV_{T} and f independence is observed, consistent with data shown in Sect. 14.3.

However, the ΔV_{OT} contribution is not negligible in SiGe25 and SiGe45 low N% devices shown in Fig. 14.13, and the following conditions are observed. First, for moderate V_{GSTR} and T stress in Fig. 14.13, for (b) SiGe25 and (d) SiGe45 devices,

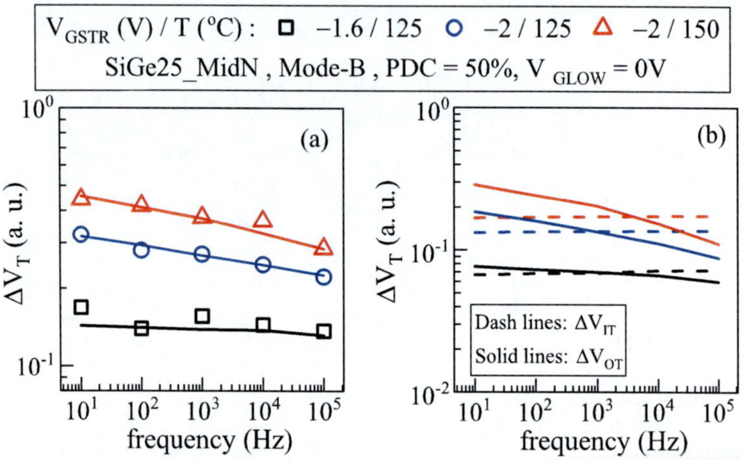

Fig. 14.12 (a) Measured and modeled fixed time (t_{STR} = 1Ks) ΔV_T and (b) underlying ΔV_{IT} and ΔV_{OT} subcomponents as a function of f at fixed PDC of 50% under Mode-B AC stress at different V_{GSTR} (=V_{GHIGH}) and T in SiGe25 p-FinFET having low N%. Symbols: experiment, lines: model calculation. Data from [25]

the ΔV_{OT} contribution is less than ΔV_{IT} for the entire f range, and therefore, the f dependence of overall ΔV_T is sort of absent.

Moreover, for high V_{GSTR} and moderate T stress in Fig. 14.13(a), (f) for SiGe25 device, although the ΔV_{OT} contribution is more than ΔV_{IT} either at only lower f or for the entire f range, however, the f dependence of ΔV_{OT} is not very strong, and therefore, the overall ΔV_T also shows weak f dependence.

However, for high V_{GSTR} and moderately high T stress in Fig. 14.13(c) for SiGe45 and for high V_{GSTR} and high T stress in Fig. 14.13(e) for SiGe25, the contribution from ΔV_{OT} is large either only at lower f or for the entire f range, and also the f dependence of ΔV_{OT} is strong. Therefore, the f dependence of overall ΔV_T is also strong.

As mentioned before, all f dependent data in Fig. 14.13 are modeled using only one parameter (K_{F30}) that is adjusted with f, all other parameters are the same as listed in Chap. 11, Table 11.1. Therefore, it is important to choose proper V_{GSTR} for the f dependent experiments, to make the stress condition similar to the operating condition, for which the f independence is always observed.

14.6 Summary

Ultra-fast NBTI measurements are done during and after Mode-A and Mode-B AC stress in Si channel GF HKMG MOSFET and RMG HKMG SOI FinFET, and during Mode-B AC stress in Si and SiGe (different Ge% and N%) channel RMG HKMG

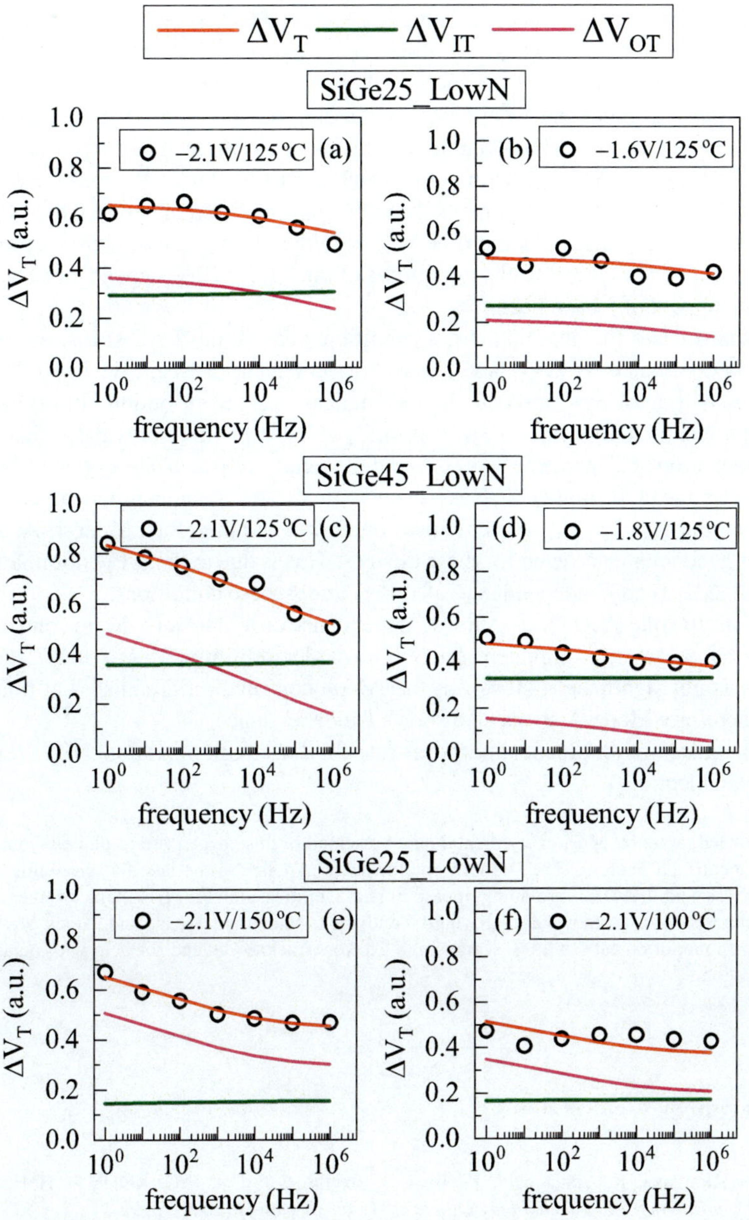

Fig. 14.13 Measured and modeled fixed time (t_{STR} = 1Ks) ΔV_T and the underlying ΔV_{IT} and ΔV_{OT} subcomponents as a function of f at fixed PDC of 50% under Mode-B AC stress in (a, b, e, f) SiGe25_LowN and (c, d) SiGe45_LowN p-FinFETs. The V_{GSTR} (=V_{GHIGH}) and T conditions are mentioned in the panels. Symbols: experiment, lines: model calculation. Data from [25]

bulk FinFETs. Experiments are done at different PDC, *f*, and V_{GLOW} for Mode-A and Mode-B stress, and also at different V_{GSTR} and T for Mode-B stress. The BAT framework can model the measured stress-recovery data under different AC conditions, using only one *f* dependent parameter that is re-adjusted between the DC and AC stress. The ΔV_T time kinetics during and after Mode-A stress is governed by ΔV_{IT}, ΔV_{HT}, and ΔV_{OT} subcomponents, while that for Mode-B stress is governed by ΔV_{IT} and ΔV_{OT} subcomponents. Features that are different between the Mode-A and Mode-B stress, such as the time kinetics slope during stress, the start time and rate of recovery after stress, the magnitude of the kink in PDC dependence near DC, and the impact of *f* have been modeled.

In most cases (devices and stress conditions) for Mode-B AC stress, the overall ΔV_T is dominated by ΔV_{IT} and shows *f* independence, except for low N% SiGe devices at higher V_{GSTR} and/or T. In the latter cases, the contribution from ΔV_{OT} can be significantly higher than ΔV_{IT}, especially at lower *f*. Therefore, due to the (yet to be understood) *f* dependence of ΔV_{OT}, the overall ΔV_T also shows *f* dependence. However, due to the higher VAF associated with the ΔV_{OT} subcomponent, its contribution reduces at low V_G (closer to use condition), and therefore, Mode-B AC stress shows *f* independence even for these devices. This is due to the ΔV_{IT} domination of overall ΔV_T (and *f* independence of ΔV_{IT}) under these conditions.

Finally, Mode-A AC stress shows *f* dependence only at lower *f* due to contribution from the ΔV_{HT} subcomponent (at low to moderately high V_{GSTR} and T, so that ΔV_{OT} is not significant). However, the ΔV_{HT} contribution diminishes at higher *f*, and therefore, Mode-A stress also shows *f* independence.

Therefore, for all practical purposes relevant for circuit operation, NBTI remains *f* independent.

Acknowledgements Most experimental data presented in this chapter are re-plotted from previously published reports. The authors acknowledge Applied Materials for providing planar MOSFETs and IBM for providing measurement facilities and FinFETs. One of the authors (Narendra Parihar) acknowledges Richard Southwick, Miaomiao Wang, and James Stathis for help with measurement facilities. Karansingh Thakor is acknowledged for help with manuscript preparation.

References

1. N. Kimizuka, K. Yamaguchi, K. Imai, T. Iizuka, C.T. Liu, R.C. Keller, T. Horiuchi, in *Symposium on VLSI Technology Digest of Technical Papers* (2000), p. 92
2. V. Huard, M. Denais, F. Perrier, N. Revil, C. Parthasarathy, A. Bravaix, E. Vincent, Microelectron. Reliab. **45**, 83 (2005)
3. S. Mahapatra, K. Ahmed, D. Varghese, A.E. Islam, G. Gupta, L. Madhav, D. Saha, M.A. Alam, in *IEEE International Reliability Physics Symposium Proceedings* (2007), p. 1
4. Y. Mitani, H. Satake, A. Toriumi, IEEE Trans. Device Mater. Reliab. **8**, 6 (2008)
5. S. Pae, M. Agostinelli, M. Brazier, R. Chau, G. Dewey, T. Ghani, M. Hattendorf, J. Hicks, J. Kavalieros, K. Kuhn, M. Kuhn, J. Maiz, M. Metz, K. Mistry, C. Prasad, S. Ramey, A. Roskowski,

J. Sandford, C. Thomas, J. Thomas, C. Wiegand, J. Wiedemer, in *IEEE International Reliability Physics Symposium Proceedings* (2008), p. 352
6. K. Joshi, S. Hung, S. Mukhopadhyay, V. Chaudhary, N. Nanaware, B. Rajamohanan, T. Sato, M. Bevan, A. Wei, A. Noori, B. McDougal, C. Ni, G. Saheli, C. Lazik, P. Liu, D. Chu, L. Date, S. Datta, A. Brand, J. Swenberg, S. Mahapatra, in *IEEE International Reliability Physics Symposium Proceedings*, 4C.2.1 (2013)
7. J. Franco, B. Kaczer, P.J. Roussel, J. Mitard, M. Cho, L. Witters, T. Grasser, G. Groeseneken, IEEE Trans. Electron Devices **60**, 396 (2013)
8. P. Srinivasan, J. Fronheiser, K. Akarvardar, A. Kerber, L.F. Edge, R.G. Southwick, E. Cartier, H. Kothari, in *IEEE International Reliability Physics Symposium Proceedings*, 6A.3.1 (2014)
9. N. Parihar, N. Goel, S. Mukhopadhyay, S. Mahapatra, IEEE Trans. Electron Devices **65**, 392 (2018)
10. N. Parihar, S. Mahapatra, in *IEEE International Reliability Physics Symposium Proceedings*, TX.5.1 (2018)
11. V. Huard, C. Ndiaye, M. Arabi, N. Parihar, X. Federspiel, S. Mhira, S. Mahapatra, A. Bravaix, in *IEEE International Reliability Physics Symposium Proceedings*, TX.4.1 (2018)
12. N. Parihar, R. Tiwari, C. Ndiaye, M. Arabi, S. Mhira, H. Wong, S. Motzny, V. Moroz, V. Huard, S. Mahapatra, in International Conference on Simulation of Semiconductor Processes and Devices (2018), p. 167
13. S. Ramey, A. Ashutosh, C. Auth, J. Clifford, M. Hattendorf, J. Hicks, R. James, A. Rahman, V. Sharma, A.St. Amour, C. Wiegand, in *IEEE International Reliability Physics Symposium Proceedings*, 4C.5.1 (2013)
14. K.T. Lee, W. Kang, E.-A. Chung, G. Kim, H. Shim, H. Lee, H. Kim, M. Choe, N.-I. Lee, A. Patel, J. Park, J. Park, in *IEEE International Reliability Physics Symposium Proceedings*, 2D.1.1 (2013)
15. J. Franco, B. Kaczer, A. Chasin, H. Mertens, L.-A. Ragnarsson, R. Ritzenthaler, S. Mukhopadhyay, H. Arimura, P.J. Roussel, E. Bury, N. Horiguchi, D. Linten, G. Groeseneken, A. Thean, in *IEEE International Reliability Physics Symposium Proceedings*, 4B.2.1 (2016)
16. G. Jiao, M. Toledano-Luque, K.-J. Nam, N. Toshiro, S.-H. Lee, J.-S. Kim, T. Kauerauf, E. Chung, D. Bae, G. Bae, D.-W. Kim, K. Hwang, in *IEEE International Electron Devices Meeting Technical Digest*, 31.2.1 (2016)
17. N. Parihar, R.G. Southwick, U. Sharma, M. Wang, J.H. Stathis, S. Mahapatra, in *IEEE International Reliability Physics Symposium Proceedings*, 2D.4.1 (2017)
18. N. Parihar, R. Southwick, M. Wang, J.H. Stathis, S. Mahapatra, in *IEEE International Electron Devices Meeting Technical Digest*, 7.3.1 (2017)
19. N. Parihar, U. Sharma, R.G. Southwick, M. Wang, J.H. Stathis, S. Mahapatra, IEEE Trans. Electron Devices **65**, 23 (2018)
20. N. Parihar, R.G. Southwick, M. Wang, J.H. Stathis, S. Mahapatra, IEEE Trans. Electron Devices **65**, 1699 (2018)
21. N. Parihar, R.G. Southwick, M. Wang, J.H. Stathis, S. Mahapatra, IEEE Trans. Electron Devices **65**, 1707 (2018)
22. N. Parihar, R. Tiwari, S. Mahapatra, in International Conference on Simulation of Semiconductor Processes and Devices (2018), p. 176
23. R. Tiwari, N. Parihar, K. Thakor, H.Y. Wong, S. Motzny, M. Choi, V. Moroz, S. Mahapatra, IEEE Trans. Electron Devices **66**, 2086 (2019)
24. R. Tiwari, N. Parihar, K. Thakor, H.Y. Wong, S. Motzny, M. Choi, V. Moroz, S. Mahapatra, IEEE Trans. Electron Devices **66**, 2093 (2019)
25. N. Parihar, U. Sharma, R.G. Southwick, M. Wang, J.H. Stathis, S. Mahapatra, in *IEEE International Reliability Physics Symposium Proceedings* (2019)
26. M. Wang, J. Zhang, H. Zhou, R.G. Southwick, R. Hsin, K. Chao, X. Miao, V.S. Basker, T. Yamashita, D. Guo, G. Karve, H. Bu, in *IEEE International Reliability Physics Symposium Proceedings* (2019)
27. N. Choudhury, U. Sharma, H. Zhou, R.G. Southwick, M. Wang, S. Mahapatra, in *IEEE International Reliability Physics Symposium Proceedings* (2020)

28. H. Zhou, M. Wang, J. Zhang, K. Watanabe, C. Durfee, S. Mochizuki, R. Bao, R. Southwick, M. Bhuiyan, B. Veeraraghavan, in *IEEE International Reliability Physics Symposium Proceedings* (2020)
29. N. Choudhury, T. Samadder, R. Tiwari, H. Zhou, R.G. Southwick, M. Wang, S. Mahapatra, in *IEEE International Reliability Physics Symposium Proceedings* (2021)
30. S. Mahapatra, N. Goel, S. Desai, S. Gupta, B. Jose, S. Mukhopadhyay, K. Joshi, A. Jain, A.E. Islam, M.A. Alam, IEEE Trans. Electron Devices **60**, 901 (2013)
31. A.E. Islam, N. Goel, S. Mahapatra, M.A. Alam, in *Fundamentals of Bias Temperature Instability in MOS Transistors* (Springer, India, 2015), pp. 181–207
32. S. Mukhopadhyay, K. Joshi, V. Chaudhary, N. Goel, S. De, R.K. Pandey, K.V.R.M. Murali, S. Mahapatra, in *IEEE International Reliability Physics Symposium Proceedings*, GD 3.1 (2014)
33. N. Choudhury, N. Parihar, N. Goel, A. Thirunavukkarasu, S. Mahapatra, IEEE J. Electron Devices Soc. **8**, 1281 (2020)
34. T. Samadder, N. Choudhury, S. Kumar, D. Kochar, N. Parihar, S. Mahapatra, IEEE Trans. Electron Devices **68**, 485 (2021)
35. N. Goel, K. Joshi, S. Mukhopadhyay, N. Nanaware, S. Mahapatra, Microelectron. Reliab. **54**, 491 (2014)
36. S. Wang, D.S. Ang, G.A. Du, IEEE Electron Device Lett. **29**, 483 (2008)
37. H. Reisinger, T. Grasser, K. Ermisch, H. Nielen, W. Gustin, C. Schlunder, in *IEEE International Reliability Physics Symposium Proceedings*, 6A.1.1 (2011)
38. T. Grasser, B. Kaczer, H. Reisinger, P.-J. Wagner, M. Toledano-Luque, in *IEEE International Reliability Physics Symposium Proceedings*, XT.8.1 (2012)
39. G. Chen, M.F. Li, C.H. Ang, J.Z. Zheng, D.L. Kwong, IEEE Electron Device Lett. **23**, 734 (2002)
40. R. Fernandez, B. Kaczer, A. Nackaerts, S. Demuynck, R. Rodriguez, M. Nafria, G. Groeseneken, in *International Electron Devices Meeting Technical Digest* (2006)
41. V. Huard, R. Chevallier, C. Parthasarathy, A. Mishra, N. Ruiz-Amador, F. Persin, V. Robert, A. Chimeno, E. Pion, N. Planes, D. Ney, F. Cacho, N. Kapoor, V. Kulshrestha, S. Chopra, N. Vialle, in *IEEE International Reliability Physics Symposium Proceedings* (2010), p. 655
42. L. Zhou, Q. Zhang, H. Yang, Z. Ji, Z. Zhang, Q. Liu, H. Xu, B. Tang, E. Simoen, X. Ma, X. Wang, Y. Li, H. Yin, J. Luo, C. Zhao, W. Wang, IEEE Electron Device Lett. **41**, 965 (2020)
43. R. Ranjan, Y. Liu, T. Nigam, A. Kerber, B. Parameshwaran, in *IEEE International Reliability Physics Symposium Proceedings*, DG.10.1 (2017)
44. M. Rafik, A.P. Nguyen, X. Garros, M. Arabi, X. Federspiel, C. Diouf, in *IEEE International Reliability Physics Symposium Proceedings*, 4A.3.1 (2018)

Index

S. Mahapatra (ed.), *Recent Advances in PMOS Negative Bias Temperature Instability*,
https://doi.org/10.1007/978-981-16-6120-4

Index

S. Mahapatra (ed.), *Recent Advances in PMOS Negative Bias Temperature Instability*,
https://doi.org/10.1007/978-981-16-6120-4

G

H

I

J

K

T